Design and manufacture of textile composites

Related titles from Woodhead's textile technology list:

3-D textile reinforcements in composite materials (1 85573 376 5)

3-D textile reinforced composite materials are obtained by applying highly productive textile technologies in the manufacture of fibre preforms. The damage tolerance and impact resistance are increased as the trend to delamination is drastically diminished due to the existence of reinforcements in the thickness direction. *3-D textile reinforcements in composite materials* describes the manufacturing processes, highlights the advantages, identifies the main applications, analyses the methods for prediction of mechanical properties and examines the key technical aspects of 3-D textile reinforced composite materials. This will enable materials scientists and engineers to exploit the main features and overcome the disadvantages in relation to laminated composite materials.

Green composites (1 85573 739 6)

Life cycle assessment is becoming increasingly important at every stage of a product's life from initial synthesis through to final disposal, and a sustainable society needs environmentally safe materials and processing methods. With an internationally recognised team of authors, *Green composites* examines polymer composite production and explains how environmental footprints can be diminished at every stage of the life cycle. This book is an essential guide for agricultural crop producers, governmental agricultural departments, automotive companies, composite producers and materials scientists all dedicated to the promotion and practice of eco-friendly materials and production methods.

Bast and other plant fibres (1 85573 684 5)

Environmental concerns have regenerated interest in the use of natural fibres for a much wider variety of products, including high-tech applications such as geotextiles, and in composite materials for automotive and light industry use. This new study covers the chemical and physical structure of these natural fibres; fibre, yarn and fabric production; dyeing; handle and wear characteristics; economics; and environmental and health and safety issues.

Details of these books and a complete list of Woodhead's textile technology titles can be obtained by:

- visiting our website at www.woodheadpublishing.com
- contacting Customer Services (e-mail: sales@woodhead-publishing.com; fax: +44 (0) 1223 893694; tel.: +44 (0) 1223 891358 ext.30; address: Woodhead Publishing Limited, Abington Hall, Abington, Cambridge CB1 6AH, England)

Design and manufacture of textile composites

Edited by

A. C. Long

The Textile Institute

CRC Press
Boca Raton Boston New York Washington, DC

WOODHEAD PUBLISHING LIMITED
Cambridge England

Published by Woodhead Publishing Limited in association with The Textile Institute
Woodhead Publishing Limited
Abington Hall, Abington
Cambridge CB1 6AH
England
www.woodheadpublishing.com

Published in North America by CRC Press LLC, 6000 Broken Sound Parkway, NW,
Suite 300, Boca Raton FL 33487, USA

First published 2005, Woodhead Publishing Limited and CRC Press LLC
© Woodhead Publishing Limited, 2005
The authors have asserted their moral rights.

British Library Cataloguing in Publication Data
A catalogue record for this book is available from the British Library.

Library of Congress Cataloging in Publication Data
A catalog record for this book is available from the Library of Congress.

Woodhead Publishing Limited ISBN-13: 978-1-85573-744-0 (book)
Woodhead Publishing Limited ISBN: 10-1-85573-744-2 (book)
Woodhead Publishing Limited ISBN-13: 978-1-84569-082-3 (e-book)
Woodhead Publishing Limited ISBN-10: 1-84569-082-6 (e-book)
CRC Press ISBN 0-8493-2593-5
CRC Press order number: WP2593

The publishers' policy is to use permanent paper from mills that operate a
sustainable forestry policy, and which has been manufactured from pulp
which is processed using acid-free and elementary chlorine-free practices.
Furthermore, the publishers ensure that the text paper and cover board used
have met acceptable environmental accreditation standards.

Project managed by Macfarlane Production Services, Markyate, Hertfordshire
(macfarl@aol.com)
Typeset by Replika Press Pvt Ltd, India
Printed by T J International Limited, Padstow, Cornwall, England

Contents

W-R Yu, Seoul National University, Korea and A C Long, University of Nottingham, UK

J Dominy, Carbon Concepts Limited, UK, C Rudd, University of Nottingham, UK

M D Wakeman and J-A E. Mânson, École Polytechnique Fédérale de Lausanne (EPFL), Switzerland

A Gokce and S G Advani, University of Delaware, USA

J Chilton, University of Lincoln, UK and R Velasco, University of Nottingham, UK

J G Ellis, Ellis Developments Limited, UK

K Van de Velde, Ghent University, Belgium

Contributor contact details

Introduction

Professor A.C. Long
School of Mechanical, Materials and
Manufacturing Engineering
University of Nottingham
University Park
Nottingham NG7 2RD
UK

Email: andrew.long@nottingham.ac.uk

Chapter 1

Professor S. Lomov and Professor I.
Verpoest
Katholieke Universiteit Leuven
Department of MTM
Kasteelpark Arenberg 44
B-3001, Heverlee
Belgium

Email:
stepan.lomov@mtm.kuleuven.ac.be
Email:
ignass.verpoest@mtm.kuleuven.ac.be

Dr F. Robitaille
Faculty of Engineering
161 Louis Pasteur
Room A306
Ottawa, Ontario
Canada K1N 6N5

Email: frobit@genie.uottawa.ca

Chapter 2

Professor P. Boisse
Laboratoire de Méchanique des
Contacts et des solides
LaMCoS, UMR CNRS 5514
INSA de Lyon
Bâtiment Jacquard
27 Avenue Jean Capelle
69621 Villeurbanne Cedex
France

Email:Philippe.Boisse@insa-lyon.fr

Professor A. C. Long
School of Mechanical, Materials and
Manufacturing Engineering
University of Nottingham
University Park
Nottingham NG7 2RD
UK

Email: andrew.long@nottingham.ac.uk

Dr F. Robitaille
Faculty of Engineering
161 Louis Pasteur
Room A306
Ottawa, Ontario
Canada K1N 6N5

Email: frobit@genie.uottawa.ca

Chapter 3

Dr P. Harrison and Dr M. Clifford
School of Mechanical, Materials and
Manufacturing Engineering
The University of Nottingham
University Park
Nottingham NG7 2RD
UK

Tel: (44) (0)115 8466134
Fax: (44) (0)115 9513800
Email: mike.clifford@nottingham.ac.uk

Chapter 4

Professor A. C. Long
School of Mechanical, Materials and
Manufacturing Engineering
University of Nottingham
University Park
Nottingham NG7 2RD
UK

Email: andrew.long@nottingham.ac.uk

Assistant Professor W-R. Yu
School of Materials Science and
Engineering
Seoul National University
San 56-1
Shillim 9 dong
Kwanak-gu
Seoul 151-742
Korea

Email: woongryu@snu.ac.kr

Chapter 5

Professor C. Rudd
School of Mechanical, Materials and
Manufacturing Engineering
University of Nottingham
University Park
Nottingham NG7 2RD
UK

Email:
christopher.rudd@nottingham.ac.uk

Professor J. Dominy
Carbon Concepts Ltd
Unit A2, Lower Mantle Close
Bridge Street Industrial Estate
Clay Cross
Derbyshire S45 9NU
UK

Email: john@carbonconcepts.co.uk

Chapter 6 and 10

Dr M. D. Wakeman and Professor
J-A. E. Månson
Laboratoire de Technologie
des Composites et Polyméres
École Polytechnique
Fédérale de Lausanne
Lausanne
Switzerand

Tel: (41) 21 693 4281
Email: martyn.wakeman@epfl.ch
Email: jan-anders.manson@epfl.ch

Chapter 7

Dr A. Gokce and Professor S. G.
Advani
Department of Mechanical
Engineering
University of Delaware
Newark
DE 19716
USA

Email: advani@me.udel.edu
Email: gokce@ME.UDel.Edu

Chapter 8

Dr I. A. Jones
School of Mechanical, Materials and
Manufacturing Engineering
University of Nottingham
University Park
Nottingham NG7 2RD
UK

Email: arthur.jones@nottingham.ac.uk

Professor A. K. Pickett
Cranfield University
School of Industrial and
Manufacturing Science
Building 61, Cranfield
Bedfordshire MK43 0AL
UK

Tel: (44) (0)1234 754034
Fax: (44) (0)1234 752473
Email: a.k.pickett@cranfield.ac.uk

Chapter 9

Professor A. R. Horrocks and
Dr B. K. Kandola
Centre for Materials Research &
Innovation
University of Bolton
Deane Road
Bolton BL3 5AB
UK

Tel: +44 (0)1024 903831
Email: A.R.Horrocks@bolton.ac.uk

Chapter 11

Dr J. Lowe
Tenax Fibers GmbH & Co. KG
Kasinostrasse 19-21
42 103 Wuppertal
Germany

Email: JLowe@t-online.de

Chapter 12

Professor J. Chilton
Lincoln School of Architecture
University of Lincoln
Brayford Pool
Lincoln LN6 7TS
UK

Email: j.chilton@lincoln.ac.uk

Dr R. Velasco
School of the Built Environment
University of Nottingham
University Park
Nottingham NG7 2RD
UK

Email:laxrv@nottingham.ac.uk

Chapter 13

Mr J. G. Ellis, OBE
Ellis Developments Limited
The Clocktower
Bestwood Village
Nottingham
NG6 8TQ
UK

Tel: 44 (0)115 979 7679
Email: julian.ellis@ellisdev.co.uk

Chapter 14

Dr K. Van de Velde
Ghent University
Department of Textiles
Technologiepark 907
B-9052 Zwijnaarde (Ghent)
Belgium

Email:
Kathleen.VandeVelde@UGent.be

Introduction

A C L O N G, University of Nottingham, UK

Textile composites are composed of textile reinforcements combined with a binding matrix (usually polymeric). This describes a large family of materials used for load-bearing applications within a number of industrial sectors. The term textile is used here to describe an interlaced structure consisting of yarns, although it also applies to fibres, filaments and yarns, and most products derived from them. Textile manufacturing processes have been developed over hundreds or even thousands of years. Modern machinery for processes such as weaving, knitting and braiding operates under automated control, and is capable of delivering high-quality materials at production rates of up to several hundreds of kilograms per hour. Some of these processes (notably braiding) can produce reinforcements directly in the shape of the final component. Hence such materials can provide an extremely attractive reinforcement medium for polymer composites.

Textile composites are attracting growing interest from both the academic community and from industry. This family of materials, at the centre of the cost and performance spectra, offers significant opportunities for new applications of polymer composites. Although the reasons for adopting a particular material can be various and complex, the primary driver for the use of textile reinforcements is undoubtedly cost. Textiles can be produced in large quantities at reasonable cost using modern, automated manufacturing techniques. While direct use of fibres or yarns might be cheaper in terms of materials costs, such materials are difficult to handle and to form into complex component shapes. Textile-based materials offer a good balance in terms of the cost of raw materials and ease of manufacture.

Target application areas for textile composites are primarily within the aerospace, marine, defence, land transportation, construction and power generation sectors. As an example, thermoset composites based on 2D braided preforms have been used by Dowty Propellers in the UK since 1987[1]. Here a polyurethane foam core is combined with glass and carbon fibre fabrics, with the whole assembly over-braided with carbon and glass tows. The resulting preform is then impregnated with a liquid thermosetting polymer via resin transfer moulding (RTM). Compared with conventional materials, the use of textile composites in this application results in reduced weight, cost savings (both initial cost and cost of ownership), damage tolerance and improved performance via the ability to optimise component shape. A number of structures for the Airbus A380 passenger aircraft rely on textile composites, including the six metre diameter dome-shaped pressure bulkhead and wing trailing edge panels, both manufactured by resin film infusion (RFI) with carbon non-crimp fabrics, wing stiffeners and spars made by RTM, the vertical tail plane spar by vacuum infusion (VI), and thermoplastic composite (glass/ poly (phenylene sulphide)) wing leading edges. Probably the largest components produced are for off-shore wind power generation, with turbine blades of up to 60 metres in length being produced using (typically) non-crimp glass or carbon fabric reinforcements impregnated via vacuum infusion. Other application areas include construction, for example in composite bridges which offer significant cost savings for installation due to their low weight. Membrane structures, such as that used for the critically acclaimed (in architectural terms) Millennium Dome at Greenwich, UK, are also a form of textile composite. Numerous automotive applications exist, primarily for niche or high-performance vehicles but also in impact structures such as woven glass/polypropylene bumper beams.

This book is intended for manufacturers of polymer composite components, end-users and designers, researchers in the fields of structural materials and technical textiles, and textile manufacturers. Indeed the latter group should provide an important audience for this book. It is intended that manufacturers of traditional textiles could use this book to investigate new areas and potential markets. While some attention is given to modelling of textile structures, composites manufacturing methods and subsequent component performance, this is intended to be substantially a practical book. So, chapters on modelling include material models and data of use to both researchers and manufacturers, along with case studies for real components. Chapters on manufacturing describe both current processing technologies and emerging areas, and give practical processing guidelines. Finally, applications from a broad range of areas are described, illustrating typical components in each area, associated design methodologies and interactions between processing and performance.

The term 'textile composites' is used often to describe a rather narrow range of materials, based on three-dimensional reinforcements produced using specialist equipment. Such materials are extremely interesting to researchers and manufacturers of very high performance components (e.g. space transportation); an excellent overview is provided by Miravette[2]. In this book the intention is to describe a broader range of polymer composite materials with textile reinforcements, from woven and non-crimp commodity fabrics to 3D textiles. However random fibre-based materials, such as short fibre mats and moulding compounds, are considered outside the scope of this book. Similarly nano-scale reinforcements are not covered, primarily because the majority of these are in short fibre or platelet forms, which are not at present processed using textile technologies.

The first chapter provides a comprehensive introduction to the range of textile structures available as reinforcements, and describes their manufacturing processes. Inevitably this requires the introduction of terminology related to textiles; a comprehensive description is given in the Glossary. Also described are modelling techniques to represent textile structures, which are becoming increasingly important for prediction of textile and composite properties for design purposes. Chapter 2 describes the mechanical properties of textiles, primarily in the context of formability for manufacture of 3D components. The primary deformation mechanisms, in-plane shear, in-plane extension and through-thickness compaction, are described in detail, along with modelling techniques to represent or predict material behaviour. Chapter 3 describes similar behaviour for pre-impregnated composites (often termed prepregs), focusing on their rheology to describe their behaviour during forming. Chapter 4 demonstrates how the behaviour described in the previous two chapters can be used to model forming of textile composite components. This includes a thorough description of the theory behind both commercial models and research tools, and a discussion of their validity for a number of materials and processes. Chapters 5 and 6 concentrate on manufacturing technologies for thermoset and thermoplastic composites respectively. Manufacturing processes are described in detail and their application to a range of components is discussed.

In Chapter 7, resin flow during liquid moulding processes (e.g. RTM) is discussed. This starts with a description of the process physics but rapidly progresses to an important area of current research, namely optimisation and control of resin flow during manufacturing. Chapter 8 describes the mechanical properties of textile composites, including elastic behaviour, initial failure and subsequent damage accumulation up to final failure. The first half of the chapter provides an excellent primer on the mechanics of composites in general, and shows how well-established theories can be adapted to represent textile composites. The second half on failure and impact builds upon this and concludes with a number of applications to demonstrate the state of the art. In Chapter 9 flammability is discussed – an important topic given the typical applications of textile composites and the flammability associated with most polymers. Chapter 10 introduces concepts associated with technical cost modelling, which is used to demonstrate interactions between the manufacturing process, production volume and component cost. Finally the last four chapters describe a number of applications from the aerospace, construction, sports and medical sectors.

References

1. McCarthy R.F.J., Haines G.H. and Newley R.A., 'Polymer composite applications to aerospace equipment', *Composites Manufacturing*, 1994 **5**(2) 83–93.
2. Miravette A. (editor), *3-D Textile Reinforcements in Composite Materials*, Woodhead Publishing Ltd, Cambridge, 2004.

1

Manufacturing and internal geometry of textiles

S L O M O V, I V E R P O E S T, Katholieke Universiteit
Leuren, Belgium and
F R O B I T A I L L E , University of Ottawa, Canada

1.1   Hierarchy of textile materials

Textiles technologies have evolved over millennia and the term 'textile' now has a very broad meaning. Originally reserved for woven fabrics, the term now applies to fibres, filaments and yarns, natural or synthetic, and most products derived from them. This includes threads, cords, ropes and braids; woven, knitted and non-woven fabrics; hosiery, knitwear and garments; household textiles, textile furnishing and upholstery; carpets and other fibre-based floor coverings; industrial textiles, geotextiles and medical textiles.

This definition introduces three important notions. First, it states that textiles are **fibrous** materials. A fibre is defined as textile raw material, generally characterised by flexibility, fineness and high ratio of length to thickness; this is usually greater than 100. The diameter of fibres used in textile reinforcements for composites (glass, carbon, aramid, polypropylene, flax, etc.) varies from 5 μm to 50 μm. Continuous fibres are called *filaments*. Fibres of finite length are called *short, discontinuous, staple* or *chopped* with lengths from a few millimetres to a few centimetres.

Fibres are assembled into *yarns* and fibrous plies, and then into textiles. The second important feature of textiles is their **hierarchical** nature. One can distinguish three hierarchical levels and associated scales: (1) fibres at the *microscopic* scale; (2) yarns, repeating unit cells and plies at the *mesoscopic* scale; and (3) fabrics at the *macroscopic* scale. Each scale is characterised by a characteristic length, say 0.01 mm for fibre diameters, 0.5–10 mm for yarn diameters and repeating unit cells, and 1–10 m and above for textiles and textile structures. Each level is also characterised by dimensionality where fibres and yarns are mostly one-dimensional while fabrics are two- or three-dimensional, and by structural organisation where fibres are twisted into yarns, yarns are woven into textiles, etc.

Textiles are **structured** materials. On a given hierarchical level one can think of a textile object as an entity and make abstraction of its internal structure: a yarn may be represented as a flexible rod, or a woven fabric as

1

a membrane. This approach is useful but the internal structure must be considered if one wishes to assess basic features and behaviour of textile objects such as the transverse compression of yarns or shear behaviour of fabrics. The diversity of textile technologies results in a large variety of available textile structures. Figure 1.1 depicts textile structures that are widely used as reinforcements for composites; these are discussed in this chapter.

The properties of a fabric are the properties of fibres transformed by the textile structure. The latter is introduced deliberately during manufacturing. Modern fibres turn millennium-old textile technologies into powerful tools for creating materials designed for specific purposes, where fibre positions are optimised for each application. Textile manufacturing methods and internal structure are two important topics that are addressed in this chapter.

1.2   Textile yarns

1.2.1   Classification

The term yarn embraces a wide range of 1D fibrous objects. A yarn has substantial length and relatively small cross-section and is made of fibres and/or filaments, with or without twist. Yarns containing only one fibre are *monofilaments*. Untwisted, thick yarns are termed *tows*. Flat tows are called *rovings* in composite parlance; in textile technology this word designates an intermediate product in spinning. *Sizing* holds the fibres together and facilitates the processing of tows; it also promotes adhesion of fibres to resin in composites. In *twisted yarns*, fibres are consolidated by the friction resulting from twist. A twist is introduced to a *continuous filament* yarn by *twisting*. For a *twisted yarn made of staple fibres* the process is called *spinning* and involves a long chain of preparatory operations. There are different *yarn spinning* processes (*ring spinning, open-end spinning, friction spinning*) leading to yarns with different internal distributions of fibres. Note that these are distinct from fibre spinning processes such as wet spinning, melt spinning or gel spinning, which are used to make individual fibres, most often from various polymers.

Fibres of different types are easily mixed when yarn spun, producing a *blend*; thermoplastic matrix fibres can be introduced among load-carrying fibres in this way. Finally, several strand yarns can be twisted together, forming a *ply* yarn.

1.2.2   Linear density, twist, dimensions and fibrous structure of yarns

The *linear density* is the mass of a yarn per unit length; the inverse quantity is called *yarn count* or *number*. Common units for linear density are given in Table 1.1. Linear density, the most important parameter of a yarn, is normalised

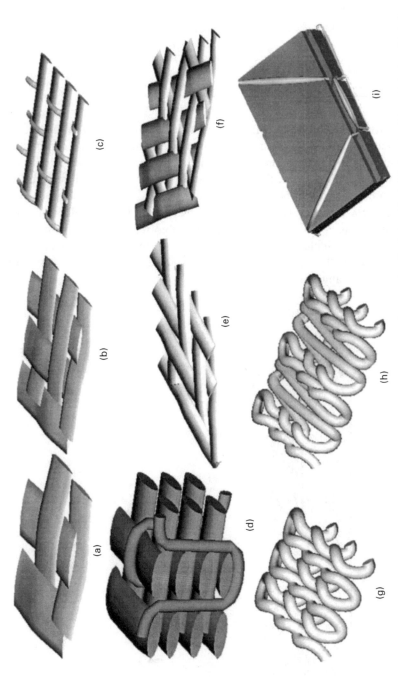

1.1 Textile structures: (a–c) 2D woven fabrics; (d) 3D woven fabrics; (e, f) 2D braided fabrics; (g, h) weft-knitted fabrics; (i) multiaxial multiply warp-knitted fabric.

Table 1.1 Units of linear density and yarn count

Unit	Definition
Tex (SI)	1 tex = 1 g/km
Denier	1 den = 1 g/9 km
Metric number	1 Nm = 1 km/kg
Glass (UK and US)	N_G = 1 pound/1 yard

by specifications and controlled in manufacturing. The *unevenness*, or coefficient of variation of linear density, is normally below 5% for continuous filament yarns while for spun yarns special measures in manufacturing keep it within a specified range. For yarns made of short and stiff fibres such as flax the unevenness can be as high as 15–20%.

The *twist K* of a yarn is a number of turns per unit length. Values of twist for yarns used in composite reinforcements are normally below 100 m^{-1}. The *twist angle* is the inclination of fibres on the surface of the yarn due to twist (Fig. 1.2). If *d* is the yarn diameter and *h* = 1/K is the length of the period of twist, the twist angle is calculated as:

$$\tan \beta = \frac{\pi d}{h} = \pi d K \qquad 1.1$$

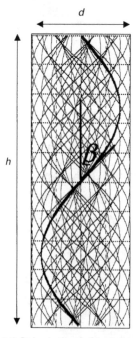

1.2 Calculation of twist angle.

The twist angle is indicative of the intensity of frictional interaction between fibres inside a yarn, and of the tensile resistance of a staple yarn. It provides an estimation of the deviation of the orientation of fibres from the yarn axis and ideal load direction in the case of textile reinforcements for composites; the effective modulus of an impregnated yarn can be estimated as proportional to $\cos^4\beta$. For example, a yarn of diameter $d = 0.5$ mm and twist $K = 100$ m^{-1} would lead to $\beta = 9°$ and $\cos^4\beta = 0.95$, resulting in a decrease in stiffness of 5% for the impregnated yarn.

Fibrous yarns do not have precise boundaries; their dimensions are arbitrary. Considering a circular yarn with similar dimensions along all radii, the diameter of a cylinder having the same average density as that yarn is:

$$d = \sqrt{\frac{4T}{\pi \rho_f V_f}}$$ 1.2

where T is the linear density of the yarn, ρ_f is the fibre density and V_f is the fibre volume fraction in the yarn (which is generally unknown). In practice, eqn 1.2 is rewritten:

$$d = C\sqrt{T}$$ 1.3

where an empirical coefficient C applies to each yarn type. Typical values of C correspond to $V_f = 0.4$ to 0.6. Measuring d presents difficulties with values depending on the pressure applied to the yarn and on image processing of yarn cross-sections. These can be avoided by extrapolating the results of transverse yarn compression tests to zero load. Remarkably, values of C obtained empirically for very different yarns (cotton, wool, polyester, aramid, glass, etc.) all lie in the range 0.03–0.04 (d in mm and T in tex), which is useful for first estimations.

Rovings and tows used in composites are normally sized, leading to better-defined surfaces without loose fibres. Roving cross-sections can be approximated by elliptical, lenticular or various other shapes with a width-to-thickness ratio of 3 to 10. Actual dimensions depend on manufacture-induced spreading; however, relationships exist between linear density and section dimensions (Fig. 1.3). Fibres are positioned randomly within cross-sections. Clusters separated by random voids are formed, and these remain present upon compaction. Fibres do not align in theoretically ideal hexagonal arrays because of fibre crimp[1].

1.2.3   Mechanical properties of yarns

During textile manufacture, yarns are subjected to different loads and constraints that define their final configuration in the textile. The behaviour of yarns under load is non-linear and non-reversible owing to their fibrous nature and inter-fibre friction. The most important yarn deformation modes in determining

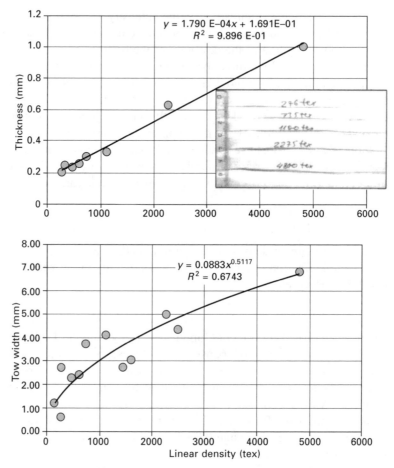

1.3 Thickness and width of glass rovings (from different manufacturers).

the internal geometry of a fabric are bending, which allows interlacing of yarns and generates yarn interaction forces in the fabric, and transverse compression which results from these interactive forces and defines the shape of yarns in the fabric. Tension and torsion are generally unimportant for unloaded fabrics.

Bending of textile yarns

Yarn bending is characterised by a bending diagram $M(\kappa)$ where M is the bending moment and κ is the curvature, with $\kappa = 1/R$ where R is the radius of curvature. Typical bending diagrams registered on the Kawabata (KES-F) bending tester[2] appear in Fig. 1.4 for a sized carbon tow; the lower cycle was

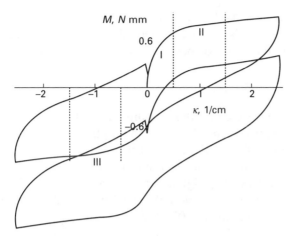

1.4 Bending diagrams for carbon roving (3.2 ktex).

shifted vertically for clarity. The initial stage of bending (stage I – κ lower than 0.2 cm^{-1}, R higher than 5 cm) features high stiffness values associated to fibres linked by sizing and friction[3]; the yarn acts as a solid rod. Curvature radii in fabrics are of the order of a few millimetres. hence stage I is not relevant to their internal structure: standard practice stipulates that bending rigidity should be determined between $\kappa = 0.5$ and $\kappa = 1.5$ cm^{-1}. The slope over stages II and III is averaged and bending rigidity is calculated as:

$$B = \frac{\Delta M}{\Delta \kappa} \qquad\qquad 1.4$$

The bending diagram also provides information on the hysteresis, typically defined as the difference between values of the bending moment at $\kappa = 1$ cm^{-1} upon loading and unloading; this is rarely used in practice but helps in understanding the physics of the phenomenon: clearly the behaviour is not elastic. There are many simple test methods for determining B using the deformation of the yarn under its own weight[4,5]. These are easier to use but their application to thick, rigid tows used in composites is limited as small deformations lead to large errors.

The bending rigidity of a yarn can be estimated from the bending rigidity of single fibres. Each fibre in a yarn containing N_f circular fibres of diameter d_f and modulus E_f has a bending rigidity of:

$$B_f = \frac{E_f d_f^4}{64} \qquad\qquad 1.5$$

Fibres in a yarn are held together by friction and/or bonding. In one extreme the yarn can be regarded as a solid rod, while in the other extreme each fibre bends independently. Lower and upper estimates of the yarn bending rigidity can be derived as $N_f B_f \leq B \leq N_f^2 B_f$; the difference between these two extremes is great as N_f is counted in hundreds or thousands. Experiments show that for rovings typical of composite reinforcements the lower estimation is valid:

$$B \approx N_f B_f \qquad\qquad 1.6$$

Figure 1.5 shows measured bending rigidity values for glass and carbon rovings and estimates from eqn 1.6. The trend for glass rovings is not linear as fibre diameters differ (13 to 21 μm) for different yarns. The carbon tows are all made of fibres with a diameter of 6 μm. The difference in the fibre diameters explains why bending rigidity of the glass rovings is higher. The outlier point in Fig. 1.5, located below the trend line, results from internal fibre crimp in an extremely thick (80K) tow.

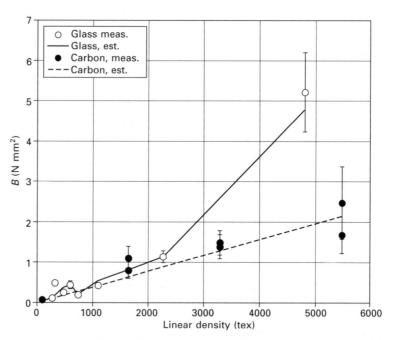

1.5 Bending rigidity of glass and carbon rovings.

Compression of textile yarns

When compressed a textile yarn becomes thinner in the direction of the force and wider in the other direction, Fig. 1.6. Changes in yarn dimensions can be expressed using coefficients of compression η:

$$\eta_1(Q) = \frac{d_1(Q)}{d_{1,0}}, \quad \eta_2(Q) = \frac{d_2(Q)}{d_{2,0}} \qquad\qquad 1.7$$

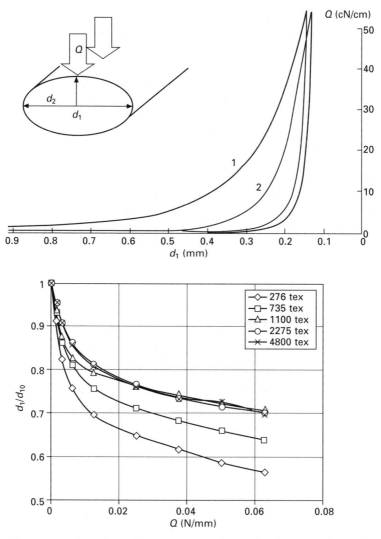

1.6 Compression of textile yarns. Top: schematic of test configuration and typical KES-F curves (two loading cycles). Bottom: compressibility of glass rovings.

where d_1 and d_2 are the yarn dimensions, Q is the load, $\eta_1 < 1$, $\eta_2 > 1$ and subscript 0 refers to the shape at $Q = 0$. The load Q is defined as a force per unit length, which is simpler to use than a pressure because of changes in yarn dimensions.

Measuring η_1 is straightforward. The Kawabata compression tester[2] is the standard tool for doing this. As shown in Fig. 1.6 the first and second compression cycles differ substantially but subsequent cycles are similar to the second. The first cycle measures forces associated to factors such as the imperfect flatness of relaxed rovings, which disappear in subsequent cycles. Latter cycles should be regarded as characteristic of yarn behaviour. Figure 1.6 also shows compression curves for glass rovings of different linear density, to be compared with zero-load thickness data in Fig. 1.3. Non-linearity is evident from Fig. 1.6. The curves may be approximated by a power law:

$$\frac{1 - \eta_1(Q)}{\eta_1(Q) - \eta_{1min}} = \left(\frac{Q}{Q^*}\right)^\alpha \qquad\qquad 1.8$$

where η_{1min} is the maximum compressibility of the yarn ($Q \to \infty$). The curve for the 276 tex glass roving can be approximated with $\eta_{1min} = 0.398$, $Q^* = 0.014$ N/mm and $\alpha = 0.612$. Measuring η_2 is a different issue for which no standard procedure exists. Published techniques[6–8] involve observing the dimensions of a yarn compressed between two transparent plates. Experimental results show that:

$$\eta_2 \approx \eta_1^{-a}, a = 0.2 \text{ to } 0.3 \qquad\qquad 1.9$$

1.3   Woven fabrics

1.3.1   Parameters and manufacturing of woven fabric

A woven fabric is produced by interlacing *warp* and *weft* yarns, identified as *ends* and *picks*. It is characterised by linear densities of warp and weft yarns, a weave pattern, a number of warp yarns per unit width P_{Wa} (ends count, inverse to warp pitch or centre-line spacing p_{Wa}), a number of weft yarns per unit length P_{We} (picks count, inverse to weft spacing p_{We}), warp and weft yarn crimp, and surface density.

Figure 1.7 represents a weaving loom. Warp yarns are wound on the warp beam, rolling off it parallel to one another under regulated tension. Each warp yarn is linked to a *harness*, a frame positioned across the loom with a set of *heddles* mounted on it. The heddle is a wire or thin plate with an eye through which a warp yarn goes. When a harness goes up or down, the warp yarns connected to it go up (warp 2) or down (warp 1). A *shed* is formed, which is a gap between warp yarns where a weft yarn may be inserted using a weft insertion device such as a shuttle, projectile, air/water jet or rapier. The weft yarn is positioned by battening from a *reed*, a grid of steel plates between which warp yarns extend. The reed is mounted on a slay which moves back and forth; the backward motion opens space for weft insertion while the forward motion battens the weft. Once a weft yarn is in position the harnesses move in opposite directions, closing the shed and locking the weft in the fabric. The fabric is moved forward by the cloth beam and the process is repeated (Fig. 1.7). In the final fabric roll the warp ends extend

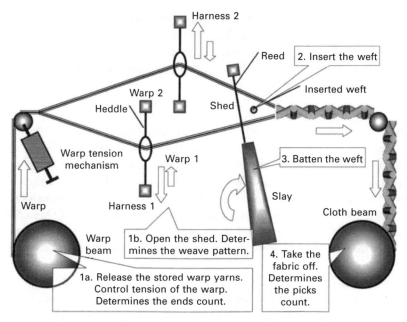

Harness 2

Reed 2. Insert the weft

Warp 2

Inserted weft

Heddle

Shed

Warp tension
mechanism

Warp 1

3. Batten the weft

Warp

Harness 1

Slay

Cloth beam

Warp
beam

1b. Open the shed. Deter-
mines the weave pattern.

4. Take the
fabric off.
Determines
the picks
count.

1a. Release the stored warp yarns.
Control tension of the warp.
Determines the ends count.

1.7 Schematic diagram of a weaving loom.

along the roll direction while the weft picks extend parallel to the roll axis. Lateral extremities are sometimes referred to as the selvedge.

The ends count, picks count and weave pattern are respectively determined by the total number of warp yarns on the beam and the fabric width, by the rotation speed of the cloth beam, and by the sequence of connection between warp yarns and harnesses and the harness motions. All motions happen within one rotation of the main shaft. The rotation speed of this shaft determines the number of weft yarns inserted per minute. Modern looms rotate at 150 to 1000 rpm. The loom speed determines the productivity A, the area of fabric produced in a given time:

$$A = \frac{nb}{P_{We}}$$ 1.10

where n and b are the loom speed and fabric width. Typical values for a composite reinforcement produced on a loom with a projectile weft insertion are $n = 300$ min^{-1}, $P_{We} = 4.0$ cm^{-1}, $b = 1.5$ m. Productivity here is then 67.5 m^2/h or, assuming a surface density of 300 g/m^2, 20 kg/h.

Weft insertion

Weaving looms are classified according to weft insertion devices: shuttle, projectile ('shuttleless'), rapier, air or water jet. Air jet insertion is unsuitable for thick glass and carbon tows. Weft insertion speeds are typically between 7 and 35 m/s, depending on the insertion technique employed. These relate to the productivity of different looms. If weft insertion takes a fraction α of the weaving cycle (typically $\alpha = 1/4$ to $1/3$), loom speed can be estimated as:

$$n = \frac{\alpha v}{b} \qquad\qquad 1.11$$

where v is the average velocity of the weft insertion.

A *shuttle* is a device shut from side to side of the loom. It carries a spool with a weft thread that is unwound during insertion. Automatic looms work on the same principle as shuttle hand looms. In order to carry enough weft yarn, the shuttle must be large. The shed must open wide with considerable warp tension to prevent it from tearing. Shuttles with masses of about 500 g prevent velocities above 10 m/s. Noisy shuttle looms consume much energy but can insert any type of weft yarn with ease, including heavy tows. They are widely used to produce composite reinforcements.

Shuttleless looms use a projectile instead of a shuttle. The projectile does not carry a spool and can be much lighter, about 50 g. A weft yarn is fixed to the projectile and unwound from the bobbin during insertion. The projectile is kept on target by a grid fixed to a reed, preventing contact with warp yarns. This allows smaller shed opening, while the lighter projectile leads to high insertion speeds (up to 20 m/s). Shuttleless looms can carry heavy yarns up to a certain limit.

Air looms use an air jet as the weft carrier. A length of weft yarn is cut off and fed into an air nozzle. An air jet carries the yarn across the loom in a tunnel attached to a reed. The tunnel helps maintaining the jet speed. On wide looms (above 1 m) additional nozzles accelerate the jet along the weft yarn path. High velocity of insertion (up to 35 m/s) ensures maximum production speed but weft linear densities are limited to 100 tex, restricting usage of these looms for composite reinforcements.

Water looms use a water jet for weft yarn insertion. This allows handling of heavier yarns, including thick tows for composite reinforcements, but excludes moisture-sensitive fibres such as aramids.

Rapier looms insert weft yarns using mechanical carriers. Yarns are stored on a bobbin and connected to the left rapier, which moves to the centre. The right rapier does the same from the other side. When they meet, the yarn is transferred mechanically from one rapier to the other and cut off before a new weaving cycle is started. Rapier looms are relatively slow (~ 7 m/s) but have no restrictions in weft thickness, and consume less energy than shuttle or projectile looms.

Shedding mechanisms

Shedding mechanisms lift and bring down warp yarns in a prescribed order, creating sheds. Groups of warp yarns are lifted and lowered by harnesses; their number determines weave complexity, with the maximum usually being 8 or 16 but sometimes as many as 32. The motion of harnesses can be effected by cams or dobby. Alternatively the motion of each warp yarn can be controlled separately, allowing any weave pattern. Such shedding mechanisms are called Jacquards.

Consider a shedding mechanism with harnesses. Figure 1.8 illustrates the way to weave a desired pattern. The pattern shows crossovers of warp (columns) and weft (rows) yarns. In the *paper-point diagram*, squares corresponding to crossovers where the warp is on top of the weft appear in black, otherwise they are white. *Top* (face) and *bottom* (back) refer to fabric on the loom. Circles in the *loom up order* diagram indicate which harness controls which warp yarns, and crosses in the shedding diagram indicate the lifting sequence of the harnesses.

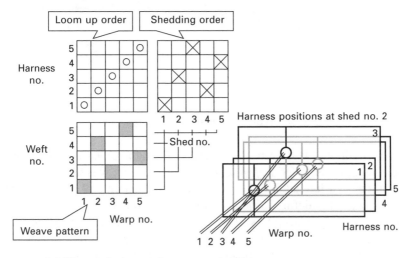

1.8 Weave design and movement of harness.

In *cam shedding mechanisms*, each cam controls the motion of one harness. To change a weave pattern one must change the cams. *Dobby shedding mechanisms* control harnesses through more complex electronic programs. They are used when the number of cams on a loom, usually eight, is insufficient for a chosen weave pattern. *Jacquard machines* control the motions of warp yarns independently. Warp yarns go through the heddles which are connected with the Jacquard machine by harness cords, lifting and lowering the heddles according to a program. Such machines are used to make 3D reinforcements.

1.3.2   Weave patterns

Figure 1.9 depicts elements of a woven pattern used for weave classification. The pattern is represented with a paper-point diagram, with black squares corresponding to crossovers where the warp yarn is on top. The minimum repetitive element of the pattern is called a *repeat*. The repeat can have a different number of warp (N_{Wa}) and weft (N_{We}) yarns.

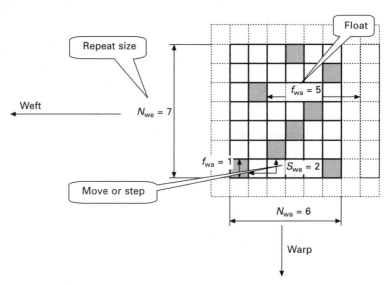

1.9 Elements of a typical weave.

   The length of a weft yarn on the face of the fabric, measured in number of intersections, is called *weft float* (f_{We}). For common weaves it is equal to the number of white squares between black squares on any weft yarn. The *warp float* (f_{Wa}) is defined similarly. Consider two adjacent weft yarns and two neighbouring warp intersections on them. The distance between them, measured in number of squares, is called *move* or *step* (s). This characterises the shift of the weaving pattern between two weft insertions. Different yarns in the pattern shown in Fig. 1.9 have different floats and steps. In common weaves these parameters take uniform and regular values, as described below.

Fundamental weaves

The family of weaves having the most regular structures and used as basis for other weaves is called *fundamental weaves*. They are characterised by a square repeat with $N_{Wa} = N_{We} = N$. Each warp/weft yarn has only one weft/warp crossing with $f = 1$ for warp/weft, and the pattern of adjacent yarns is regularly shifted with s being a constant.

The *plain weave*, Fig. 1.10, is the simplest fundamental weave with repeat size $N = 2$ and step $s = 1$. Each weft yarn interlaces with each warp yarn. *Twill weaves*, Fig. 1.11, are characterised by $N > 2$ and $s = \pm 1$. Constant shift by one position creates typical diagonals on the fabric face. If the step is positive, diagonals go from the lower left to the upper right of the pattern and a *right* or *Z-twill* is formed. If the step is negative, the twill is called *left* or *S-twill*. Different twill patterns are designated as f_{Wa}/f_{We} where f_{Wa} is the number of warp intersections on a yarn (warp float) and f_{We} is the number of weft intersections on a yarn (weft float). If there are more warp intersections than weft ones on the fabric face, the twill is a *warp twill*. The term *weft twill*

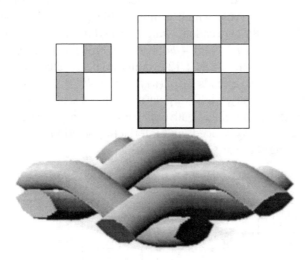

1.10 Plain weave.

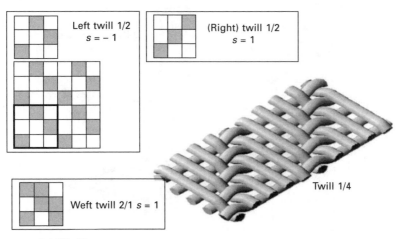

1.11 Twill weaves.

for the opposite case is normally omitted. A full designation for a twill weave reads *(right)|left* or *warp|(weft) twill* f_{Wa}/f_{We} where terms in brackets can be omitted and vertical lines mean 'or'. Twill weaves are often characterised by a propensity to accommodate in-plane shear, resulting in good drape capability (see section 2.2).

Satin weaves, Fig. 1.12, are characterised by $N > 5$ and $| s | > 1$. Sparse positioning of the interlacing creates weaves in which the binding places are arranged with a view to producing a smooth fabric surface devoid of twill lines, or diagonal configurations of crossovers. The terms *warp satin*, or simply satin, and *weft satin*, or sateen, are defined similarly as for twills. N and s cannot have common denominators, otherwise some warp yarns would not interlace with weft yarns, which is impossible (Fig. 1.12). The most common satins (5/2 and 8/3) are called *5-harness* and *8-harness*. In composites literature the terms *3-harness* and *4-harness* (or *crowfoot*) can be found. Such satins are actually twills (Fig. 1.12). Satins are designated as *N/s*; a full designation of a satin weave reads *warp|(weft) satin N/s*, or *N harness* and *s step satin|sateen*. Rigorously, each warp yarn interlaces with each weft yarn only once, each weft yarn interlaces with each warp yarn only once, interlacing positions must be regularly spaced, and interlacing positions can never be adjacent – both along the warp and weft.

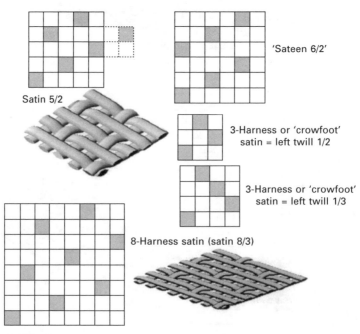

Satin 5/2

'Sateen 6/2'

3-Harness or 'crowfoot' satin = left twill 1/2

3-Harness or 'crowfoot' satin = left twill 1/3

8-Harness satin (satin 8/3)

1.12 Satin weaves.

Modified and complex weaves

Fundamental weaves can serve as a starting point for creating complex patterns, to create some design effect or ensure certain mechanical properties for the fabric. Figure 1.13 shows modified weaves created by doubling warp intersections in plain (resulting in *rib* and *basket* weaves) and twill patterns. Such weaves are identified by a fraction f_{Wa}/f_{We}, where f_{Wa} is the number of warp intersections on a warp yarn and f_{We} is the number of weft intersections on a yarn (Fig. 1.13a–c). Complex twills can have diagonals of different width identified by a sequence $f_{Wa1}/f_{We1}, \ldots f_{WaK}/f_{WeK}$, (Fig. 1.13d). A twill pattern can also be broken by changing diagonal directions (sign of s), creating a *herringbone* weave (Fig. 1.13e). If an effect of apparent random interlacing is desired the repeat can be disguised (*crepe* weave) by combining different weaves in one pattern (Fig. 1.13f).

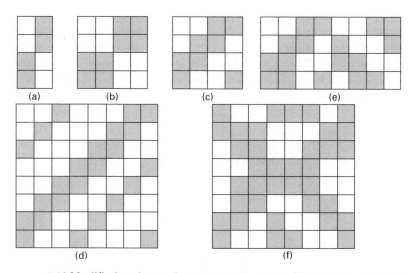

1.13 Modified and complex weaves: (a) warp rib 2/2; (b) basket 2/2; (c) twill 2/2; (d) twill 2/3/1/2; (e) herringbone; (f) crepe (satin 8/3 + twill 2/2).

Tightness of a 2D weave

The *weave tightness* or *connectivity* is determined by the weave pattern and quantifies the freedom of yarns to move. This is not to be confused with fabric tightness, which is the ratio of yarn spacing to their dimensions. The weave tightness characterises the weave pattern, providing an indication on fabric properties as a function of weave type. Two simple ways to characterise weave tightness follow:

$$\text{Tightness} = \frac{N_{\text{liaison}}}{2N_{Wa}N_{We}} \qquad 1.12$$

where N_{liaison} is the number of transitions of warp/weft yarns from one side of the fabric to another and the denominator is multiplied by 2 so as to have a value of 1 for plain weaves, and:

$$\text{Tightness} = \frac{2}{f_{\text{Wa}} + f_{\text{We}}} \qquad\qquad 1.13$$

Equation 1.12 will be used in the remainder of the chapter. Lower weave tightness values indicate less fixation of the yarns in a fabric, less fabric stability and better fabric drapability. As the fabric is less stable it tends to be distorted easily as shown by differences in fabric resistance to needle penetration (Fig. 1.14a). Higher tightness means also higher crimp, which deteriorates fabric strength (Fig. 1.14b).

Multilayered weaves

Conventional weaving looms allow the production of multilayered weaves used for heavy apparel and footwear. Multilayer integrally woven reinforcements are often called *3D* or *warp-interlaced 3D* weaves. A multilayer weave is shown in Fig. 1.15. The weave is called *1.5-layered satin* as the pattern is similar to satin on the fabric face and the fabric has two weft layers and one warp layer – 1.5 being an average. The weave pattern is more complex. Positions where warp yarns appear at the fabric face above the upper weft layer (Arabic figures) are black. Positions between the upper and the weft layers lower (Roman figures) appear as crosses and positions at the back of the fabric appear in white.

The weaving pattern does not reveal the weave structure clearly. The spatial positioning of yarns is created by stopping the fabric upon insertion of a lower weft and only resuming after insertion of an upper weft, hence inserting two wefts at the same lengthwise position in the fabric. The spatial weave structure is better revealed by a section in the warp direction, (Fig. 1.15b). A fabric with L weft layers can have warps occupying $L + 1$ levels, level 0 being the fabric face and level L being the back. Each warp is coded as a sequence of level codes and the entire weave is coded as a matrix.

In composite reinforcements warps are often layered as shown in Fig. 1.16. Warp paths are coded as *warp zones*, identifying sets of warp yarns layered over each other. The 1.5-layered satin of Fig. 1.15 has four warp zones, each occupied by one yarn. Yarns going through the thickness of a fabric are called *Z-yarns*. Multilayered composite reinforcements are termed *orthogonal* when Z-yarns go through the whole fabric between only two columns of weft yarns (Fig. 1.16a), *through-thickness angle interlock* when Z-yarns go through the whole fabric across more than two columns of weft yarns (Fig. 1.16b) or *angle interlock* when Z-yarns connect separate layers of the fabric (Fig. 1.16c).

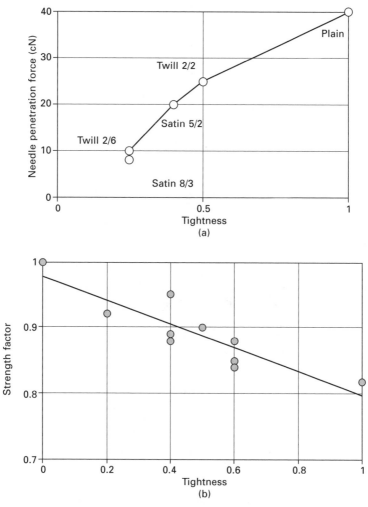

1.14 Influence of fabric tightness on the fabric properties:
(a) resistance of cotton fabrics (warp/weft 34/50 tex, 24/28 yarns/cm)
to needle penetration; (b) strength factor of yarns in aramid fabrics
(ratio of the strength of the fabric per yarn to the strength of the
yarn).

Matrix coding of weaves allows the analysis of their topology. Consider
for example the front warp yarn in Fig. 1.16b with level codes $\{w_i\} = \{0, 2,
4, 2\}$. This coding means that the yarn path goes from the top of the fabric
(level 0) on the left, to beneath the second weft (level 2), to the bottom of the
fabric (level 4), then back to level 2 before the pattern repeats. Consider
crimp intervals, which are yarn segments extending between two crossovers.
Over the first crimp interval the yarn interacts with weft yarns in layers

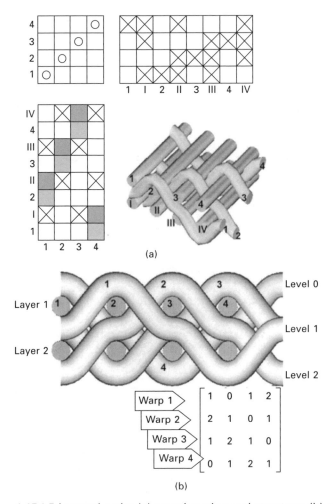

(a)

(b)

1.15 1.5-Layered satin: (a) weaving plan and structure; (b) coding.

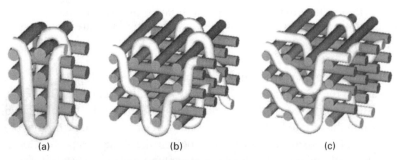

(a)   (b)   (c)

1.16 Types of multilayered weaves: (a) orthogonal; (b) through-the-thickness angle interlock; (c) angle interlock.

$l_1^1 = 1$ and $l_1^2 = 2$, where the subscript and superscript identify the crimp interval and each of its ends. The yarn is above its supporting weft at the left end (1) of the crimp interval ($P_1^1 = 1$) and below its supporting weft at the right (2) end ($P_1^2 = -1$). For all the crimp intervals of this yarn:

$$l_1^1 = 1, l_1^2 = 2, P_1^1 = +1, P_1^2 = -1$$
$$l_2^1 = 3, l_2^2 = 4, P_2^1 = +1, P_2^2 = -1$$
$$l_3^1 = 4, l_1^2 = 3, P_1^1 = -1, P_1^2 = +1$$
$$l_4^1 = 2, l_1^2 = 1, P_1^1 = -1, P_1^2 = +1$$

1.14

These values can be obtained from the intersection codes $\{w_i\}$, using the following algorithm:

$$w_i = w_{i+1} \Rightarrow l_i^1 = l_i^2 = \min(w_i + 1, L), P_i^1 = P_i^2 = \begin{cases} +1, w_i < L \\ -1, w_i = L \end{cases}$$
$$w_i < w_{i+1} \Rightarrow l_i^1 = w_i, l_i^2 = w_{i+1} + 1, P_i^1 = +1, P_i^2 = -1$$
$$w_i > w_{i+1} \Rightarrow l_i^1 = w_i + 1, l_i^2 = w_{i+1}, P_i^1 = -1, P_i^2 = +1$$

1.15

Similar descriptions of weft yarns are obtained from intersection codes and crimp interval parameters for warp yarns. For a weft yarn i at layer l, the first warp yarn with $l_i^1 = l$ or $l_i^2 = l$ (supported by weft yarn i at layer l) is found from the crimp interval parameters list; this is the left end of the first crimp interval on the weft yarn. The support warp number is thus found, with a weft position sign inverse to that of the warp. Then the next warp yarn supported by weft (i, l) is found; this is the right end of the first weft crimp interval and the left end of the second. This is then continued for all crimp intervals.

1.3.3   Geometry of yarn crimp

The topology and waviness of interlacing yarns are set by the weave pattern. The waviness is called *crimp*. The term also characterises the ratio of the length of a yarn to its projected length in the fabric:

$$c = \frac{l_{\text{yarn}}}{l_{\text{yarn}} - l_{\text{fabric}}}$$

1.16

Crimp is caused primarily by out-of-plane waviness; reinforcements do not feature significant in-plane waviness because of the flat nature of their rovings. Typical crimp values range from less than 1% for woven rovings to values of well over 100% for warp-interlaced multilayered fabrics. The wavy shape of a yarn in a weave may be divided in intervals of crimp, say between intersections A and B; Fig. 1.17. Considering a warp yarn, let p and h be distances

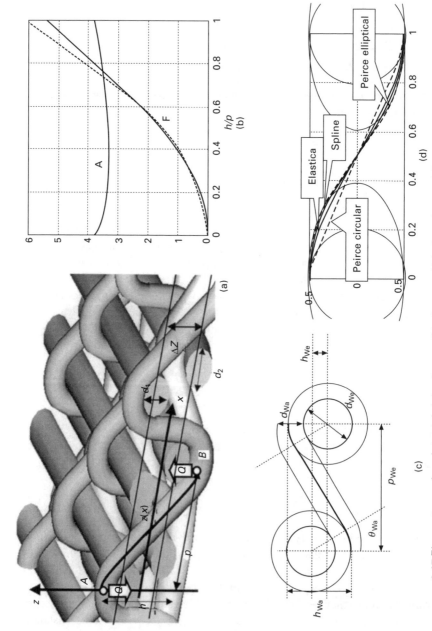

1.17 Elementary crimp interval: (a) scheme; (b) characteristic functions; (c) Peirce's model; (d) comparison between various models.

between points A and B along the warp direction x and thickness z, with p defined as the yarn spacing. The distance h is called the crimp height. If this is known one may wish to describe the middle line of the yarn over the crimp interval:

$$z(x) : z(0) = h/2; \; z'(0) = 0; \; z(p) = -h/2; \; z'(p) = 0 \qquad 1.17$$

The actual shapes of warp and weft yarn cross-sections can be very complex, even if dimensions d_1 and d_2 known are known (subscripts 'Wa' and 'We', Fig. 1.17c). Two simplified cases will be considered here. In the first, cross-section shapes are defined by the curved shape of interlacing yarns. This is acceptable when fibres move easily in the yarns, for example with untwisted continuous fibres. In the second case cross-section shapes are fixed and define the curved shape of interlacing yarn. This is representative of monofilaments or consolidated yarns with high twist and heavy sizing.

Elastica model

In the first approach one can consider yarn crimp in isolation, find an elastic line satisfying boundaries represented by eqn 1.17 and minimise the bending energy:

$$W = \frac{1}{2} \int_0^p B(\kappa) \frac{(z'')^2}{(1 + (z')^2)^{5/2}} \, dx \to \min \qquad 1.18$$

where $B(\kappa)$ is the yarn bending rigidity which depends on local curvature. A further simplification using an average bending rigidity leads to the well-known problem of the elastica[9] for which a solution can be written using elliptical integrals. Calculations are made easier with an approximation of the exact solution:

$$\frac{z}{h} = \underline{\frac{1}{2}(4\bar{x}^3 - 6\bar{x}^2 + 1)} - A\left(\frac{h}{p}\right)\bar{x}^2(\bar{x} - 1)^2\left(\bar{x} - \frac{1}{2}\right), \; \bar{x} = \frac{x}{p} \qquad 1.19$$

where function $A(h/p)$ is shown in Fig. 1.17b; the value $A = 3.5$ provides a good approximation in the range $0 < h/p < 1$. The first, underlined term of eqn 1.19 is the solution to the linear problem. This cubic spline very closely approximates the yarn line.

As the yarn shape defined by eqn 1.19 is parameterised with the dimensionless parameter h/p, all properties associated with the bent yarn centreline can be written as a function of this parameter only. This allows the introduction of a *characteristic function F* for the crimp interval, used to define the bending energy of the yarn W, transversal forces at the ends of the interval Q and average curvature κ, (Fig. 1.17b):

$$W = \frac{1}{2} B(\kappa) \int_0^p \frac{(z'')^2}{[1 + (z')^2]^{5/2}} \, dx = \frac{B(\kappa)}{p} F\left(\frac{h}{p}\right) \qquad 1.20$$

$$Q = \frac{2W}{h} = \frac{2B(\kappa)}{ph} F\left(\frac{h}{p}\right) \qquad 1.21$$

$$\bar{\kappa} = \sqrt{\frac{1}{p} \int_0^p \frac{(z'')^2}{[1 + (z')^2]^{5/2}} \, dx} = \frac{1}{p} F\left(\frac{h}{p}\right) \qquad 1.22$$

If a spline approximation of eqn 1.19 and linear approximation of eqn 1.20 are used,

$$W = \frac{1}{2} B(\kappa) \int_0^p (z'')^2 \, dx = 6 B(\kappa) \frac{h^2}{p^3} \qquad 1.23$$

and $F \approx 6(h/p)^2$ (dotted line, Fig. 1.17b). The difference with the exact $F(h/p)$ is negligible.

Peirce type models

In the second approach, yarn cross-sections take set shapes. The simplest model for this case, proposed by Peirce[10] and still widely used, is based on two assumptions: weft cross-sections are regarded as circular when considering crimp intervals for warp yarns, and warp yarns are straight when not in contact with the weft (Fig. 1.17c). The vertical distance between centres of the weft yarns is the weft crimp height h_{We}. As the warp and weft yarns are in contact the following geometric constraint holds:

$$h_{Wa} + h_{We} = d_{Wa} + d_{We} = D \qquad 1.24$$

The yarn spacing p, yarn length l and contact angle θ must satisfy the following relations:

$$p_{We} = (l_{Wa} - D\theta_{Wa})\cos \theta_{Wa} + D \sin \theta_{Wa}$$
$$h_{Wa} = (l_{Wa} - D\theta_{Wa})\sin \theta_{Wa} + D(1 - \cos \theta_{Wa}) \qquad 1.25$$

The same equations may be written for weft yarns:

$$p_{We} = (l_{We} - D\theta_{We})\cos \theta_{We} + D \sin \theta_{We}$$
$$h_{We} = (l_{We} - D\theta_{We})\sin \theta_{We} + D(1 - \cos \theta_{We}) \qquad 1.26$$

The system 1.24–1.26 providess five equations for six unknowns h_{Wa}, h_{We}, l_{Wa}, l_{We}, θ_{Wa} and θ_{We}. If one of the crimp heights is given or some relation between them is assumed then all parameters can be determined. Peirce's model can represent non-circular cross-sections in a similar manner.

Mixed model

The two above models have their advantages and limitations. Figure 1.17(d) compares yarn mid-line shapes obtained from the different models. Differences are small, justifying a mixed model where average characteristics of crimp intervals are obtained from (1.20–1.22), taking advantage of the single parameter h/p involved in these equations, while yarn shapes over crimp intervals are defined by Peirce-type models for an assumed cross-section shape. This approach is taken in the following section.

1.3.4   Balancing yarn crimp

Plain weave, incompressible yarns

Consider a plain weave fabric made of incompressible yarns. Yarn spacing p, bending rigidity B and vertical dimensions d are given. Crimp height h is needed to derive a full description of the yarn geometry from the previous equations. This can be calculated by minimising bending energy of all yarns in the repeat:

$$W_\Sigma = \frac{B_{Wa}}{p_{We}} F\left(\frac{h_{Wa}}{p_{We}}\right) + \frac{B_{We}}{p_{Wa}} F\left(\frac{h_{We}}{p_{Wa}}\right) \rightarrow \min \qquad 1.27$$

Using the approximation from eqn 1.23 this is easily solved:

$$h_{Wa} = D\frac{B_{We}p_{We}^3}{B_{Wa}p_{Wa}^3 + B_{We}p_{We}^3}, \; h_{We} = D\frac{B_{Wa}p_{Wa}^3}{B_{Wa}p_{Wa}^3 + B_{We}p_{We}^3} \qquad 1.28$$

A few special cases are illustrated in Fig. 1.18(b). If $B_{Wa} = B_{We}$ and $p_{Wa} = p_{We}$ then $h_{Wa} = h_{We} = D/2$, which is typical of a balanced fabric. If $B_{Wa} \gg B_{We}$ and $p_{Wa} = p_{We}$ then $h_{Wa} = 0$ and $h_{We} = D$. Rigid warp yarns stay straight and compliant weft yarns wrap around them, which is typical of quasi-unidirectional woven fabrics. If $p_{Wa} \gg p_{We}$ and $B_{Wa} = B_{We}$ then again $h_{Wa} = 0$ and $h_{We} = D$. Long segments of weft yarn extending between crossovers are easier to bend than short segments of warp yarns.

Plain weave, compressible yarns

Consider the same problem with compressible yarns. Compressibility of warp and weft yarns is given by experimental diagrams. Subscripts 0, 1 and 2 respectively designate the relaxed state, yarn dimension in the vertical direction and in-plane dimensions; q is a compression force per unit length.

$$d_1^{Wa} = d_{10}^{Wa}\eta_1^{Wa}(q), d_2^{Wa} = d_{20}^{Wa}\eta_2^{Wa}(q) \qquad 1.29a$$

$$d_1^{We} = d_{10}^{We}\eta_1^{We}(q), d_2^{We} = d_{20}^{We}\eta_2^{We}(q) \qquad 1.29b$$

Equation 1.27 applies with constraint including dimensions depending on yarn interaction forces:

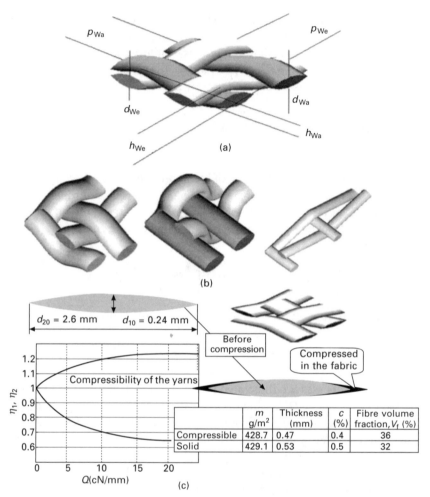

1.18 Balance of crimp in plain weave: (a) scheme; (b) three special cases; (c) change in yarn dimensions for typical fabric with compressible yarns.

$$h_{Wa} + h_{We} = d_{10}^{Wa} \eta_1^{Wa} \left(\frac{Q}{d_2^{Wa}} \right) + d_{10}^{We} \eta_1^{We} \left(\frac{Q}{d_2^{We}} \right) \qquad 1.30$$

where the transversal forces Q are calculated from eqn 1.21, giving:

$$Q = \frac{B_{Wa}}{p_{We} h_{Wa}} F \left(\frac{h_{Wa}}{p_{We}} \right) + \frac{B_{We}}{p_{Wa} h_{We}} F \left(\frac{h_{We}}{p_{Wa}} \right) \qquad 1.31$$

Equation 1.30 assumes that the zone over which Q acts has dimensions equal to in-plane yarn dimensions; a better description of the yarn contact zone would improve the model. The non-linear system (eqn 1.29–1.31) can be solved iteratively by setting:

$$d_1^{Wa} = d_{10}^{Wa}, d_2^{Wa} = d_{20}^{Wa}, d_1^{We} = d_{10}^{We}, d_2^{We} = d_{20}^{We}, h_{We} = h_{Wa}$$

$$= (d_1^{Wa} + d_1^{We})/2 \qquad\qquad 1.32$$

Transversal forces are computed from eqn 1.31, yarn dimensions are obtained from eqn 1.29 and the minimum problem of eqn 1.27 is solved for h_{Wa} and h_{We}. Then convergence is checked by comparing with values from eqn 1.32 and the above steps are repeated as necessary.

Figure 1.18(c) shows an example of changes in yarn dimensions for a glass plain weave with tow linear density of 600 tex, tow bending rigidity of 0.44 N mm^2 and ends/picks count 36/34 yarns/cm. The compression diagram in Fig. 1.18(c) was measured using the KES-F tester. Results show that the yarns are compressed considerably, changing the fabric thickness by 8%.

1.4   Braided fabrics

1.4.1   Parameters and manufacturing of a 2D braided fabric

In braiding, three or more threads interlace with one another in a diagonal formation, producing flat, tubular or solid constructions. Such fabrics can often be used directly as net-shape preforms for liquid moulding processes such as resin transfer moulding (RTM). This section discusses 2D flat or tubular braids.

Braiding process

The principle of braiding is explained in Figs 1.19 and 1.20 for a maypole machine. Carriers move spools in opposite directions along a circular path. Yarn ends are fixed on a mandrel and interlace as shown in Fig. 1.19. Interlaced yarns move through the convergence zone of the machine, towards the mandrel which takes the fabric up the loom. The yarns follow helical paths on the mandrel and interlace each time spools meet. Producing a thin, tight braided lace or a circular tube does not require a mandrel. On the other hand, 2D braids produced on shaped mandrels can be used as reinforcements for composite parts. Braiding over shaped mandrels allows the introduction of curvature, section changes, holes or inserts in the reinforcement without need to cut the yarns, as shown in the photograph in Fig. 1.19.

In the process described above spools must travel to and from the inner and outer sides of the circular path to create interlacing. Fig. 1.20 shows the necessary spool motions for a part of the path. Carriers with odd and even numbers move from left to right (clockwise) and from right to left (counter-clockwise) respectively. The notched discs are located on the circumference

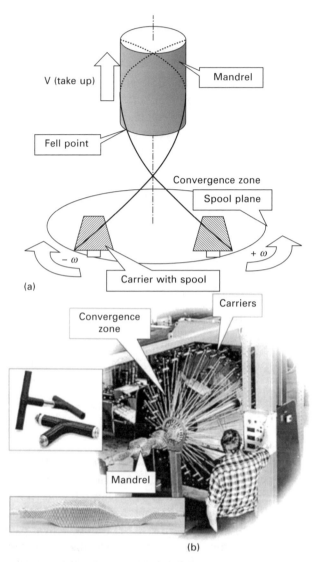

1.19 Scheme and photograph of a maypole braider: diagonal interlacing of the yarns, where only two carriers are shown. Insets: non-axisymmetric mandrel and braided products with holes and inserts.

of the machine, spool carrier axles are located in these notches and adjacent disks rotate in opposite directions. Consider carrier 1; once taken by a disk to a position where notches superpose with those of the neighbouring disk the carrier is transferred to that disk by centrifugal force and a rotating guide. Because of the difference in rotation directions, spool 1 is taken toward the

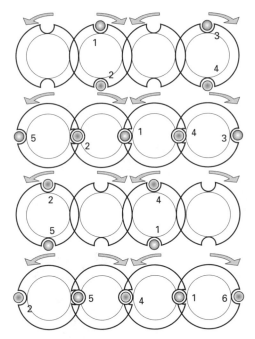

1.20 Horn gears and movements of the yarn carriers.

outside of the machine by the second disk. The same happens to its counterpart 2 carried by the notched disks in the opposite direction. Hence carriers 1 and 2 cross each other and change sides, creating interlacing. Figure 1.21(a) shows carrier paths producing a tubular braid. If the sequence of disks is interrupted, the carriers create a flat braid as shown in Fig. 1.21(b). In modern braiders notched disks are replaced by horn gears and carrier feet are directed along slots in a steel plate covering the gears.

Additional yarns extending along the braid axis can be incorporated, producing what are commonly described as triaxial braids. They do not travel on carriers but are fed from stationary guides situated at the centres of horn gears, Fig. 1.21(c). Such *inlays* or *warp yarns* are supported by the braided yarns and are almost devoid of crimp. They are common in reinforcements.

Braid patterns and braiding angle

Patterns created by braiding are similar to weaves and can be described in these terms (Fig. 1.22). Braids are identified in the same way as twills, by floats lengths for the two interlacing yarn systems. Three patterns have special names: *diamond* (1/1), *regular* or *plain* (2/2) and *hercules* (3/3) braids. A *plain* weave and a *plain* braid correspond to different patterns.

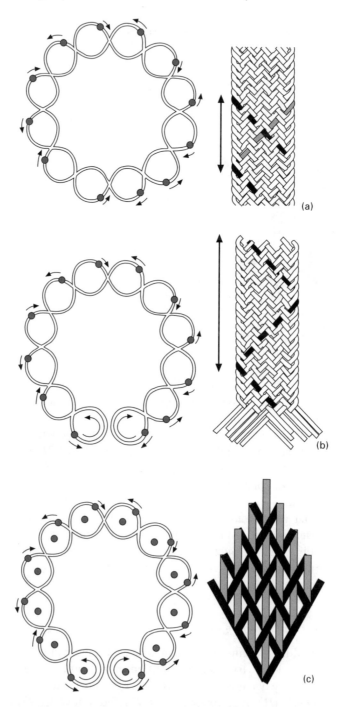

1.21 Carrier paths and braided patterns: (a) circular braid; (b) flat braid; (c) braid with inlays (triaxial braid).

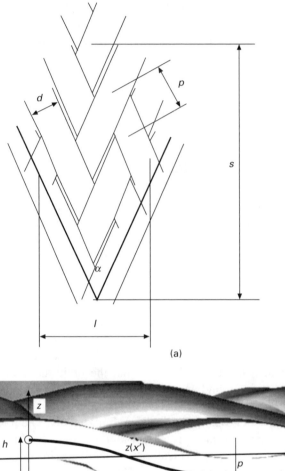

(a)

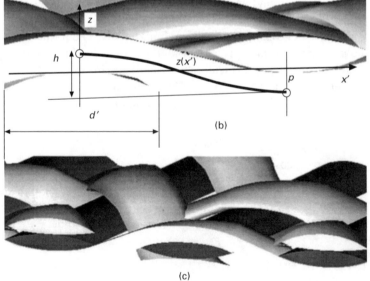

(b)

(c)

1.22 Geometry of a braid unit cell: (a) dimensions of the unit cell;
(b) geometry of crimp; (c) yarn crimp in a braid with inlays (darker
yarns).

The repeat of a braid is defined as the number of intersections required for a yarn to leave at a given point and to return to the equivalent position further along the braid (arrows, Fig. 1.21). This measure is called *plait*, *stitch* or *pick* and equals the total number of yarns in a flat braid or half their number in tubular braids. The definition differs from that of repeat in weaves; to avoid confusion the term *unit cell* is used to identify the basic repetitive unit of a pattern. The most important parameter of a braid is the *braiding angle* between interlaced yarns (Fig. 1.22), calculated from the machine speed and mandrel diameter D_m:

$$\alpha = 2 \arctan \frac{\pi D_m n}{v} \qquad\qquad 1.33$$

where n and v are the rotational speed (s^{-1}) and take-off speed. The angle depends on the mandrel diameter; if a complex mandrel is used the take-up speed must be constantly adjusted to achieve a uniform angle. By varying the take-up speed variable angles can be induced, allowing stiffness variations for the composite part. The practical range of braiding angles is between $20°$ and $160°$.

Equation 1.33 implies that the mandrel is circular (or at least axisymmetric), but as shown in Fig. 1.19 this is not a requirement for the process. An average braid angle for a non-axisymmetric mandrel can be obtained by replacing the term πD_m in eqn 1.33 with the local perimeter of the mandrel. However, it has been observed that the braid angle changes significantly with position around any section of a complex mandrel. Models have been proposed recently to predict this behaviour[11,12].

A braider can be set vertically or horizontally. The former is common for the production of lace or general tubular braids, and the latter for braiding over long mandrels. Typical carrier numbers are up to 144. Rotation and take-up speeds can reach 70 rpm (depending on carrier numbers) and 100 m/min respectively. Productivity as mass of fabric per unit time can be estimated by the equation:

$$A = \frac{TvN}{\cos^2 \dfrac{\alpha}{2}} = \frac{2\pi D_m TNn}{\sin \alpha} \qquad\qquad 1.34$$

where T and N are the yarn linear density and number of carriers. The productivity can reach several hundreds of kilograms per hour; braiding is more productive than weaving by an order of magnitude.

1.4.2   Internal geometry of 2D braids

The internal geometry of braids is governed by the same phenomenon mentioned for woven fabrics namely yarn crimp. The non-orthogonal interlacing of the yarns results in certain peculiarities. Consider a braided unit cell. Let

n be a float in the braid pattern identified as n/n; $n = 2$ in Fig. 1.22. Unit cell dimensions are given by the braiding angle, stitch length per unit cell s and line length per unit cell l:

$$l = s \tan \frac{\alpha}{2} \qquad\qquad 1.35$$

The shape of yarn centrelines can be modelled in a similar way as for woven fabrics (Fig. 1.22b). On a section of the fabric including the centre of a yarn (axis x'), spacing p and dimensions d' of support cross-sections are given by:

$$p = \frac{s}{4n \sin \frac{\alpha}{2}} = \frac{l}{4n \cos \frac{\alpha}{2}}, d' = \frac{d}{\sin \frac{\alpha}{2}} \qquad\qquad 1.36$$

where d is the yarn width. A function $z(x')$ is constructed for a given crimp height h using an elastica or Peirce-type model and crimp balance is computed using the algorithm previously formulated. Observation of triaxial braids reveals mostly straight and crimp-free inlays extending along the machine direction. This is expected for balanced braids, while unbalanced braids are quite rare in practice. The presence of inlays changes the crimp height of interlacing yarns but the algorithms for the calculation of the mid-line shape and crimp balance remain unchanged, (Fig. 1.22c).

When the braided yarns are not perpendicular the yarn paths will be twisted, so that yarn cross-sections appear rotated. This also applies to woven fabrics subjected to shear deformation (for example during fabric draping).

1.4.3   3D braided fabrics

3D braiding presents some similarities with 2D braiding. In 2D maypole braiding only two sets of carriers rotate around the braiding axis in opposite directions, creating a single textile layer. In contrast, 3D braids can be regarded as multiple layers of interlaced yarns which are connected more or less extensively by individual yarns extending through the thickness. It is possible to braid preforms where the level of interlacing between different layers is such that it becomes impossible to discern the distinct layers in the final preform. Carrier paths may be defined over concentric circles or Cartesian arrays, which are square or rectangular.

Two examples of 3D braiding processes are described by Byun and Chou[13], known as the four-step and two-step processes. The four-step process (Fig. 1.23a) uses a framework of yarn carriers in a rectangular or circular array. As the name suggests, the process consists of four steps, each involving alternate movements of the rows and columns of yarn carriers. Between cycles the yarns are 'beaten up' into the structure and the braid is hauled off by one pitch length. The two-step process (Fig. 1.23b) involves a large number of axial yarns arranged in the required preform geometry, with a smaller number

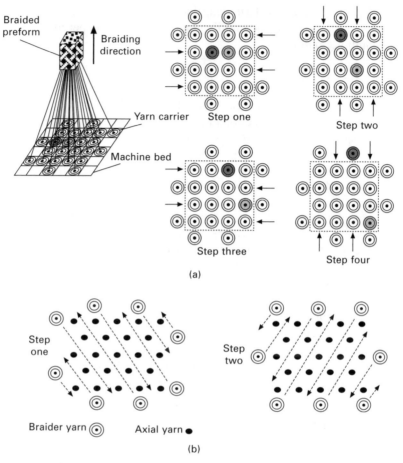

1.23 Schematic illustration of the (a) four-step and (b) two-step 3D braiding processes.

of yarn carriers arranged around them. The carriers are moved through the array of axial yarns in two alternate directions.

Such processes can be used to produce preforms featuring yarns that extend along many directions. Yarn directions are not limited to a plane, and this constitutes the main advantage of the process. However, any increase in out-of-plane properties is achieved at the expense of in-plane properties. 3D braids generally feature axial yarns. Preforms of various shapes can be produced, for composite parts of which the geometry is very different to the shells associated to parts based on 2D textiles such as weaves or non-crimp fabrics. Such applications are still relatively rare; rocket engines constitute a classic example while biomedical engineering is one area of emergence.

1.5   Multiaxial multiply non-crimp fabrics

The above textile reinforcements offer economical advantages while being easier to process than prepregs. However, compromises are made on the performance of the composite because of crimp that causes fibre misalignments. Non-crimp fabrics combine unidirectional crimp-free fibre layers by assembling them together by stitching – sewing or knitting – and/or bonding by chemical agents.

1.5.1   Terminology and classification

European standard EN 13473 defines a multiaxial multiply fabric as

> A textile structure constructed out of one or more laid parallel non-crimped not-woven thread plies with the possibility of different orientations, different thread densities of single thread plies and possible integration of fibre fleeces, films, foams or other materials, fixed by loop systems or chemical binding systems. Threads can be oriented parallel or alternating crosswise. These products can be made on machines with insertion devices (parallel-weft or cross-weft) and warp knitting machines or chemical binding systems.

The definition stresses that fibrous plies are laid up with threads. Stitching the plies together by warp-knitting can be achieved in such a way that the stitches pierce the plies between the laid yarns, resulting in an open preform architecture; Fig. 1.24a. On the other hand wide threads (flat tows) laid close together form continuous fibrous plies bound by warp-knitting with piercing sites positioned according to the needle spacing, without any connection to tow positioning. Needles pierce the fibrous plies and distort them locally; Fig. 1.24b. Such a preform construction is close to ideal unidirectional uniform plies. The latter case is the most common, and is the subject of this section.

A fabric is characterised by the nature of the fibres (E/G/C/A glass, carbon or aramid), surface density of each ply (g/m^2), fibre direction in each ply given as an angle with respect to the machine direction in the range [–90°, 90°], type of binding agent, nature of the binding agent (polyamide/poly (ethylene)/poly(ether sulphone)/poly(ethylene terephthalate/polypropylene; PA/PE/PES/PET/PP, chemical binder) and surface density of the binding agent. For example the designation stipulated by the standard for a 3-ply +45°/–45°/0° glass fabric stitched by a polyester yarn is:

$$[G, 235, +45° \text{ // } G, 235, -45° \text{ // } G, 425, 0°][PES, 12, L]$$

with a total surface density of 907 g/m^2. Ply weight depends on yarn weight and placement density. In general yarn weight is characterised by a number

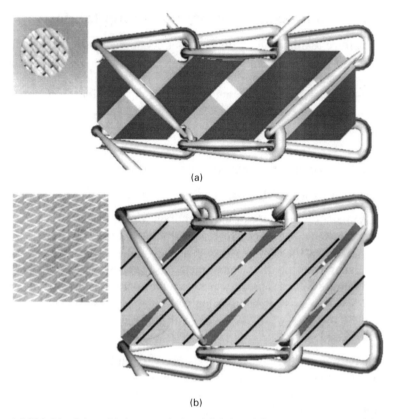

(a)

(b)

1.24 Multiaxial multiply warp knitted fabrics: (a) open structure, with stitching sites in between the tows; (b) continuous plies, with stitching penetrating the tows forming 'cracks' in the plies.

of filaments ranging from 3K to 80K for carbon tows. Surface densities usually range from 100 g/m² upwards. Higher yarn weights normally translate into higher ply surface density values for a closely covered surface. Low surface density values may be reached with thick tows by spreading, using a special unit. Yarns with higher fibre count present the economical advantages of cheaper raw material and faster production. Low ply surface weights are desirable as lighter fabrics have better drapability. Current technology achieves 150 g/m² with 12K yarns in the weft or 24K in 0° layers.

1.5.2   Manufacturing

Producing a multiaxial multiply fabric involves laying plies or *weft insertion*, and stitching or *knitting* them together. Hence the machine has two main parts, a weft insertion device and a knitting unit.

Weft insertion

Weft insertion carriages travelling across the machine lay several tows simultaneously at a desired angle across the fabric. The tows are fixed to the needles of the moving chains. The length of the weft insertion part of a typical machine is about 10 m and allows up to three layers with angles from 20° to 90° or –20° to –90°. The speeds of weft insertion carriages dictate the fibre angles. Measurements show that the orientation of oblique fibres can deviate from the prescribed value by several degrees.

Weft inserted tows together with optional 0° plies are stitched by a warp-knitting device. A powder binder (e.g. epoxy) can be added beforehand to facilitate composite processing. A foil or non-woven ply may be added below the plies. A warp-knitting device as described below stitches the plies by an action of needles, which are collected in a needle bed and move upwards, piercing the plies.

Walking needle

Stitching of ply construction by warp-knitting is optimised to decrease misalignment and disturbance of the weft yarns by the knitting needles. In older knitting units needles move in one direction only, standing at a certain angle to the plies. The needles move linearly upwards to catch the warp knitting yarn and then down to pull the yarn through the ply, building a stitching mesh. Plies move constantly forward. The relative movement of plies and needles in the horizontal direction causes misalignment during penetration, the extent of which depends on the stitching length and other parameters. This limitation was reduced drastically by the development of the *walking needle* device, introducing a new needle movement where the needle is placed at 90° to the textile and moves in the vertical direction as well as in the direction of ply movement during penetration. The relative horizontal motion between the textile and needle is minimal. The needle head is designed to penetrate the ply construction by pushing away the filaments of reinforcing yarns, minimising the damage.

Warp knitting

The principle of warp knitting is illustrated in Fig. 1.25. An actual warp knitted fabric or a knitted web, whereby warp knitting is used to stitch unidirectional fibrous plies, is an assembly of loops formed by the interaction of needles and guides. Yarns are fed through all guides and all needles are fixed on a needle bar, moving simultaneously and forming a loop around every needle in every cycle.

Consider a warp yarn about to be looped, going thorough the guide at the beginning of the knitting cycle (Fig. 1.25a). Knitting needles are in their

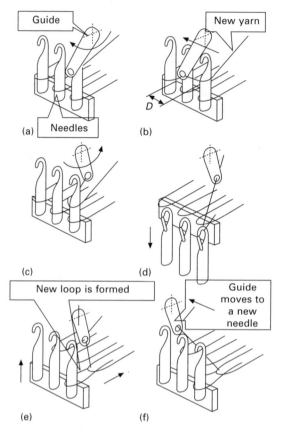

1.25 Principle of warp knitting, showing interaction of the knitting needles and guides.

upper position. Loops forming the edge of the fabric are on the stems of the needles. The guide carries the new yarn around a needle, forming a new loop (Fig. 1.25b,c). The needle moves down, connecting the new loop with the old one sitting on its stem (Fig. 1.25d,e); in the case of ply stitching the loop is pulled through the plies. The textile is pulled and a new cycle starts. The knitting action creates loop connections in the machine (*course*) direction. Movements of the guides across the needle bed create connections in the cross (*wale*) direction, forming a warp-knit pattern; Fig. 1.26. The positions of the gaps between the needles, where guides pass in subsequent knitting cycles, can be used to code the pattern by the so-called Leicester notation. Consider a diagram of the guides' movement, called a *lapping diagram* (Fig. 1.26). Rows of dots represent needle positions in plane view. The numbering of needles assumes that the pattern mechanism is on the right side. As the guides position themselves in the spaces between needles, the positions

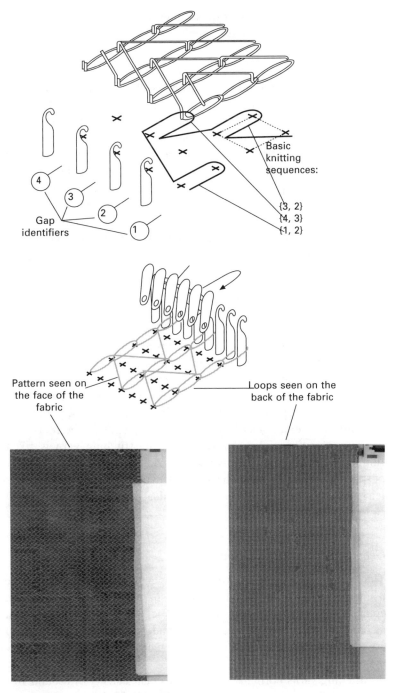

Basic
knitting
sequences:

{3, 2}
{4, 3}
{1, 2}

Gap
identifiers

Pattern seen on
the face of the
fabric

Loops seen on the
back of the fabric

1.26 Warp-knit pattern formation and coding.

between vertical columns of dots represent lateral shifts of the guides. The pattern is a sequence of these numbers:

$$s_1-S_1/s_2-S_2/ \ldots / s_N-S_N \qquad\qquad 1.37$$

where s_i and S_i are the positions of the guide forming the ith loop and N is the number of knitting cycles in the pattern. Positions s_i and S_i refer to gaps between needles where the elementary movement of guides start. A pair s_i–S_i represents an overlap motion (guides move behind needles) while S_i/s_{i+1} is an underlap motion (guides move in front of needles). The following consider patterns where overlaps extend by one needle only, i.e. $|s_i-S_i| = 1$.

When fibrous plies are stitched, the pattern is seen on the face of the fabric (Fig. 1.26). These yarns are laid on the face of the fabric by the guides. On the back of the fabric chains of loops are seen, extending in the machine direction. These loops are formed on separate needles and interconnect through the knitting action of the needles.

1.5.3   Internal geometry

Positions of the stitching sites

A stitching yarn pierces the fibrous plies at positions defined by the needle spacing in the needle bed (spacing in cross direction A) and by the speed of the material feeding in the knitting device (spacing in machine direction B). The value of A is also expressed by the *machine gauge*, a number of needles per inch. In a warp-knitted fabric devoid of fibrous plies, yarn tension leads to significant deviations of the actual loop spacing in the relaxed fabric. In the case of multiaxial multiply fabrics the stitching yarn is fixed by the fibrous plies, resulting in fairly regular spacing. Figure 1.27 gives an example of spacing variability for such a fabric. Deviations in the positions of stitching lead to deviations of the loops from the ideal machine direction; the deviation can be as high as 15°.

Geometry of the stitching loop

The loop geometry can be assumed using a geometric approach or described using a mechanical approach where the shape is calculated from equilibrium equations applied to the looped yarns. The latter approach is unsuitable here as loop positions are set by the plies, and large fluctuations of loop shapes are observed in practice. Hence the geometric approach is considered more appropriate.

Stitching yarns normally have a low linear density of around 10 tex with tyically 15 filaments, and low twist (<100 m^{-1}). They are easily compressible, which explains the large observed variations of their dimensions. Stitch yarn

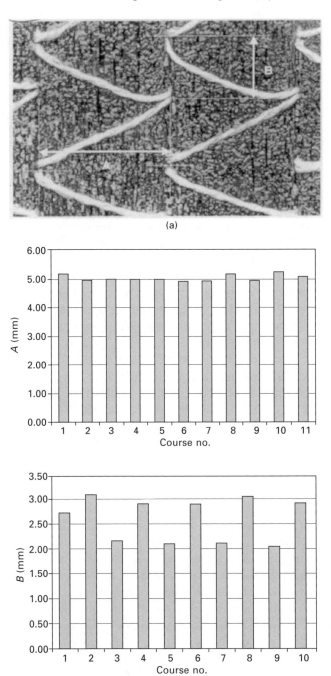

(a)

(b)

1.27 Spacing of the stitching loops for a carbon 0/–45/50 fabric:
(a) definition of *A* and *B* and (b) measured values for adjacent wales.

shapes are observed to vary within the repeating pattern, with cross-sections ranging from circular to flattened (elliptical) and with varying dimensions[14].

The geometry of the yarn centreline is defined by identifying a set of anchor points A–M along the loop and approximating the centreline between the anchor points by straight lines or circular arcs. In Fig. 1.28 h is the total thickness of the plies; arcs AB, CD, FE, GH, IJ and KL have constant radii and diameter d_0. The loop width w depends on knitting tension. Considerable tension produces narrow loops with $w \approx 3d_0$; lesser tension increases w to 4 to 5 times d_0.

Interactions of the stitching with fibrous plies

Stitching causes deviations to fibre orientations. These deviations can result in linear channels (Fig. 1.29a), or be localised near stitching sites as *cracks* (Fig. 1.29b). Localised cracks have a rhomboidal shape of width b and length l. A channel is formed when localised cracks touch or overlap. The direction of cracks and channels corresponds to that of fibres in the ply. The cracks/channels are also evident inside the fabric, where they provide routes for resin flow. In a composite they create resin rich zones which can play an important role in the initiation of damage.

Crack and channel dimensions feature much scatter (Fig. 1.29c). Table 1.2 shows measured crack and channel dimensions for various fabrics. Crack widths are approximately proportional to the thickness of the stitch yarn, so that an empirical coefficient k can be used to calculate $b = kd_0$, where d_0 is the compacted stitch diameter. Fibres in a powdered fabric are more difficult to displace, explaining the low values of k and l/b for the latter fabric. In non-powdered fabrics channels are wider than cracks, with rough k averages of 4 (cracks) and 7 (channels) and l/w values for cracks of about 20. Dimensions are also influenced by knitting tension.

When large channels extend along the machine direction, stitch yarn sections on the fabric face or loops on the back can sink in the channels. This phenomenon plays a major role in nesting of non-crimp fabric layers in laminates. When this sinking occurs the layers can come closer, closing any gap between them that might be introduced by stitched yarns laying on the surface of the fabric. This is evidenced by compression curves of multiple layers of fabric such as a biaxial 'B' –45°/45° and quadriaxial 'Q' 0°/–45°/ 90°/45°; Fig. 1.30. In the former, stitch yarns extend across the fibre direction and cannot sink in cracks. The face of the latter textile has wide channels into which stitches sink deeply (Fig. 1.30a), allowing nesting. Hence the thickness per layer of the compressed fabric 'Q' decreases with an increase in number of layers while it stays constant for fabric 'B'.

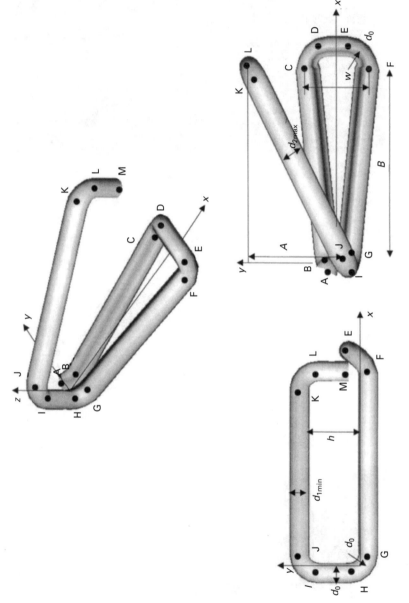

1.28 Geometry of the stitching loops, showing positions of anchor points.

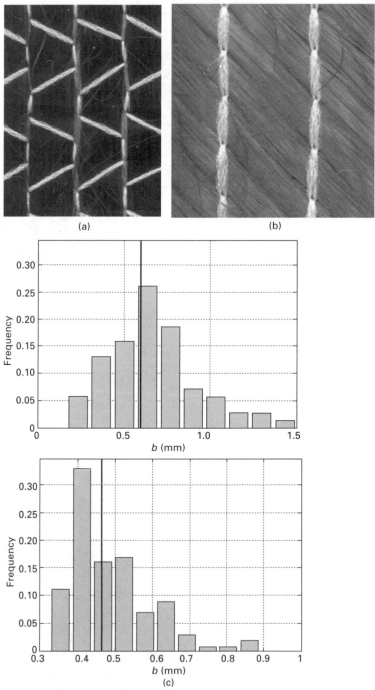

1.29 Channels and cracks in the fibrous layers: (a) channels in a 0°
(face) ply; (b) crack in a 45° (back) ply; (c) distribution of channel/
crack widths.

Table 1.2 Dimensions of channels and cracks in multiaxial multiply carbon fabrics

Fabric	Stitching		Crack/channel	w (mm)	l (mm)	k	l/w
	Material, linear density	d_0 (mm)					
-45/-45, 12K	PES, 7.6 tex	0.088	Face cracks	0.276	5.05	3.13	18.3
			Back cracks	0.434	7.15	5.39	16.5
0/90, 24K	PES, 7.6 tex	0.088	Face channels	0.62	n/a	7.04	n/a
			Back channels	0.36	n/a	4.09	n/a
0/-45/90/45	PES, 7.6 tex	0.088	Face channels	0.658	n/a	7.47	n/a
			Back cracks	0.483	7.28	5.48	15.0
45/-45, 12K	PES, 6 tex	0.071	Face cracks	0.28	7.43	3.94	26.5
			Back cracks	0.27	7.85	3.80	29.1
0/90, 12K, powdered	PA, 10 tex	0.107	Face channels	0.18	n/a	1.68	n/a
			Back cracks	0.53	3.46	4.95	6.53

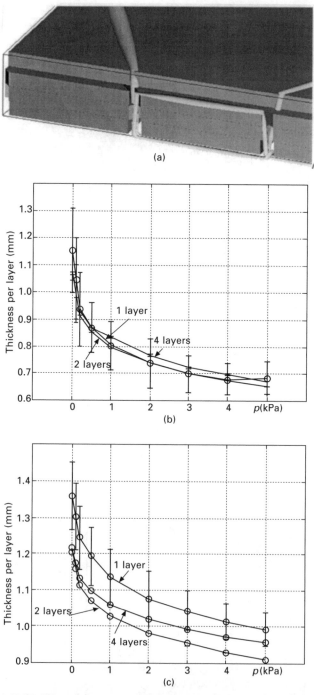

1.30 Sinking of the stitching in channels in the fibrous plies: (a) fabric 'Q' (see text), photo and schematic, and compression curves for (b) biaxial fabric 'B' and (c) quadriaxial fabric 'Q'.

1.6   Modelling of internal geometry of textile preforms

Textiles are hierarchically structured fibrous materials. As discussed in the classic paper by Hearle *et al.*[15] this description of the nature of textiles allows efficient construction of mathematical models for the geometry and the mechanical behaviour of textile structures. In spite of the generally recognised usefulness of this approach, it has not been used to its full strength for the creation of textile structure models. During the 1930s the first serious mechanical treatments of the structure of textile materials were published by Peirce[10], Pozdnyakov[16] and Novikov[17]. Since then, numerous publications have been proposed on the mechanical behaviour of textiles, culminating in a comprehensive treatment on *Mechanics of Flexible Fibre Assemblies* in 1980[18]. In the following years the ideas and approaches outlined in this book were pursued further. Textile mechanics at the outset of the 21st century includes models of the internal geometry of textile structures such as continuous-filament and staple yarns, random fibre mats, woven and knitted fabrics. The hierarchical description of textile structures is implemented using the minimum energy principle as introduced by Hearle and Shanahan[19] and de Jong and Postle[20]. This allows the decomposition of a structure into a set of structural elements, leading to models that are physically sound and computationally feasible. It should be noted that the principle of minimum energy is heuristic when applied to non-conservative mechanical systems such as textiles.

Following the hierarchical approach one can consider a description of the internal geometry of a fabric on two different levels: the *yarn paths mode* describing the spatial configuration of yarns and the *fibre distribution mode* describing spatial distribution properties of the fibrous assembly. In the former case the fabric is regarded as an agglomeration of yarns – slender curved bodies. Their spatial positions are defined by description of yarn midlines and cross-section at each point of the midlines. The internal fibrous structure of the yarns is not considered. In the latter case the fibrous structure of yarns should be defined. The definition of the spatial positions of yarns states whether any point in the unit cell lies inside a yarn or not; if so, parameters of the fibrous assembly at this point are determined. The yarn path mode is sufficient for the calculation of simpler data related to the internal geometry, such as surface density, fabric thickness and local inter-yarn porosity. The fibre distribution mode is needed when detailed information on the fibrous structure is required.

1.6.1   Yarn path mode

Figure 1.31 illustrates the description of yarn spatial configuration. The midline of the yarn shown in Fig. 1.31b is given by the spatial positions of the

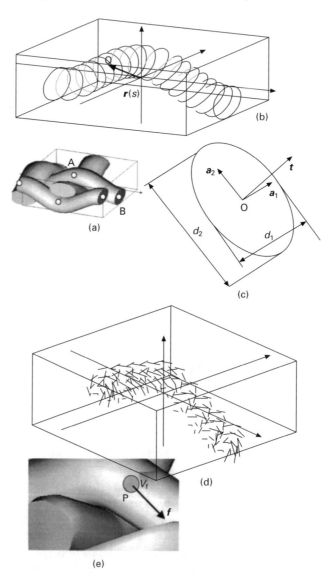

1.31 Models of textile structures: (a) yarn spatial positions; (b) yarn path mode; (c) parameters of a cross-section; (d) fibre distribution mode (fibre directions for a twisted yarn); (e) parameters of a fibrous assembly.

centres of the yarn cross-sections O: $r(s)$ where s and r are coordinates along the midline and the radius-vector of point O. Let $t(s)$ be the tangent to the midline at point O. The yarn cross-section is normal to t and defined by its dimensions $d_1(s)$ and $d_2(s)$ along axes $a_1(s)$ and $a_2(s)$, which are defined in the cross-section and rotate around $t(s)$ if the yarn twists along its path.

Because of this rotation the system $[a_1a_2t]$ may differ from coordinate system along the path.

A definition of the spatial position of a yarn with a given cross-section shape therefore consists of the five functions $r(s)$, $a_1(s)$, $a_2(s)$, $d_1(s)$, $d_2(s)$. Models defining these functions for different textile structures were described in previous sections. Two approaches are possible for definition of the midline path $r(s)$: a simpler geometric and a more complex mechanical one.

Geometric approach

In the geometric approach, midline anchor points are defined from topological information for the textile structure. These points can be located at yarn contacts or crossovers for some textiles, as the distance between contacting yarns is known from cross-section dimensions. For more complex textiles in may be necessary to introduce additional points to represent some curvatures. Yarn midlines are represented by smooth lines going through these points; typically Bézier curves are used[21–23].

The main advantage of the geometric approach is that it can represent any textile structure under one single format while producing simple and portable definitions. As a result, downstream models of the physical properties of textiles and their composites can be used with these geometric definitions, regardless of the textile manufacturing process. Furthermore, the format is equally well suited to simpler closed-form physical property models as it is to more computationally intensive ones. The philosophy behind the geometric approach consists in creating geometric models that are ensured to be appropriate, devoid of interference for example, with the view of refining the assumed initial geometry through mechanical methods, and this for any textile and manufacturing processes. Figure 1.32 shows diverse examples, with anchor points forming 3D linear segments represented as vectors on the left and the resulting geometric models on the right. Any interferences resulting from user input were corrected by a geometric algorithm where sections are modified arbitrarily and yarn V_f remains constant. Another major advantage is that it allows explicit consideration of important practical phenomena such as the statistical variation of dimensions inherent within actual textiles.

Although the geometric approach method offers the advantages of simplicity, speed and certainty over aspects such as interference, the models that it produces clearly depend on user input. The anchor points are predefined and assumptions are made about yarn interaction or crimp height, for example. Symmetry considerations can assist but only to a certain extent, and in its basic form predictive geometry determination is not possible. For an accurate description, dimensions must be measured from real fabrics.

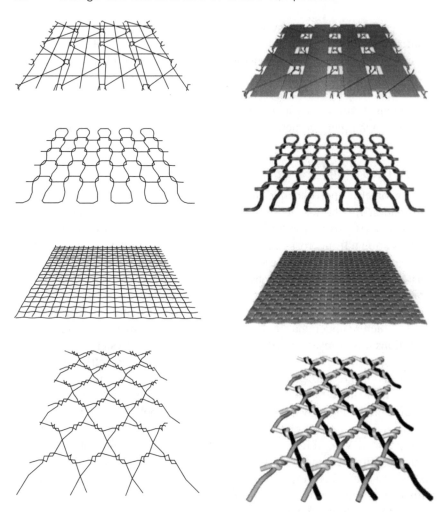

1.32 Geometric approach to textile modelling for a range of textiles, showing vectorial description of yarn paths (left) and resulting textile models (right).

Mechanical approach

A mechanical approach to the modelling of yarn paths requires a mathematical description of the interaction of contacting yarns. In relaxed fabrics this involves bending and compression of yarns; twisting may also be present. Therefore, it also requires appropriate experimental data for the yarns in these deformation modes. If this is available, the model must account for contacts between interacting yarns and describe the equilibrium of the final configuration. The minimum energy principle is normally used for this description. Examples of such models were shown above for weaves and braids.

Computational efficiency is critical in the mechanical approach and therefore the definition of fabric geometry should use the principle of hierarchy and structure decomposition. The repeat is a natural structural unit for fabrics and contains a number of yarns $\{Y\}$ that contact others. The contact regions occupy certain zones on the yarns. Characteristic points provide natural boundaries for *structural elements* of the repeat (Fig. 1.31a). In a woven fabric the structural elements consist of intervals of warp and weft yarns located between subsequent crossovers, where characteristic points are located.

A yarn Y consists of a set of structural elements $\{e\}_Y$, each characterised by the coordinates of its end points A_{Ye} and B_{Ye}, the dimensions of the contact regions near the end-points, and the forces acting on the contact regions. The exact shape and dimensions of the contact regions on a structural element and the forces acting on them are determined by yarn interactions in the structure.

The basic assumption of the decomposition routine is that the spatial positions of the end-points of structural elements play a central role in calculating the geometry of that element. Supposing that the end positions of the structural elements for all yarns in a repeat are known and fixed, contact regions develop near the end-points due to yarn interactions, and internal forces arise in the contact zones. Geometric constraints imposed on the structure by the end-point positions determine the local deformations of yarns at contact regions and the shape and dimensions of these regions, which in turn determine the contact forces.

In general cases a set of parameters q_1, q_2, q_3, ... are designated that determine the complex mechanical behaviour of structural elements. Relating to the assumption formulated in previous paragraphs, these parameters consist of the positions of end-points of the structural elements, $\{q_i\} = \{A, B\}$.

Consider now the problem of computing the spatial position of yarns in the fabric repeat space. Let $r_Y(s)$ be the parametric representation of the centreline of yarn Y. According to the minimum energy principle the set of $r_Y(s)$ for all yarns in the fabric repeat should satisfy:

$$\sum_{(Y)} W[r_Y(s)] \to \min \qquad\qquad 1.38$$

If $r_Y(s)$ is split into separate functions for each structural element e on yarn Y this takes the form:

$$\sum_{(Y)} \sum_{(e)} W[r_{Ye}(s)] \to \min \qquad\qquad 1.39$$

which can be recast as:

$$\sum_{(Y)} \sum_{(e)} W[r_{Ye}(s; q_1, q_2, q_3...)] \to \min \qquad\qquad 1.40$$

This global minimisation problem can be reformulated as a series of minimisation problems for structural elements:

$$\min_{r(s)} \sum_{(Y)} \sum_{(e)} W[r_{Ye}(s; q_1, q_2, q_3...)]$$

$$= \min_{q_1, q_2, q_3...} \sum_{(Y)} \sum_{(e)} \min_{r(s)} \{W[r_{Ye}(s \mid q_1, q_2, q_3...)]\} \qquad 1.41$$

Each minimisation problem should be solved with parameters $\{q_i\}$ fixed, yielding the solution:

$$\min_{r(s)} \{W[r(s \mid q_1, q_2, q_3 ...)]\} \Rightarrow r(s \mid q_1, q_2, q_3...) \qquad 1.42$$

which is used to calculate the energy of the structural element:

$$r(s \mid q_1, q_2, q_3 ...) \Rightarrow W(q_1, q_2, q_3 ...) \qquad 1.43$$

The function W, which depends only on parameters of the structural element, is termed the *characteristic function* of the structural element. The minimisation problem (1.40), which has the *functions* $r_Y(s)$ as arguments, is reduced to the following minimisation problem:

$$\sum_{Y} \sum_{e} W_{Ye}(q_1, q_2, q_3...) \rightarrow \min \qquad 1.44$$

where the arguments are a set of *scalar parameters* $\{q_i\}$. This leads to a system of non-linear algebraic equations instead of differential or integral equations, which would result from eqn 1.38. A practical application of this appears in section 1.3.3 for a simple weave. Note that the energy functions involved in eqn 1.44 may represent bending, compression and torsion and accounting for non-linearity.

Mechanical models aim to be predictive and to determine the structure of fabrics from independent measurements of yarn properties. In some cases this cannot be achieved as yarns undergo mechanical loading during fabric manufacture and the resulting deformations are not relaxed, leading to deviations between reality and analysis based on equilibrium in the relaxed state. This is especially true for cases when manufacturing tensions are extreme, say for example in 3D weaving. For such cases one must rely on a geometric model coupled with physical measurements taken from the fabric.

Using yarn path mode

The yarn path mode (mesoscopic geometric model) is a necessary step in building models at the microscopic scale (fibre distribution mode). Mesoscopic models also have inherent value: they define yarn volumes and therefore inter-yarn porosity, as well as local fibre directions of yarns. Information on the *local direction of the yarns* is essential for most predictive models of the physical properties of textiles, say for example in micro-mechanical models of composites to calculate the homogenised stiffness matrix of the unit cell, or in permeability models for composites processing. One approach to the former consists in defining ellipsoidal inclusions with stiffness properties equivalent to those of elements of the impregnated yarn, and then using the solution of Eshelby for homogenising the stiffness of agglomerated inclusions[24]. Both approaches rely on a representation of the local, microscopic fibrous structure as a unidirectional array of fibres, the equivalent stiffness of which is obtained from empirical formulae[25] or finite element analysis results[26]. These issues are discussed further in Chapter 8.

The second use of the yarn path mode is related to numerical modelling. The geometry of *yarn volumes* can be easily transferred to mesh generators; (Fig. 1.33). Yarn segments can be assembled into solid models of fabrics and meshed. The geometry of the empty volumes extending between the yarns is far more intricate; this must also be meshed for problems of flow and heat transfer through textiles and for all problems related to textile composites. Although commercial solid modellers can sometimes process this, good results often require very high mesh densities. Other solution techniques that do not use conformal meshing, or that simplify the problem, constitute important topics towards the routine use of such tools as a support to industrial design of textile applications. Figure 1.34 shows a 2.5D mesh used for calculating in-plane permeability tensors of textile reinforcements based on a simplified, efficient technique[27].

1.6.2   Fibre distribution mode

Yarn path mode geometric modelling describes yarn volumes but says nothing about the fibrous structure of yarns, unit cells or fabrics. The fibre distribution mode provides such information. Consider a point P (Fig. 1.31e) and a fibrous assembly in the vicinity of this point. The assembly can be characterised by different parameters including physical and mechanical properties of the fibres near point P (which may vary throughout the fabric), fibre volume fraction V_f and direction f. Length distribution of fibres near the point, average curvature and other parameters can be specified if the point is in a yarn. A fibre distribution mode model should provide such information for any point P.

Using fibre distribution mode

The ability to query the fibrous structure near arbitrary points can be used in many applications dealing with mesh representations of heterogeneous media. In *finite element* representations of fabrics the fibre distribution mode is used to determine mechanical properties for each element. Figure 1.33(c) illustrates

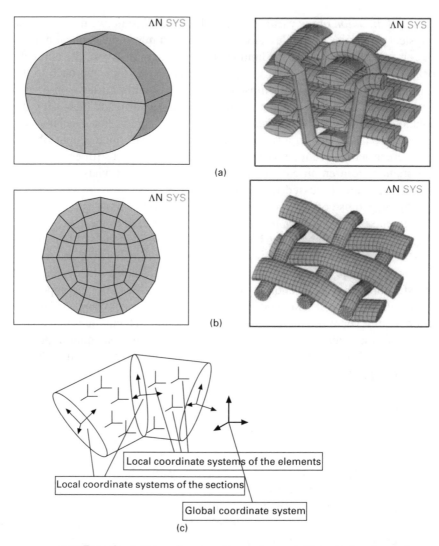

1.33 Transformation of yarn path mode model into finite element description: (a) yarn volumes; (b) mesh on the yarns; (c) local coordinate systems, representing fibre directions.

local coordinate systems for each element, aligned with the average fibre direction; this allows correct representation of anisotropy. Mechanical properties of a dry fabric include the tensile stiffness along the fibres and the compression stiffness through thickness (see Chapter 2). For composite materials, the homogenised stiffness of unidirectional fibre reinforced composite and local fibre volume fraction are required. The same is done in *cell models* of composite materials[28]. In such models a unit cell is subdivided into sub-cells (quasi-

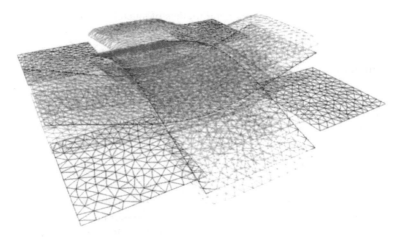

1.34 Mesh representing local yarn and resin/free volume mid-planes for a single crossover of a plain weave. Each element has an associated thickness (free volumes) or permeability (yarn). This model is used for the 'streamsurface' permeability prediction method[27].

finite elements), the effective stiffness of which is calculated from average properties of the local fibrous assembly or matrix.

In *models of flow* through the reinforcement, a regular mesh or lattice is defined in the unit cell. Nodes of the lattice can lie outside or inside the yarns. In the former case the Navier–Stokes equation governs the fluid flow near the node. In the latter case the flow is described by the Brinkmann equation. The local permeability of the media is calculated using a unidirectional description of the fibrous assembly near the node. These equations are solved simultaneously for all the nodes in the unit cell, using, for example, the lattice Boltzmann method[29] or commercial software[30].

1.6.3   Implementation of textile hierarchical model

Table 1.3 shows the hierarchy of structural elements in textile composites, and the modelling problems associated with each scale. Unnecessary mixture of hierarchical levels should be avoided: yarn data should be used to predict behaviour of fabrics as opposed to fibre data, and so forth. Each hierarchy level is occupied by models which use data from that level and the next lower to predict properties of the defined structure at the next upper level.

Object oriented programming provides a powerful tool for the construction virtual textiles. Associated data fields and methods are outlined in Table 1.4. The results of geometric modelling serve as crucial input to models of composites processing and mechanical behaviour. The main advantage is that local variations in textile geometry, and hence in physical properties,

Table 1.3 Hierarchy of structure and models of a textile composite

Structure	Elements	Models
Yarn (tow)	Fibres	Fibre distribution in the yarn and its change under load/strain Mechanical properties of the yarn
Fabric (woven, knitted...)	Yarns	Geometry of yarns in the fabric and its change under load/strain Mechanical behaviour of the fabric repeat under complex loading
Composite unit cell	Fabric	Mechanical properties (stiffness matrix/non-linear law; strength)
	Matrix	Permeability tensor
Composite part	(Deformed/draped) unit cells	Behaviour under loading Flow of the resin Behaviour in the forming process

can be accounted for explicitly. A composite material, as the final product in the sequence 'fibre to yarn to textile to preform to composite', is included in the hierarchical description of the textile, taking full advantage of the versatility of the approach.

Fabrics are usually assembled into laminates to form reinforcements with appropriate thickness. The layers of the laminate are not precisely positioned one against another, causing a geometric and mechanical phenomenon of nesting. Nesting plays an important role in determining the physical properties of dry laminated reinforcements and composite parts. It causes statistical distribution of the laminate properties, both within one part and between different parts. A study of the nesting effect is therefore important for the correct representation of the internal geometry in predictive models of processing and performance and for the assessment of the statistical characteristics of properties, determining processing parameter windows and confidence intervals for the performance indicators. Such an approach is possible using a hierarchical model of the type described here[31].

Table 1.4 Implementation of virtual textile model via OOP approach

Object	Data		Methods
	Group	Fields	
Fibre	General	Linear density (tex) Diameter (mm) Density (g/cm^3)	Compute mass for a given length
	Mechanics	Elastic constants Tenacity (MPa) Ultimate elongation	
Yarn	General	Yarn type: monofilament, continuous filament or spun Linear density (tex)	Compute volume of the given yarn length
	Geometry of the cross-section	Assumed shape: elliptical/ lenticular/rectangular Dimensions of cross-section in free state d_{01}, d_{02} (mm)	Determine, whether the given point (x, y) lies inside the yarn
	Compression	Type of compression behaviour: no compression, specified compression law	Compute compressed yarn dimensions under a given force per unit length
		Compression coefficient $\eta_1 = d_1/d_{10}/$ function of compressive force Q per unit length	Compute compression of two intersecting yarns for given normal force and intersection angle
		Flattening coefficient $\eta_2 = d_2/d_{20}/$ function of compressive force Q per unit length	
	Bending	Bending curve 'torque–curvature' (linear for constant bending rigidity) $M(\kappa)$	Compute bending rigidity value B for a given curvature

(Contd)

Table 1.4 (Continued)

Object	Data		Methods
	Group	Fields	
	Friction	Friction law yarn-yarn in the form $F = fN^n$, where N is a normal force	Compute friction force for a given normal force
Yarn with fibre data	Inherits data and methods of yarn. Adds the following and replaces (**bold**) some of yarn methods		
	General	Twist (1/m) Twist direction (S or Z)	Compute the twist angle Compute linear density from fibre data
	Fibre	Fibre data (*fibre* object) Number of fibres in cross-section Fibre distribution in the yarn	Compute fibrous content and fibre direction in the vicinity of the given point (x, y).
	Compression		Compute compressed yarn dimensions **and fibre distribution in it**
	Bending		Compute bending resistance from fibre data
Yarn path	Fibrous assembly	Implements yarn path mode Holds a reference to yarn with fibre data object	
	Yarn path	Array of descriptions of dimensions and orientation of yarn cross-sections	Compute yarn volume, length and mass Compute average fibre volume fraction
Fabric	Fibrous assembly	Generic description of a fabric Array of *yarn path* objects	Build the yarn descriptions (abstract). Compute overall fabric parameters: dimensions of unit cell, areal density, fibre volume fraction Compute local fibre data at given point

Table 1.4 (Continued)

Object	Data		Methods
	Group	Fields	
			Visualise the fabric
			Export data for micro-mechanical and permeability analysis
Woven fabric		Inherits from *fabric* object. Adds and replaces (**bold**) the following	
	Topology	Weave coding	
	Yarns	References to *yarn with fibre data* objects for all yarns in fabric repeat	
	Spacing	Ends/picks count	
	Fibrous assembly		Build the yarn descriptions

References

1. Kurashiki T., Zako M. and Verpoest I., 'Damage development of woven fabric composites considering an effect of mismatch of lay-up', *Proc. 10th European Conf. on Composite Materials*, Brugge, June 2002 (CD edition).

2. Kawabata, S., *The Standardisation and Analysis of Hand Evaluation*, Osaka, Textile Machinery Society of Japan, 1975.

3. Hearle J.W.S., Grosberg P. and Baker S., *Structural Mechanics of Fibres, Yarns and Fabrics*, New York, Wiley Interscience, 1969.

4. Abbot G.M., 'Yarn-bending and the weighted-ring stiffness test', *J. Textile Institute*, 1983 **74**(5) 281–286.

5. Lomov S.V., Truevtzev A.V. and Cassidy C., 'A predictive model for the fabric-to-yarn bending stiffness ratio of a plain-woven set fabric', *Textile Research J.*, 2000 **70**(12) 1088–1096.

6. Grishanov S.A., Lomov S.V., Harwood R.J., Cassidy T. and Farrer C., 'The simulation of the geometry of two-component yarns. Part I. The mechanics of strand compression: simulating yarn cross-section shape', *J. Textile Institute*, 1997 **88 part 1**(2) 118–131.

7. Harwood R.H., Grishanov S.A., Lomov S.V. and Cassidy T., 'Modelling of two-component yarns. Part I: The compressibility of yarns', *J. Textile Institute*, 1997 **88 Part 1** 373–384.

8. Potluri P., Wilding M.A. and Memon A., 'A novel stress-freezing technique for studying the compressional behaviour of woven fabrics', *Textile Research J.*, 2002 **72**(12) 1073–1078.

9. Love A.E.H., *A Treatise on the Mathematical Theory of Elasticity*, New York, Dover Publications, 1944.

10. Peirce F.T., 'The geometry of cloth structure', *J. Textile Institute*, 1937 **28**(3) T45–T96.

11. Long A.C., 'Process modelling for liquid moulding of braided preforms', *Composites Part A*, 2001 **32**(7) 941–953.

12. Kessels J.F.A. and Akkerman R., 'Prediction of the yarn trajectories on complex braided preforms', *Composites Part A*, 2002 **33**(8) 1073–1081.

13. Byun J.H. and Chou T.W., 'Process–microstructure relationships of 2-step and 4-step braided composites', *Composites Sci. Technol.*, 1996 **56** 235–251.

14. Lomov S.V., Belov E.B., Bischoff T., Ghosh S.B., Truong Chi T. and Verpoest I, 'Carbon composites based on multiaxial multiply stitched preforms. Part 1. Geometry of the preform', *Composites Part A*, 2002 **33**(9) 1171–1183.

15. Hearle J.W.S., Konopasek M. and Newton A., 'On some general features of a computer-based system for calculation of the mechanics of textile structure', *Textile Research J*, 1972 **42**(10) 613–626.

16. Pozdnyakov B.P., *Fabric Resistance to Tension in Different Directions*, Moscow-Leningrad, GIZLegProm., 1932, 70 (in Russian).

17. Novikov N.G., 'A fabric structure and its design with the geometrical technique', *Textilnaya Promishlennost*, 1946 **6**(2) 9–17 (in Russian).

18. Hearle J.W.S., Amirbayat J. and Twaites J.J. (editors), *Mechanics of Flexible Fibre Assemblies*, Alphen aan den Rijn, Sijthoff and Nordhof, 1980.

19. Hearle J.W.S. and Shanahan W.J., 'An energy method for calculations in fabric mechanics', *J Textile Institute*, 1978 **69**(4) 81–110.

20. de Jong S. and Postle R., 'A general energy analysis in fabric mechanics using optimal control theory', *Textile Research J.*, 1978 **48**(3) 127–135.

21. Robitaille F., Clayton B.R., Long A.C., Souter B.J. and Rudd C.D., 'Geometric modelling of industrial preforms: woven and braided textiles', *Proc. Institution of Mechanical Engineers Part L*, 1999 **213** 69–84.
22. Robitaille F., Clayton B.R., Long A.C., Souter B.J. and Rudd C.D., 'Geometric modelling of industrial preforms: warp-knitted textiles', *Proc. Institution of Mechanical Engineers Part L*, 2000 **214** 71–90.
23. Goktepe O. and Harlock S.C., 'Three-dimensional computer modelling of warp knitted structures', *Textile Research J*, 2002 **72**(3) 266–272.
24. Huysmans G., Verpoest I. and Van Houtte P., 'A poly-inclusion approach for the elastic modelling of knitted fabric composites', *Acta Materials*, 1998 **46**(9) 3003–3013.
25. Chamis C.C., 'Mechanics of composite materials: past, present and future', *J. Composites Technol. Res.*, 1989 **11**(1) 3–14.
26. Carvelli V. and Poggi C., 'A homogenization procedure for the numerical analysis of woven fabric composites', *Composites Part A*, 2001 **32**(10) 1425–1432.
27. Robitaille F., Long A.C. and Rudd C.D., 'Geometric modelling of textiles for prediction of composite processing and performance characteristics', *Plastics Rubber and Composites*, 2002 **31**(2) 66–75.
28. Vandeurzen P., Ivens J. and Verpoest I., 'Micro-stress analysis of woven fabric composites by multilevel decomposition', *J. Composite Materials*, 1998 **32**(7) 623–651.
29. Belov E.B., Lomov S.V., Verpoest I., Peters T., Roose D., Parnas R.S., Hoes K. and Sol V., 'Modelling of permeability of textile reinforcements: lattice Boltzmann method', *Composites Sci. Technol.*, 2004 **64**(7–8) 1069–1080.
30. Long A.C., Wong C.C., Sherburn M. and Robitaille F., 'Modelling the effect of fibre architecture on permeability for multi-layer preforms', *Proc 25th SAMPE Europe Int. Conf., Paris*, March/April 2004, pp. 325–330.
31. Lomov S.V., Verpoest I., Peeters T., Roose D. and Zako M., 'Nesting in textile laminates: geometrical modelling of the laminate', *Composites Sci. Technol.*, 2003 **63**(7) 993–1007.

2

Mechanical analysis of textiles

A C L O N G, University of Nottingham, UK,
P B O I S S E, INSA Lyon, France and
F R O B I T A I L L E, University of Ottawa, Canada

2.1   Introduction

This chapter describes the mechanical behaviour of textile reinforcements, with the primary aim of understanding their behaviour during forming and consolidation processes. A number of deformation mechanisms are available. However, during typical composites manufacturing processes, it is generally agreed that the most important of these are in-plane shear and tensile behaviour and through-thickness compaction. Of these mechanisms, the ability of fabrics to shear in-plane is their most important feature during forming, although given their low shear stiffness, in-plane tensile behaviour represents the largest source of energy dissipation. Compaction behaviour defines the fibre volume fraction that can be obtained after manufacturing. Other properties such as fabric bending and ply/tool friction are not considered here, primarily because these have received relatively little attention elsewhere and little data are available.

This chapter introduces a number of experimental methods for characterising the deformation of textiles. These methods have been developed within research studies, usually to obtain material data for manufacturing simulation (see Chapter 4). One important consideration here is that none of these tests is standardised – and in fact nearly all published studies use slightly different test methods and specimen dimensions. This issue is being addressed at present as part of a round-robin exercise[1]. Several standard tests are used in the wider textiles community (e.g. BS ISO 4606:1995, BS 3356:1990, BS 3524-10:1987). Of particular relevance here is the 'Kawabata Evaluation System for Fabrics (KES-F)', a series of test methods and associated testing equipment for textile mechanical behaviour including tensile, shear, bending, compression and friction[2]. However, while this system has been used widely for clothing textiles, its application to reinforcement fabrics has been limited[3]. This is probably because KES-F provides single point data at relatively low levels of deformation, coupled with the limited availability of the (expensive) testing equipment.

As discussed in Chapter 1, the geometry of textile reinforcements can be described at a number of length scales. Individual fibres represent the microscopic scale, with large numbers of fibres (typically several thousand) making up the tow or yarn. The scale of the yarns and of the fabric repeating unit cell is the mesoscopic scale. Finally, the fabric structure constitutes the macroscopic scale. The macroscopic mechanical behaviour of fabrics depends on phenomena at smaller scales, and in particular it is dependent on geometric and contact non-linearities. Such considerations will be used throughout this chapter to develop predictive models for textile mechanical behaviour.

2.2   In-plane shear

2.2.1   Characterisation techniques

As mentioned above, in-plane (or intra-ply) shear is generally considered to be the primary deformation mechanism during forming of textile reinforcements to three-dimensional geometries. Characterisation of this mechanism has therefore received a great deal of attention. The objectives are usually twofold: to measure the non-linear mechanical response of the material during shear, and to characterise the limit of deformation. At the simplest level, mechanical behaviour is of interest for ranking of materials in terms of ease of forming. More recently such data have been required for mechanical forming simulation software based on finite element analysis (see Chapter 4). Here shear force should be normalised by a representative length to eliminate sample size effects. The limit of forming is often characterised by measurement of the 'locking angle', which represents the maximum level of shear deformation that can be achieved before fabric wrinkling occurs. In practice this limit varies widely and is highly dependent on the test method employed. The primary use of locking angle data is to identify areas of wrinkling within kinematic draping codes for fabric forming.

The majority of published studies have characterised resistance to intra-ply shear using two approaches (Fig. 2.1). Bias extension tests, involving uniaxial extension of relatively wide samples in the bias direction, are favoured by a number of researchers[4-6], as the testing procedure is relatively simple. However the deformation field within the sample is non-uniform, with maximum shear observed in the central region and a combination of shear and inter-yarn slip observed adjacent to the clamped edges. In addition the shear angle cannot be obtained directly from the crosshead displacement, so that the test must be monitored visually to measure deformation. Nevertheless this test can provide a useful measure of the locking angle, which in this case represents the maximum shear angle achieved during the test. Above this angle deformation occurs entirely by inter-yarn slip, indicating that the energy required to achieve shear deformation has reached a practical limit. Although this may not indicate the exact angle at which wrinkling would occur, its measurement is repeatable as the bias extension test is not affected significantly by variability in boundary conditions. Bias extension testing is discussed in more detail in Chapter 3, in the context of forming of pre-impregnated composites.

The other popular test method for shear resistance of textile reinforcements is the picture frame test[7-10]. Here the fabric is clamped within a frame hinged at each corner, with the two diagonally opposite corners displaced using a mechanical testing machine. Cruciform specimens are used typically, with the corners of specimens removed adjacent to the bearings in the corners. Samples must be mounted such that the fibres are parallel to the sides of the picture frame prior to testing. Any small misalignment will lead to tensile or compressive forces in the fibre directions, resulting in large scatter in measured force readings. Nevertheless the picture frame test has proved popular as it produces uniform shear deformation (if performed with care).

A number of picture frame test results obtained at Nottingham are reported here as examples of typical behaviour. The picture frame shearing equipment used is illustrated in Fig. 2.1, where the distance between the clamps (l) is 145 mm. Crimped clamps are used typically to ensure that the fabric does not slip from the grips during testing. The apparatus is operated using a Hounsfield mechanical testing machine, which monitors axial load versus crosshead displacement. Here the results are converted into shear force versus shear angle using the following relationships. The shear force (F_s) can be obtained from the measured force in the direction of extension (F_x) using:

$$F_s = \frac{F_x}{2 \cos \Phi} \qquad\qquad 2.1$$

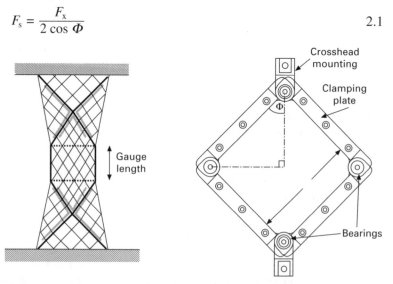

2.1 Characterisation tests for in-plane shear of biaxial fabrics – bias extension (left) and picture frame shear (right).

where the frame angle Φ is determined from the crosshead displacement D_x and the side length of the shear frame (l) using:

$$\Phi = \cos^{-1}\left[\frac{1}{\sqrt{2}} + \frac{D_x}{2l}\right]$$
2.2

The shear angle (reduction in inter-yarn angle) is given by:

$$\theta = \frac{\pi}{2} - 2\Phi$$
2.3

For tests reported here, a pre-tensioning rig was used to position dry fabrics within the picture frame. This consists of a frame within which fabric samples are clamped prior to mounting in the picture frame. Two adjacent sides of the pre-tensioning frame are hinged, to which a measured force is applied to impart tension to the fabric. This device serves two purposes, to align the material within the rig and to enhance repeatability. For the results reported here, a tension of 200 N was applied to fibres in each direction. All tests were conducted at a crosshead displacement rate of 100 mm/min, although experiments at other rates for dry fabric have produced almost identical results[11]. A minimum of six tests were conducted for each fabric, with error bars produced using the t-distribution at 90% confidence limit.

Initial yarn width and pitch (centreline spacing) values were measured using image analysis from digital images oriented normal to the fabric. Video images were also used to estimate the locking angle, although here the procedure was more effective with the camera placed at an oblique angle to the plane of the fabric. Samples were marked with horizontal lines, which buckled when wrinkling occurred. A range of glass fibre reinforcements were tested, including woven fabrics and non-crimp fabrics (NCFs). Descriptions of the fabrics tested are given in Table 2.1, which also includes locking angles averaged for at least four samples.

As shown in the table, for woven fabrics shear angles of at least 55° were achieved before wrinkling was observed. Skelton[5] proposed a lower limit when adjacent yarns come into contact (i.e. yarn width is equal to yarn pitch). Application of this model to the fabrics described in Table 2.1 is of limited use, as the predicted locking angles are typically 20° lower than those measured. Here, fabric locking occurred some time after adjacent yarns came into contact. For the light plain weave (P150), the relatively large spacing between the yarns permitted large shear angles, and in fact only three of the six samples wrinkled before the end of the shear test. The locking angle was lower for the heavier plain weave (P800). Woven fabrics can be ranked for locking angle using the ratio between initial yarn (tow) pitch and width, which is sometimes referred to as the *fabric tightness*. For woven fabrics a high ratio appears to indicate a high locking angle. The same procedure appears to apply to non-crimp fabrics, although ratios here are

Table 2.1 Fibre architecture descriptions and measured locking angles for woven and non-crimp glass fabrics characterised in shear. In each case, tow properties and spacings were identical in warp and weft

Fabric ID	Style	Fibre angles (°)	Surface density (g/m²)	Tow linear density (tex)	Tow pitch (mm)	Tow width (mm)	Tow pitch/width	Locking angle (standard deviation)
P150	Plain weave	0/90	147	150	2.04	1.22	1.67	≈68° (–)
P800	Plain weave	0/90	800	1600	4.01	3.09	1.30	60° (1.4)
S800-01	4-Harness satin	0/90	800	270	0.68	0.62	1.10	55° (1.7)
S800-02	4-Harness satin	0/90	788	1450	3.70	2.80	1.32	61° (1.6)
T800	2:2 Twill weave	0/90	790	2500	6.30	4.40	1.43	62° (1.7)
Ebx936	Tricot 1&1 NCF	±45	936	388	0.82	0.70	1.17	62° (–)
Ebx318*	Pillar NCF	±45	318	99	0.62	0.47	1.32	37°/64° (–)

*Locking angle for this material depends on direction of testing. Value for fabric sheared parallel to stitching is significantly lower than value when sheared perpendicular to stitch.

generally lower than those for woven fabrics with similar locking angles. In addition, for these materials the locking angle can depend on the direction of testing, with different behaviour observed when the fabric is sheared parallel or perpendicular to the stitching thread (as discussed below).

Typical shear compliance curves for woven fabrics are shown in Fig. 2.2. Figure 2.2(a) compares the behaviour of fabrics with similar surface densities

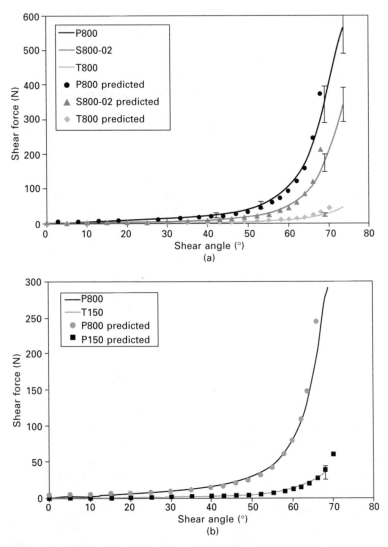

2.2 Experimental and predicted shear compliance curves for woven fabrics. (a) Effect of weave pattern: plain weave (P800), 4-harness satin weave (S800-02) and 2:2 twill weave (T800). (b) Effect of surface density for plain weave: P150 (150 g/m^2) and P800 (800 g/m^2).

but different fibre architectures. The plain weave (P800) requires the highest force to achieve a particular shear angle, while the twill weave (T800) is the most compliant. Clearly this is related to the ratio between yarn width and pitch as described above, although difference in weave style is also significant (as discussed in the next section). Similar observations can be made from Fig. 2.2(b), which compares the behaviour of a two plain weaves with different surface densities. For all fabrics tested, two distinct regions may be identified within the shear compliance curve. The initial resistance to shear is relatively low, and is likely to be caused by friction at the yarn crossovers as suggested by Skelton[5]. Once adjacent yarns come into contact, the resistance increases significantly as the yarns are compressed together. This is the region where wrinkling is usually observed. If the test were continued, the curve would tend towards an asymptote corresponding to maximum yarn compaction (i.e. close packing of the filaments within each yarn).

Figure 2.3 shows typical shear compliance curves obtained for NCFs with both tricot and pillar warp-knit stitching threads. The tricot warp-knit resembles a 'zigzag' pattern, whereas the pillar warp-knit is similar to a chain stitch. In both cases it is apparent that the compliance is lower when the fabric is sheared parallel to the stitching direction. Testing in this direction results in a tensile strain within the stitch, which causes an increase in shear force. The effect is more pronounced for the pillar warp-knit, as the majority of the stitching thread is aligned with the applied force. Here testing parallel to the stitch results in a linear increase in force until the stitching thread snaps. After this point the force is reduced until inter-yarn compaction occurs. The directionality exhibited by NCFs during shear can result in non-symmetric fibre patterns during forming, as described in Chapter 4.

2.2.2   In-plane shear modelling

Given the large number of reinforcements available, it is desirable to develop a model for fabric shear behaviour, both to predict resistance to shear and the change in fabric geometry. Broadly, this must capture the evolution of the fabric geometry during shear, and include contributions from each deformation mechanism to the overall shear behaviour. From the experimental analyses above, it may be concluded that the primary mechanisms are shear at yarn crossovers and friction between yarns, both at crossovers and between parallel yarns. NCFs also dissipate energy via relative displacement and stretching of the stitching threads. The geometry of the fabric can be relatively complex, so that precise determination of forces or energy dissipated via any individual mechanism may be difficult to determine. For example in woven fabrics parallel yarns are only in contact with each other over a fraction of their lengths, and crossover contact may act over relatively complex curved surfaces. Hence to produce an analytical model, a number of simplifications are required.

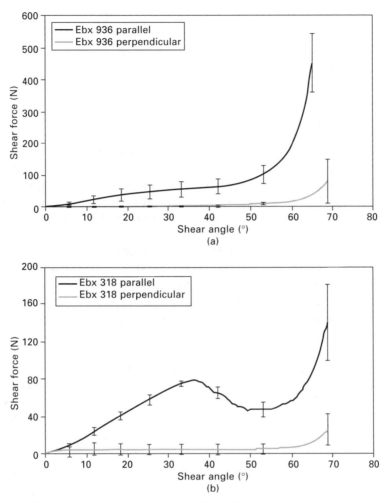

2.3 Shear compliance curves for ±45° NCFs tested parallel and perpendicular to the stitch. (a) tricot 1&1 warp-knit (Ebx936); (b) pillar warp-knit (Ebx318).

Skelton[5] suggested that resistance to in-plane shear for woven fabrics was a result of friction at yarn crossovers, and developed an expression for shear stiffness as a function of the number of crossovers. However, this implied a linear relationship between shear force and shear angle, which is not the case for large deformations (as illustrated above). Kawabata *et al.*[12] performed a more detailed study for a plain woven fabric. In this work the torque required to rotate a single yarn crossover was expressed as an empirical function of contact force and shear angle. Yarn paths were modelled using a saw tooth pattern (straight lines between crossovers), which allowed the contact force to be derived from the tensile forces in the yarns. This model is described in detail in section 2.3.4 in the context of in-plane tensile behaviour. However, it is of limited use for modelling of fabric shear, as it relies on specialised equipment to determine the parameters for the yarn interaction model.

McBride and Chen[13] modelled the yarn paths using sinusoidal curves, which were allowed to evolve during shear deformation. Bias extension tests were performed for a number of plain weave reinforcements, with image analysis used to study the evolution in yarn spacing and yarn width during shearing. Yarn width was found to remain constant until adjacent yarns came into contact, after which the yarns were compacted. For all materials, fabric thickness (measured using a fabric micrometer) remained constant prior to wrinkling. These results may be explained by the fact that an increase in thickness would require elongation of the yarns, resulting in forces significantly higher than those measured above.

Souter[11] developed a general model for shear resistance of woven fabrics. A combination of inter-yarn friction at rotating crossovers and compaction between adjacent yarns was used to predict shear resistance. Yarn cross-sections were assumed to be lenticular based on microscopy, with yarns in intimate contact at crossovers and following straight paths between crossovers. The distance between yarn crossovers along each yarn remained constant (i.e. inter-yarn slip was neglected), whereas the yarn width reduced once yarns came into contact. Fabric thickness was assumed constant during shearing, based on experimental observations by the authors and supported by those from McBride and Chen[13].

The resistance to shearing at each yarn crossover was modelled by considering the torque (T_c) required to rotate a yarn crossover, defined by Kawabata et al.[12] as:

$$T_c = \mu_c \times F_c \times R_{eff} \qquad\qquad 2.4$$

where μ_c is the coefficient of friction (taken as 0.3), F_c is the contact force between the yarns and R_{eff} is the effective radius of rotation of the contact area (i.e. the radius of a circle with the same area as the yarn crossover). Yarn width was assumed to remain constant until adjacent yarns came into contact (i.e. the yarn width is equal to the yarn pitch). After this point the yarn width was reduced to prevent adjacent yarns passing through each other.

Inter-yarn contact force was calculated by resolving the contributions from yarn tension and compaction into the direction normal to the crossover. The resulting contact force depends on the weave style, as woven fabrics can consist of a number of contact geometries as illustrated in Fig. 2.4. Compaction was modelled here using the model from Cai and Gutowski[14], who developed the following equation for the fibre bundle pressure:

$$\sigma_b = \frac{e_b F_{11}}{(F_{bb} F_{11} - F_{b1}^2)} \qquad\qquad 2.5$$

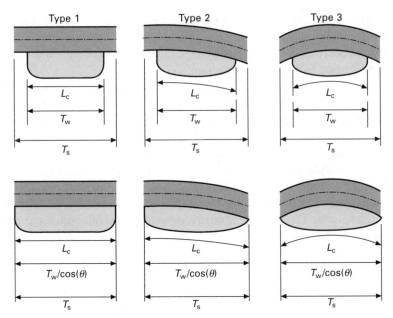

2.4 Three different types of tow crossover: before shearing (top) and during shearing (bottom). L_c = *contact length; T_w = tow width; T_s = tow spacing.*

where the bulk strain in the yarn is defined as:

$$e_b = 1 - \sqrt{\frac{v_f}{v_{f0}}} \qquad\qquad 2.6$$

In the above, v_f and v_{f0} are the current and initial fibre volume fractions, and F_{11}, F_{b1} and F_{bb} are terms of the yarn compliance tensor, which are functions of fibre modulus, fibre waviness (wavelength divided by amplitude for a notional sinusoidal fibre path) and maximal fibre volume fraction. Parameters determined by McBride[15] for compaction of unidirectional yarns were used in the present study.

The torque required to shear the fabric is obtained by summing the torque for each crossover (T_c). The shear force for each specimen can then be calculated from the number of unit cells it contains (N):

$$F_s = \frac{1}{l \cos \theta} \sum_{i=1}^{N} T_{c_i} \qquad\qquad 2.7$$

This method of analysis was applied to the four different woven fabrics described in Table 2.1. Results from the model are included in Fig. 2.2, showing good agreement with experimental data. In particular, the model represents the two observed regions within the shear compliance curve very clearly. Before yarns come into contact, the resistance to shear is small as the yarn contact force is low. Once yarns contact the shear force increases significantly as the contact force includes an additional (and substantial) contribution from yarn compaction.

The above modelling approach has recently been extended to NCFs[16]. The primary differences here are that the yarns are assumed to be perfectly straight, so that significant energy is dissipated via shearing and compaction between parallel yarns. In addition the stitch contributes a significant resistance to shearing via stitch tension and inter-stitch friction. These were modelled using a simple geometric representation for a tricot stitch (Fig. 2.5). Results from this model are shown in Fig. 2.6, which illustrates the difference in shear behaviour predicted for two ±45° tricot-stitched fabrics with different stitch lengths. The model is able to represent the asymmetric shear behaviour typical of NCFs, and also predicts the effect of the stitch geometry on this behaviour.

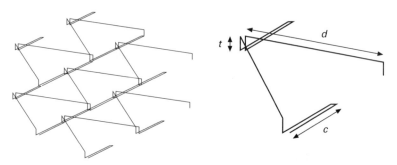

2.5 Tricot stitch pattern for NCF model (left), and stitch unit-cell (right). It is assumed that fabric thickness (t) remains constant, and dimensions c and d evolve as the fabric is sheared.

This section has presented approaches based on analytical modelling to predict the shear response of textile reinforcements. An alternative here is to use 3D finite element analysis to predict fabric shear behaviour. This approach has many attractions, eliminating many of the simplifying assumptions required by analytical models. The approach described in section 2.3.5 has recently been applied to this problem. Such techniques are very useful in the design of reinforcement materials, allowing behaviour to be predicted prior to fabric manufacture and providing useful information to the fabric designer. To assess the effects of fabric mechanics on forming, predictions from these approaches can be used directly as input data for simulations of fabric forming[17], which may allow fabric formability to be optimised for a particular component.

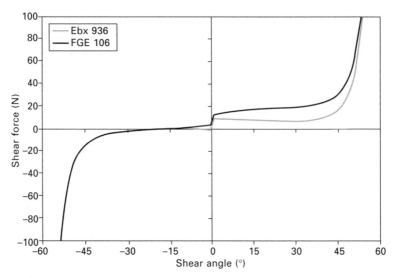

2.6 Predicted shear compliance curves for two ±45° tricot stitched NCFs with different stitch lengths (Ebx 936 c = 2 mm; FGE 106 c = 2.5 mm). Positive shear angles represent shearing parallel to the stitch, whereas negative values represent shearing perpendicular to the stitch.

2.3   Biaxial in-plane tension

2.3.1   Introduction to tensile behaviour of textiles

This section concentrates on tensile behaviour of textile reinforcements when loaded in the fibre directions. Whereas NCFs exhibit relatively linear behaviour, as the fibres remain largely parallel to the plane of the fabric, woven fabrics are well known to have non-linear mechanical properties. Hence woven fabrics form the focus of this section. The diameter of individual fibres within the yarns is very small (5–7 μm for carbon and 5–25 μm for glass) compared with their length. Consequently they can only be submitted to a tensile stress in the fibre direction h_1:

$$\sigma = \sigma^{11}\, h_1 \otimes h_1 \quad \sigma^{11} \geq 0 \qquad\qquad 2.8$$

Fibres are assembled into yarns. Different yarn structures can be obtained according to the fibre arrangement within the yarn. This section considers yarns that are simply juxtaposed (untwisted). This permits relative sliding of the fibres if yarns are subjected to bending, so that the stress state in the yarn defined by eqn 2.8 remains valid.

From the stress state, it is convenient to define the tension in the yarn:

$$T^{11} = \int_{A_y} \sigma^{11} \mathrm{d}S, \quad T = T^{11} h_1 \otimes h_1 \qquad\qquad 2.9$$

where A_y is the sum of the surfaces of the fibres constituting the yarn.

If two yarn networks in directions h_1 and h_2 are considered (Fig. 2.7a), the stress state for the domain defined by these two yarn networks is of the form:

$$\sigma = \sigma^{11} h_1 \otimes h_1 + \sigma^{22} h_2 \otimes h_2 \qquad 2.10$$

and the tensor of the tensions that can be transmitted:

$$T = T^{11} h_1 \otimes h_1 + T^{22} h_2 \otimes h_2 \qquad 2.11$$

$$T^{11} = \int_{A_{y1}} \sigma^{11} \mathrm{d}S, \; T^{22} = \int_{A_{y2}} \sigma^{22} \mathrm{d}S, \; T^{11} \geq 0, \; T^{22} \geq 0 \qquad 2.12$$

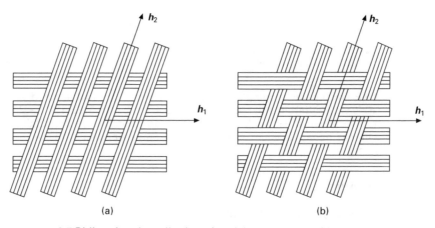

2.7 Bidirectional textile domains: (a) non-woven; (b) woven.

In the above, superscripts 11 and 22 indicate loading parallel to the two yarn directions. When the warp and weft yarns are woven, T^{11} and T^{22} interact because of the interlaced structure. Yarns within woven fabrics are 'wavy' or crimped (Fig. 2.8). Under tensile loads, yarns tend to straighten and to become flattened. In the extreme case, where transverse yarns are free to displace, loaded yarns become totally straight, with transverse yarns becoming highly crimped (Fig. 2.8, $T^{22} = 0$). In intermediate cases, an equilibrium state is reached where the two directions show undulation variations. It is clear that this phenomenon is biaxial and that the two directions interact. The tensile behaviour of woven fabrics is non-linear at low tensions, even if the yarn is linear in tension (Fig. 2.9). As will be shown in the following sections this response depends on the ratio between warp and weft strains. This non-linear phenomenon is observed in the macroscopic tensile behaviour of fabrics, but it is due to geometrical non-linearities at the mesoscopic (repeating unit-cell) scale and is amplified by yarn compaction at the microscopic scale.

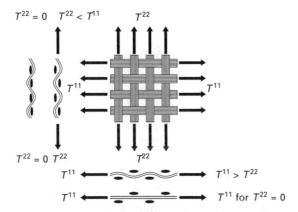

$T^{22} = 0$ $T^{22} < T^{11}$ T^{22}

T^{11} T^{11}

$T^{22} = 0$ T^{22} T^{22}

T^{11} $T^{11} > T^{22}$

T^{11} T^{11} for $T^{22} = 0$

2.8 Tension interactions due to interlacing in woven fabrics.

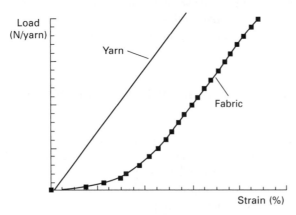

Load
(N/yarn)

Yarn

Fabric

Strain (%)

2.9 Non-linear tensile behaviour for woven fabrics based on linear elastic yarns.

2.3.2   Tensile behaviour surfaces

For a woven domain subjected to in-plane tension (Fig. 2.7), the stress state can be described by the tension tensor, which is of the form:

$$T = T^{11}(\varepsilon_{11}, \varepsilon_{22}) \, h_1 \otimes h_1 + T^{22}(\varepsilon_{11}, \varepsilon_{22}) \, h_2 \otimes h_2 \qquad 2.13$$

The tensions in warp and weft directions, T^{11} and T^{22}, depend on both axial strains, ε_{11} and ε_{22}, because of warp and weft interactions as described above. For a woven domain made of n_{cell} elementary woven cells submitted to in-plane biaxial tension, the dynamic equation can be written in the following form:

$$\sum_{p=1}^{n_{\text{cell}}} {}^P\varepsilon_{11}(\eta)^P T^{11}\,{}^P L_1 + {}^P\varepsilon_{22}(\eta)\,{}^P T^{22}{}^P L_2 - T_{\text{ext}}(\eta) = \int_{\Omega} \rho \ddot{u}\, \eta \, \mathrm{d}V \quad 2.14$$

$$\forall \; \eta/\eta = 0 \text{ on } \Gamma_{\mathrm{u}}$$

$T_{\text{ext}}(\eta)$ is the virtual work due to exterior loads:

$$T_{\text{ext}}(\eta) = \int_{\Omega} f\eta\,dV - \int_{\Gamma_t} t\eta\,dS \qquad\qquad 2.15$$

In the above, f and t are the prescribed volume and surface loads, respectively. Γ_u and Γ_t are boundaries of the fabric domain (Ω) with prescribed displacements and surface loads, respectively. L_1 and L_2 are yarn segment lengths in the two yarn directions for an individual cell, and ρ is the material density.

In eqn 2.13 the tensile mechanical behaviour of the fabric reinforcement is defined by relations between the yarn tensions, which in turn are functions of both yarn strains. Woven fabric tensile behaviour is defined by two biaxial mechanical surfaces, which define the warp or weft loads as a function of the biaxial strains. These are identical if the fabric is balanced, i.e. if the yarns and their paths are the same in the warp and weft directions. Three different analysis techniques are presented for this behaviour in the following sections: experimental measurement, simplified analytical models and 3D finite element analysis.

2.3.3   Experimental analysis of the biaxial tensile behaviour

In this section biaxial tensile experiments are described, the primary aim of which is to determine the two tension surfaces linking warp and weft tensions to the corresponding strains. A number of specialist biaxial tensile devices have been designed for fabric testing[18, 19]. The device used in this study is operated using a conventional tensile/compressive testing machine, and needs no servo-control between the two tensile axes. It is an extension of devices developed for sheet metal, and works via two deformable lozenges[20, 21] as illustrated in Fig. 2.10. When the machine crosshead is raised, it compresses the whole system. This generates a displacement in both the warp and weft directions within a specimen set in the middle of the device. A cruciform specimen is used, where only the square central region (50 mm wide) is of interest. The arms of the cruciform have the transverse yarns removed, so that these regions are uncrimped. This is essential to avoid the use of systems that allow transverse deformation (as required in the Kawabata device, for example[18, 19]).

Various ratios (denoted k) between strains in the warp and weft directions can be imposed by adjusting the dimensions of the testing fixture in one direction. In addition a regulation system allows the inter-yarn angle to be modified from 90° (i.e. effects of fabric shear on tensile behaviour can be measured). Load cells, positioned directly behind the specimen, give the total load in each direction. Strain measurements are performed using either

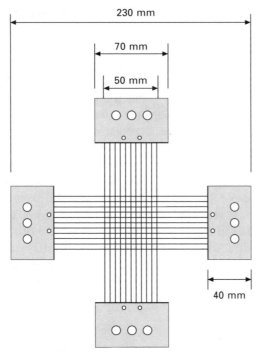

230 mm

70 mm

50 mm

40 mm

2.10 Biaxial tensile testing device and cruciform specimen.

optical method[22, 23] or extensometers. The two methods lead to results in good accordance; optical strain measurements are particularly useful as they allow the full strain field across the central woven region to be assessed. Such analyses have confirmed that the strain field is uniform and homogeneous, even in the inner corners that pose problems for materials with higher in-plane shear stiffness (as is the case for metals[20]).

Results from tests on three fabrics are presented here to illustrate typical biaxial behaviour. These materials are used to produce preforms for the RTM (resin transfer moulding) process. Data are presented for several strain ratios $k = \varepsilon_1/\varepsilon_2$ (1 is the direction under consideration, which may be warp or weft). The fabric density in one direction is the number of yarns per unit fabric length; the crimp characterises undulation and is defined as:

$$\text{Crimp } (\%) = \frac{\text{yarn length} - \text{fabric length}}{\text{fabric length}} \times 100 \qquad 2.16$$

Balanced glass plain weave

This fabric is approximately balanced (the properties in the warp and weft directions are almost identical). The yarn density is 0.22 yarns/mm and the crimp is 0.4%. The individual yarns have slightly non-linear tensile behaviour because fibre cohesion in the yarn is not particularly efficient and the filaments are not rigorously parallel. The tenacity of the glass yarns is 350 N and their stiffness is 38 KN. Tension versus strain curves for the fabric are highly non-linear at low forces and then linear for higher loads (Fig. 2.11a). This non-linearity is a consequence of non-linear geometric phenomena occurring at lower scales, more precisely straightening at the mesoscale and compaction at the microscale. This non-linear zone depends on the prescribed strain ratio, which illustrates the biaxial nature of the fabric behaviour. The non-linear zone is largest for tests in which displacement in the other direction is free (i.e. uniaxial tests, $k = 0$). Here the yarns tend to become completely straight under very low loads. Once the yarns in one direction become straight, the behaviour of the fabric is very similar to that of the yarn alone. The value of the strain corresponding to this transition is representative of the fabric crimp in the loading direction. The experimental results are highly reproducible[24, 25]. From the tension–strain curves for different strain ratios it is possible to generate the experimental biaxial behaviour surface (Fig. 2.11b).

Balanced carbon 2 × 2 twill weave

This balanced fabric is made of yarns composed of 6000 high-strength carbon fibres. In both warp and weft directions the yarn density 0.35 yarns/mm. Their tenacity is 420 N, and their uniaxial tensile behaviour is approximately linear (Fig. 2.12a). The fabric crimp is 0.35%. Biaxial tensile curves for different strain ratios are presented in Fig. 2.12(a), and the tension surface is shown in Fig. 2.12(b). Although the yarns exhibit linear behaviour, the tensile force versus strain behaviour of the fabric is highly non-linear at low loads and then linear at higher loads. The non-linearity decreases as k is increased, in a similar manner to the glass plain weave fabric. Failure has been reached for $\varepsilon = 0.8\%$ ($k = 1$)[25] while the non-linear zone extends to approximately 0.3%. Consequently, when a woven material is loaded, for instance during forming, a major part of the fabric is in a state where its behaviour is highly non-linear; it is, then, important to describe this non-linearity in the mechanical behaviour.

Highly unbalanced glass plain weave

The stiffness of the weft yarn of this fabric is 75 kN per yarn and that of the warp is 8.9 kN per yarn. Consequently the fabric is very unbalanced. The behaviour of the fabric is thus significantly more non-linear than that of the individual yarns. The two warp and weft behaviour curves (and surfaces) are different (Figs 2.13a&b). The weft, which is much more rigid, has a behaviour that is influenced very little by the warp strain. The tension surface is obtained

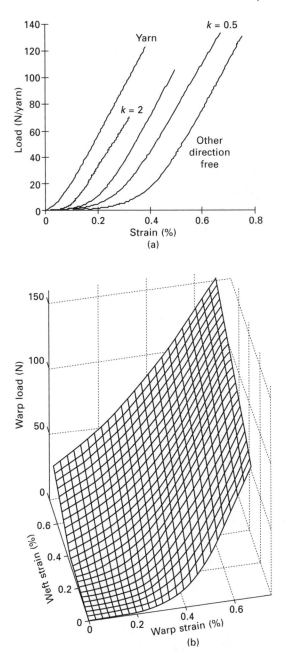

2.11 Results of biaxial tests for a glass plain weave: (a) load–strain curves; (b) tension surface.

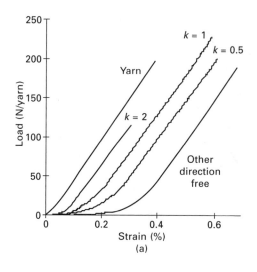

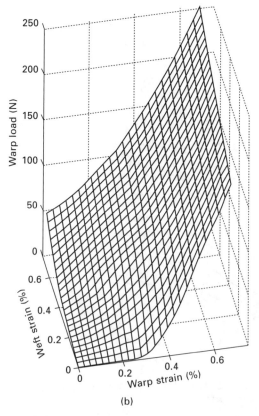

2.12 Results of biaxial tests for a carbon 2 × 2 twill weave: (a) load–strain curves; (b) tension surface.

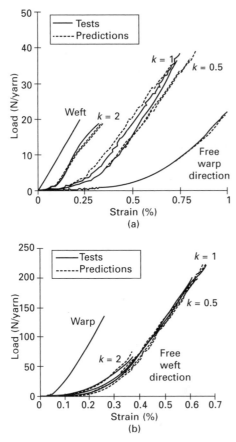

2.13 Results of biaxial tests for an unbalanced glass plain weave:
(a) weft load–strain curve; (b) warp load–strain curve.

by a straight translation in the warp strain direction. In contrast, the warp
strain depends strongly on the weft strain.

2.3.4   Analytical models

The biaxial experiments described above are rather difficult to perform and
it is convenient to use simplified analytical models to predict the mechanical
behaviour of the fabric from parameters related to the weave structure and
the mechanical characteristics of the yarns. Such models can be very fast and
can allow the influence of material parameters to be examined. Many models
have been proposed in the textile science literature[18, 26–29]. Here the well-
known Kawabata model[18] is described, and a second model that is consistent
with the geometry of the woven fabric is presented.

Kawabata model

In this model, yarns within a plain weave are described by straight lines as shown in Fig. 2.14. From the geometry of the mesh the following equation can be derived for the strain in each yarn direction:

$$\varepsilon_{f\alpha} = \{4[e_{0\alpha} + (-1)^{\alpha}w_{\alpha}]^2 + L_{0\alpha}^2 (1 + \varepsilon_{\alpha})^2\}^{1/2} (L_{0\alpha}^2 + 4e_{0\alpha}^2)^{-1/2} - 1$$

2.17

The angle between each yarn and x_3 is:

$$\cos\theta_{\alpha} = 2[e_{0\alpha} + (-1)^{\alpha}w_{\alpha}]\{L_{0\alpha}^2(1 + \varepsilon_{\alpha})^2 + 4[e_{0\alpha} + (-1)^{\alpha}w_{\alpha}]^2\}^{-1/2}$$

2.18

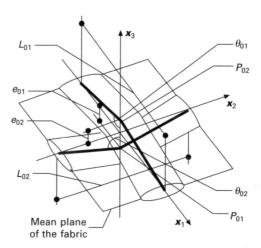

2.14 Geometric representation of a plain weave in Kawabata's model.

In the above, α (= 1 or 2) refers to the yarn direction, and w_{α} is the yarn displacement in the x_3 direction. Other symbols are defined in Fig. 2.14. Equilibrium conditions and the projection of the loads onto the fabric mean plane leads to:

$$T_f^1 \cos\theta_1 = T_f^1 \cos\theta_2 \quad T^{\alpha\alpha} = T_f^{\alpha} \sin\theta_{\alpha} \quad F_c = 2T_f^{\alpha}\cos\theta_{\alpha} \quad 2.19$$

where F_c is the compressive force between yarns along x_3. The tensile behaviour of each yarn is given by:

$$T_f^{\alpha} = g(\varepsilon_f^{\alpha})$$

2.20

A large number of studies have been published concerning the mechanics of yarn compression[14, 18, 29–32], and this is discussed in the context of fabric compaction in section 2.4. Here an empirical compaction law is used, which is different from that proposed by Kawabata[18]:

$$\phi(F_c) = A[2 - \exp(- BF_c) - \exp(- CF_c)] = w_1 - w_2 \qquad\qquad 2.21$$

The compaction coefficients A, B and C are identified using an inverse method[33, 34] from the biaxial test at $k = 1$.

The system of equations given above can be reduced to one non-linear equation depending on transverse displacement, and can be solved by dichotomy if strains are imposed. If loads are known, it is necessary to solve the whole system; an iterative method (e.g. Broyden) can be used efficiently. Good agreement between the model and experimental results for different values of k is demonstrated in Fig. 2.15. Additional results for the fabrics presented in section 2.3.3 are given by Buet-Gautier & Boisse[25].

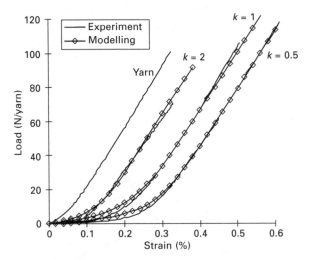

2.15 Comparison between Kawabata's model and experimental data for a glass plain weave.

This model is simple and efficient. However, the 3D geometry of the yarns built on the straight segments penetrate through each other, and in order to obtain good results as shown in Fig. 2.15 the values of parameters A, B and C are not consistent with physical values for the fabric. The goal of the next model is to avoid penetration of the yarns (to be geometrically consistent) and to work with physical values.

A geometrically consistent yarn model for fabric tensile biaxial behaviour

Here yarn longitudinal sections are curved lines with surfaces that are in consistent contact with each other at yarn crossovers. Yarn cross-sections are defined by the intersection of two circular arcs with different diameters. Between yarn crossovers, the yarns are assumed to follow straight paths.

Consequently, the yarn paths are represented by circular arcs and straight segments (Fig. 2.16). The geometry is defined in the general case (unbalanced fabric) by a set of seven material parameters. It is convenient to measure for the warp ($\alpha = 1$) and weft ($\alpha = 2$) direction, the crimp e_α, the interval between two yarns a_α and the yarn width c_α. The last parameter required is the thickness of the fabric w. It can be shown that the solution of this system can be reduced to two non-linear equations giving $\varphi_{\alpha 1}$:

$$\tan \varphi_{\alpha 1} \left\{ 4 \left[L_\alpha - \frac{1}{2} \frac{(a_{\alpha'} - c_{\alpha'})}{\cos \varphi_{\alpha 1}} \right]^2 a_{\alpha'}^{-1} - c_{\alpha'}^2 \right\}^{1/2} = c_{\alpha'} \qquad 2.22$$

where $\alpha' = 3 - \alpha$. If the fabric is balanced, only three parameters (crimp, yarn width and interval between two yarns) are necessary to define the whole geometry. For the balanced glass plain weave presented in Fig. 2.11, the geometric measurements are $c_\alpha = 3.80$ mm, $a_\alpha = 4.55$ mm and $e_\alpha = 0.40\%$. The geometric model leads to $\varphi_{\alpha 1} = 0.134$, $c_\alpha = 2.30$ mm, $b_\alpha = 0.130$ mm and $w = 0.710$ mm. These calculated quantities agree well with their physical measurements, for example the fabric thickness w. The 3D geometry obtained

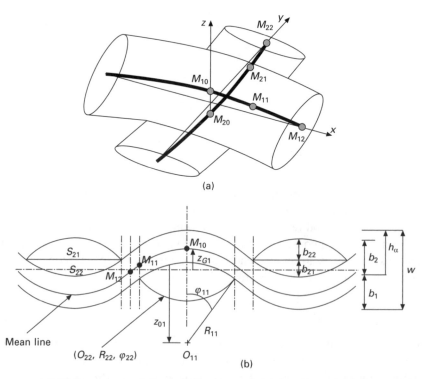

2.16 Geometry of the consistent model (a) and section through the model along one yarn direction (b).

can be used directly for generation of finite element meshes, as will be described in the following section.

The relations between the two strains and tensions in the warp and weft directions are obtained from equilibrium in the deformed configuration:

$$T_f^\alpha = PR_{\alpha 1} R_{\alpha' 1} \varphi_{\alpha 1} \quad T^{\alpha\alpha} = T_f^\alpha \cos \varphi_{1\alpha} \qquad 2.23$$

where P is the contact pressure between crossing yarns. Axial strains in the yarn direction (which are assumed to be small) and the strains in the plane of the fabric are respectively:

$$\varepsilon_f^\alpha = \frac{L_\alpha - L_{\alpha 0}}{L_{\alpha 0}} \quad \varepsilon_{\alpha\alpha} = \frac{1}{2} \left(\frac{a_{\alpha^2} - a_{\alpha_0^2}}{a_{\alpha_0^2}} \right) \qquad 2.24$$

where the subscript 0 indicates the initial value (prior to loading). The tensile behaviour of the yarns is given by eqn 2.20. Transverse compaction stiffness is defined by relating the parameter $r = |z_2 - z_1|$, which describes the relative displacement between yarn paths, with the loads on the mesh. As suggested above, this stiffness depends both on the compaction state and the yarn tensions:

$$r = A \left[2 - e^{-BSP} - e^{-\frac{C}{2}(T_f^1 + T_f^2)} \right] \qquad 2.25$$

where S is the yarn contact surface area. Considering the difficulty in performing a simple test to measure compaction of the yarn, the coefficients A, B and C are again determined by an inverse method using the result of the biaxial test for $k = 1$.

If a strain state of the fabric (ε_{11}, ε_{22}) is prescribed, the equations in the deformed state give the two warp and weft yarn tensions (T^{11}, T^{22}). It can be shown[21, 35] that this set of equations can be reduced to a non-linear system of two equations that can be solved using an iterative (Newton) method. The tension surface obtained in the case of the balanced plain weave glass fabric is compared with the experimental surface in Fig. 2.17. Both surfaces are in good agreement, particularly at low strains where non-linearity is most pronounced.

2.3.5   3D finite element analysis of the unit cell under tension

3D finite element (FE) analysis is an alternative to the two previous approaches to determine the biaxial tensile surfaces for fabric reinforcements. This gives information on local phenomena that influence the global fabric mechanical behaviour. It can be used before fabrication to define the geometrical characteristics and the types of yarn for a fabric in order to obtain the desired

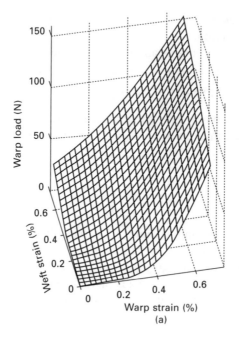

(a)

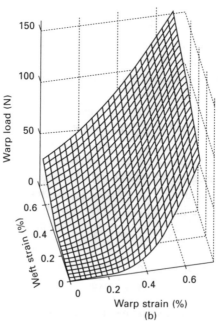

(b)

2.17 Tensile behaviour surface for balanced glass plain weave:
(a) experimental and (b) geometrically consistent model prediction.

mechanical properties. This is particularly interesting for complex fabrics manufactured in small quantities and used in highly demanding structural applications such as aerospace components. In addition the FE approach allows the influence of material parameters (e.g. crimp, yarn density, weave pattern) on the biaxial mechanical properties of the fabric to be analysed in detail[36].

The yarn is composed of a large number of fibres with small cross-sections that are very flexible and can slide in relation to each other. Consequently yarn mechanical behaviour is highly non-linear, and exhibits very low rigidity in all directions other than parallel to the fibres. Furthermore, the transverse stiffness depends on yarn compaction and tension. The equivalent material used in 3D FE analysis must account for this behaviour. This can be achieved using an elastic orthotropic model where the principal frame is defined with direction 1 parallel to the yarn, with very small shear moduli (G_{12}, G_{13}, G_{23}) and Poisson ratios (v_{12}, v_{13}, v_{23}), and with low transverse moduli (E_2, E_3) in comparison with the longitudinal modulus (E_1). Because local geometric non-linearities are of primary importance, the simulation must be performed using large strain theory. Since the axial (1 direction) and transverse moduli are very different, it is important to specify the mechanical characteristics in the appropriate directions during the analysis. Therefore, the yarn stiffnesses are prescribed in material directions, i.e. linked to the finite elements in their current position. In each of these material directions, hypoelastic behaviour is considered.

The longitudinal modulus may be constant or strain dependent, and can be determined by tension tests on single yarns. As already noted, yarn compaction plays a major role in the deformation of woven fabrics. In the current FE-based approach, local values of the mechanical properties are needed, so that to model compaction the transverse modulus E_3 is required. When the yarns are under tension, it can be observed that it is more difficult to compress them. It is therefore necessary to take this tension into account. Here, yarn compaction is represented by the evolution of the transverse modulus, which increases if transverse and longitudinal strains increase:

$$E_3 = E_\varepsilon + E_0 | \varepsilon_{33}^n | \varepsilon_{11}^m \qquad 2.26$$

where E_0, m and n are three material parameters. E_ε is the transverse Young's modulus in the unloaded state, which is almost zero for the fabric yarns described in section 2.3.3. Equation 2.26 is consistent with results given by global yarn compression models for the evolution of the yarn thickness[30, 31]. When loading starts, the transverse stiffness increases because the voids between fibres are reduced. Finally, the stiffness tends to an asymptotic value, which is very large in comparison with the value at low strain. As for the analytical models presented above, an inverse method[33, 34, 36] was used to determine the parameters E_0, m and n, using the results for a biaxial test with

$k = 1$. In this 3D simulation of the woven unit cell, a numerical difficulty comes from the very large difference between weak (or zero) values of some moduli (particularly shear modulus) and the modulus in the direction of the yarn. This can result in spurious zero energy modes. The hourglass control method, developed for finite elements with reduced integration, is used here to avoid this problem[37, 38]. A master–slave approach is used to model contact between the yarns. It has been observed that contact non-linearities are important for the twill weave, but less so for the plain weave.

Figure 2.18 presents the unit cell mesh for the carbon 2 × 2 twill weave (30 000 degrees of freedom), as described in section 2.3.3 The tensile behaviour surface obtained from computations under biaxial tensile loads with different strain ratios is also shown. This is in good agreement with the surface determined by experiments (Fig. 2.12); similarly good agreement between FE predictions and experimental data was obtained for the other fabrics described earlier[21]. Curves obtained from FE analysis are included in the graphs in Fig. 2.13 (dotted lines) for the unbalanced fabric, again demonstrating good agreement.

The 3D FE analyses show that yarn compaction is extremely important in most cases. This is particularly evident under equal biaxial strains (i.e. $k = 1$), where the yarn compaction (logarithmic) strain reaches a maximum value of 40% for the glass plain weave (Fig. 2.19). In contrast, under uniaxial loading ($k = 0$), compaction is negligible: the stretched yarn becomes straight while the undulation of the free yarn increases (Fig. 2.19a).

The 2.5D carbon fabric presented in Fig. 2.20 is used in aerospace structural components. The weave pattern is such that the weft yarns cross two warp layers, so avoiding delamination between the different plies. The FE model presented in Fig. 2.20 has 53 000 degrees of freedom. The warp yarns are initially straight. This explains why biaxial tests give the same response in the warp direction for different strain ratios, so that the tensile behaviour surface is flat (Fig. 2.20b). On the other hand, the response in the weft direction is non-linear and depends on the strain ratio (Fig. 2.20c). Given the complexity of such fabric reinforcements, 3D FE simulations can facilitate fabric design so that the appropriate type of yarn and weave pattern for a given application can be determined before fabric manufacture. Results of computations for other fabrics are given by Boisse and coworkers[21, 36].

2.4   Compaction

2.4.1   Importance of compaction in the processing textile preforms and composites

The widespread use of textile preforms and reinforcements for composites manufacturing results from the easy handling of large quantities of fibres

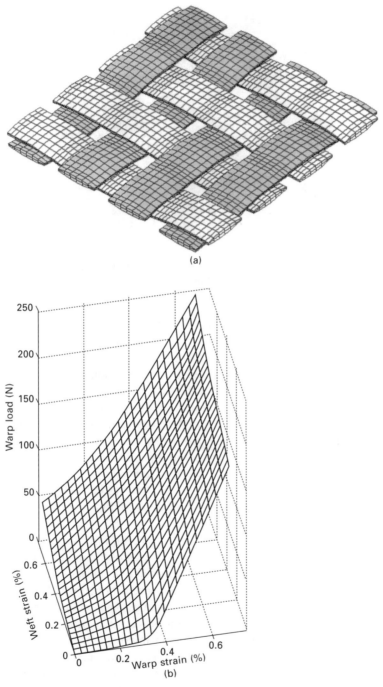

2.18 3D finite element analyses for biaxial tensile behaviour of carbon 2 × 2 twill weave. (a) FE mesh and (b) predicted tensile behaviour surface.

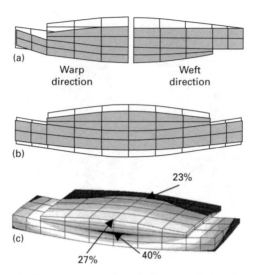

Warp
direction

Weft
direction

23%

27%

40%

2.19 Yarn compaction during tensile loading of a plain weave, for strain ratios of (a) $k = 0$ and (b, c) $k = 1$.

offered by these material forms, which make fast and economical production possible. Despite lower costs, stringent imperatives on processing and performance are nonetheless present when manufacturing composites from textiles. Preforms must be processed in a configuration that eases manufacturing and maximises the properties of the final part.

In RTM (section 5.3.1), the injection of the resin through the preform is eased by low fibre volume fractions. These must be kept within a certain range, above a minimum value in order to prevent movement of the preform in the mould upon injection and below a maximum value to ensure that resin flows at sufficiently high rate under practicable injection pressures. Final parts must feature high fibre volume fractions as this ensures minimal resin usage and high specific properties, as well as limiting the possible occurrence of resin-rich areas.

Similar considerations apply to VI (vacuum infusion, section 5.3.2). In this process the vacuum level and the design of the mould and gates affect the thickness at each point of the part, in its final state and at all times during processing. In turn the thickness affects preform permeability, flow, filling dynamics and pressure distribution. Thickness and compaction behaviour of the textile reinforcement have a major influence on process kinetics for VI.

Compaction behaviour also plays an important role in the stamping of textile-reinforced thermoplastic composites. Here the reduction in thickness is an integral part of the manufacturing cycle, and must be timed accordingly. Initially, any air must be evacuated by flowing along and/or between solid fibres. Then the viscosity of the polymer (present as fibres or in partly

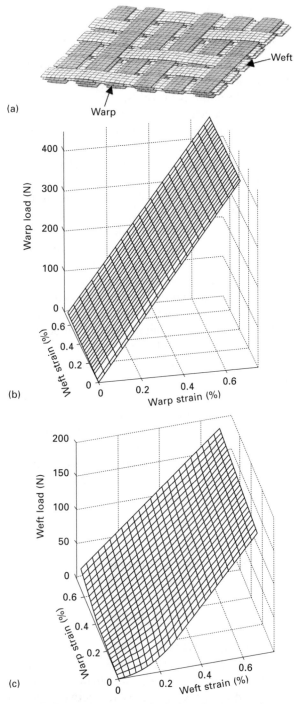

2.20 3D finite element analysis for biaxial tensile behaviour of a 2.5D fabric: (a) FE mesh and (b, c) predicted tensile behaviour surfaces.

consolidated form) is lowered throughout the preform, so that thorough consolidation can take place under increased compaction forces. Critically, these operations must be completed within a minimal cycle time. The thermal behaviour of the material throughout the cycle depends on the imposed compaction level. Knowledge of the fibre volume fraction in a part being processed is also critical to minimising voids.

Textiles in their unloaded state have a low fibre volume fraction (typically between 10 and 25%[39]) which must be increased during processing. The final fibre volume fraction and the rates at which it is increased are generally crucial to the economic production of quality parts. Hence control of the compaction is essential.

Predictive simulation tools for composites manufacturing processes are becoming increasingly available (see Chapters 4 and 7). Accurate predictions require accurate data, including compaction data. However, the usefulness of such compaction data extends beyond process modelling and control. Knowledge of the compaction behaviour is equally important to the specification of manufacturing equipment and textile reinforcements for RTM. Process engineers need to know what forces are involved in bringing a preform to a target fibre volume fraction, and which textiles may be most amenable to this. Similarly, when selecting a vacuum pump for VI moulding one needs to know what preform thicknesses can be achieved for a given port set-up and part dimensions, so that the manufacturing operation can be designed appropriately and operated with confidence.

The precise scope of this section is detailed below, followed by a description of the phenomena covered here. Further information on the effect of diverse parameters that relate to industrial situations is presented, and practical recommendations are made. While the information applies to any manufacturing process where textile reinforcements are used, throughout this section the concepts and their practical consequences are discussed for three processes, namely RTM, VI and thermoplastic stamping.

2.4.2   Scope of the section

This section describes the behaviour of textile reinforcements subjected to compaction forces normal to their plane. The focus is on the relation between the force F applied to the textile and its thickness h, which can be converted respectively to a pressure P by dividing by the projected area A, and to a fibre volume fraction v_f.

In manufacturing, pressure is typically applied between the two rigid surfaces of a mould (RTM, stamping, etc.) or between a lower rigid tool and a flexible membrane (VI, diaphragm forming, etc.). In this section the pressure refers to the force applied on a textile compacted between two platens of finite area. The pressure is obtained by dividing the value of the force by the area of the platens, assuming that the textile sample has the same dimensions. Textiles are made of yarns, which in turn are made of fibres. As it is transmitted through the textile, the externally applied force is actually split into numerous small forces that are transmitted at the numerous points of contact between individual fibres. Therefore, the concept of an average pressure that is constant over the surface of the platens is an approximation. A more detailed investigation would reveal higher pressures on the yarns and no pressure in the empty gaps separating them (Fig. 2.21). The pressure distribution on any given plane cutting through the material would reflect the geometry of the preform on that plane, with pressure levels on individual yarns being significantly higher than the average applied pressure. Here the compaction pressure is defined as the applied force divided by the area of the platens, including the yarns and the empty volumes defined between them.

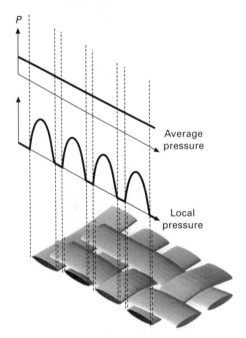

2.21 Definitions of local and average pressure during textile compaction.

The other variable recorded in a compaction test is the preform height (thickness), which is subsequently expressed as a fibre volume fraction. The reason for this is that preforms can be made of any number of textile layers. A given preform will often feature varying numbers of textile layers in different locations, as well as diverse combinations of different textiles. Using a single relationship between the applied pressure and the resulting fibre volume fraction is simpler and more general. Also, the fibre volume fraction of a part is often more critical to designers than the thickness. However, although the concept of a fibre volume fraction applied to a single layer of textile compacted between two rigid platens is unambiguous, real preforms are generally made of more than one textile layer. In such cases the volume, or thickness, occupied by each layer is not precisely defined and individual layers typically run into each other. Fibres from a given layer can lodge themselves between fibres from another layer, and at a larger scale the peaks and valleys formed, say, by a woven textile can either fill each other or not, (Fig. 2.22). This phenomenon is known as nesting and affects the way in which the applied force is transmitted through textile layers.

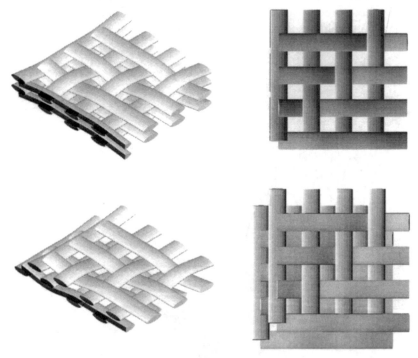

2.22 Illustration of nesting between fabric layers – no nesting (top) and maximum nesting (bottom).

Experiments have shown that the compaction behaviour of multilayer stacks of a given textile differs from that of a single layer of the same material[39]. While a single relation between P and v_f for a given textile may be used as a guideline, in simulation one should ensure that the data represent the actual lay-up envisaged for production, especially when a high-fibre volumes fraction is required. Furthermore, properties such as in-plane permeability of preforms or bending stiffness of composites depend on the relative thickness of individual textile layers. Preforms are sometimes built with random mats located at their core or distributed through the thickness, and enclosed between layers of directional reinforcements. Such preforms will show high in-plane permeability without major reductions in their specific bending stiffness if the compaction of the random mats is controlled. The reader should note that European Standard BS EN ISO 5084:1997 'Textiles – Determination of thickness of textiles and textile products', which supersedes BS 2544, involves a static test where the textile is subjected to one nominal load only. The evolution of v_f as a function of P is not supplied by the test.

Compaction data such as those presented here are usually gathered from tests where platens are parallel and textiles are flat and normal to the compaction axis. A typical rig used to perform such tests is illustrated in Fig. 2.23[40].

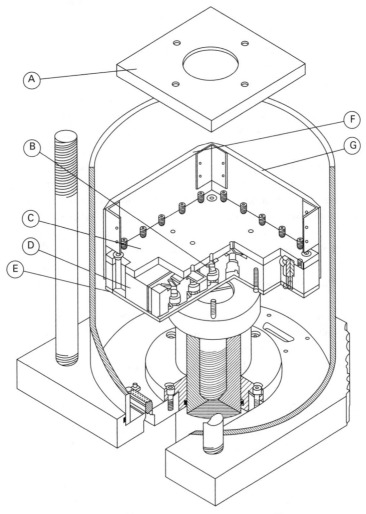

2.23 Testing fixture for reinforcement compaction[40].

Textile layers are compacted between platens A and C. Other items B to G in the figure are used to perform compaction tests on samples saturated in fluid, as discussed later in the text. The two platens are mounted to the ram and fixed traverse of a universal testing machine and the ensemble is loaded, usually at a constant displacement rate in the order of 1 mm/min. Calibration tests performed with an empty rig allow the subtraction of the deformation of the rig from the total deformation at a given force, ensuring that only the compaction of the textile layers is recorded. The sample is compacted either to a predetermined load or to a predetermined fibre volume fraction, depending on the information sought. Upon reaching this state, a number of things may happen. The displacement will be held constant to record the relaxation at a constant fibre volume fraction, or the force will be held constant to record compaction creep, or the sample will be unloaded at the same rate in order to record a compaction hysteresis.

The surfaces of production tools are rarely flat in their entirety, and they are not parallel where changes in thickness occur. Therefore at most points of the tool surfaces the compaction direction is not parallel to the axis of the press. In simulations these factors are accommodated by projecting the press displacement on the normal to the tool surface for every point of that surface, and by obtaining the corresponding pressure level. One may note that assuming that a textile compacts along the normal to the tool surfaces is not rigorously correct, as the actual relative motion of the surfaces includes a lateral component to any normal that is not parallel to the axis of the press. No published experimental data document this phenomenon, and hence it is usually neglected.

The term compaction can be used in the context of individual tows or yarns. Tow compaction happens when textiles are compacted, and is also observed when textiles are sheared (section 2.2) and submitted to tension (section 2.3). Tow compaction is discussed in more detail in Chapter 1. While it may be possible to infer tow compaction properties from experiments performed on textiles through intelligent use of textile models, the focus here is industrial production. Basic compaction cycles are described, and phenomena such as behaviour upon impregnation, relaxation and unloading are discussed. The important aspect of repeatability and statistical distribution of the data is covered, along with implications for mould design.

2.4.3   Basic compaction and relaxation curves

Figure 2.24 shows a typical compaction and unloading cycle for a stack of textile reinforcements. In this case the distance between platens was reduced at a constant rate of 0.5 mm/min, held constant for 1 min and returned to its initial value at the same rate. The curve shows the evolution of the compaction pressure P as a function of the average fibre volume fraction v_f for six layers of a woven textile; v_f varies between 0.40 and 0.60 and the maximum pressure

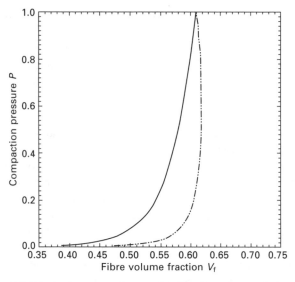

2.24 Typical compaction and unloading curves for a stack of plain weave reinforcements, tested at a constant displacement rate compaction test (pressure in MPa).

is 1.0 MPa or 10 bar; such high compaction pressures represent a practical maximum. The compaction part of the cycle can be characterised by an initial fibre volume fraction $v_{f,o}$ which relates to the unloaded textile, a representative fibre volume fraction $v_{f,r}$ read at a predetermined pressure level, and an average stiffness defined over part of the loading curve where pressure levels are significant. The curve shows that compaction pressure P builds up rapidly with v_f and that over the range of fibre volume fractions used in production parts, a small increase in v_f requires a large increase in pressure.

As one aims at maximising the fibre volume fraction in commercial parts, textiles are often compacted to levels where P increases very rapidly with v_f, making knowledge of the compaction behaviour critical in selecting a press. In processes such as RTM the press must equilibrate the compaction pressure as well as the pressure of the resin injected in the tool. Slower injection rates lead to longer production times but allow marginally higher fibre volume fractions for a given press; such aspects must be considered carefully when designing production equipment. The resistance to changes in v_f with fluctuations in P is advantageous in VI as small fluctuations in vacuum levels do not affect the reproducibility of local part thicknesses. However, it should be pointed out that v_f is more sensitive to P at the lower levels of pressure involved in that latter process. Finally, it is generally easier to control the stamping of thermoplastics when conducted under a membrane as opposed to a positive displacement press; this is important especially as achievement of low void content requires accurate timing of heating and compaction.

The dotted line in Fig. 2.24 corresponds to the relaxation step where the distance between platens is held constant, and to load removal. The most prominent feature is that the two curves for loading and unloading do not superimpose, hence the behaviour is not elastic. A practical consequence is that if v_f is overshot, say in preform stamping, the properties of the final assembly will be altered. For example, if a RTM preform targeted at a fibre volume fraction of 50% reached a value of 55% during its production (preforming) it would produce lower compaction pressures upon closure of the actual RTM mould (constant thickness), possibly resulting in a displacement of the preform upon resin injection. However, the same phenomenon can have fortuitous consequences. In VI compaction pressure is highest when the preform is dry, and it is progressively reduced behind the flow front. Pre-compacted preforms show reduced spring back for the same compaction pressure, hence higher fibre volume fractions may be reached in the final parts. More generally, energy is absorbed when a textile is compacted, its structure is altered, and in most cases it will not return to its initial height.

Figure 2.25 illustrates a typical time relaxation curve. The curve shows the evolution of P with time t when the distance separating the platens is held constant after the initial compaction. Once a certain maximum pressure level is reached and platen travel is stopped the pressure reduces. If after a certain

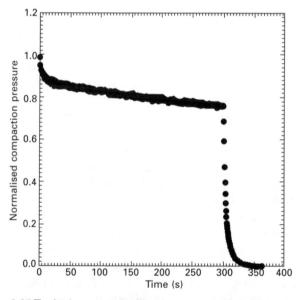

2.25 Typical pressure relaxation curve, where the pressure is measured as a function of time with the reinforcement held at a constant thickness.

time the pressure is returned to its initial maximum level the fibre volume fraction will increase further. This is why some pressure-controlled preforming presses can be observed to re-clamp at more or less regular intervals. The practical benefits of this are limited as relatively long relaxation periods lead to only small increases in v_f, especially at higher fibre volume fractions. In spite of this, relaxation is an important phenomenon as changes in relative thickness for a multilayer preform affect permeability, and reduced overall compaction pressure may render a preform more prone to movement during injection.

Figure 2.26 shows the evolution of applied load and resulting platen displacement with time for a compaction test conducted at a constant loading rate. Such tests are conducted using instrumented machines, the control of which is fairly remote to that of industrial presses. The curve is informative as its first section shows that compaction slows down as it progresses, reflecting the stiffening behaviour of compacted textiles. The second section of the curve shows the effect of an imposed constant load. The fibre volume fraction typically increases by 1–2% in 60 s, while relaxation curves typically show reductions in pressure of 15–25% in 5 min. This reinforces previous remarks stating that for fibre volume fractions representative of actual parts, time does increase v_f but only marginally. The third section corresponds to load removal and confirms previous observations. The curve features a hysteresis, hence the textile is irreversibly modified as it is compacted.

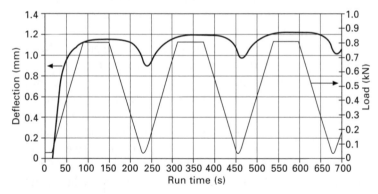

2.26 Evolution of applied load and resulting displacement (bold line) with time for a constant loading rate compaction test.

2.4.4   Effects of variable parameters

The compaction and relaxation behaviour of textile reinforcements is affected by preform construction and production parameters. For example, both compaction and relaxation are altered when the number of layers present in a preform is changed. This section documents such effects. In order to provide a usable summary of general trends, the effect of parameters on a number of variables are considered.

For compaction the following quantities were introduced previously: the initial fibre volume fraction $v_{f,o}$ of the unloaded textile, the representative fibre volume fraction $v_{f,r}$ read at a predetermined pressure level, and the average stiffness defined over a set range of pressure levels. Another variable that will be commented upon is the rate at which the rigidity of the textile increases as the textile is compacted; this will be referred to as the stiffening index. For relaxation, a reduction in compaction pressure is observed at a constant thickness and therefore the reduction in pressure observed after a period of 5 min will be commented upon; this will be referred to as the pressure decay.

The parameters investigated for both compaction and relaxation are the number of layers (NOL) and saturation by a fluid (SAT). In the latter case, experimental observations show that the compaction behaviour of a set number of layers of reinforcement differs if the reinforcement is compacted dry or saturated with resin. Considering VI for example, this means that when the resin front reaches a given point of the preform, the compaction behaviour at that point changes, effectively becoming less stiff. This can result in a local reduction in thickness at the flow front, a phenomenon usually visible to the naked eye. Additional parameters will also be commented upon. For compaction, the effect of the number of cycles (NOC) applied to a preform will be discussed. This applies to numerous practical situations such as RTM operations where preforms are often manufactured in a separate operation to the actual moulding.

The observed effects of the NOL on compaction are as follows. Different authors have reported different trends, which may be explained by the different textiles and pressure levels used. For some textiles compacted to higher pressures, it has been shown that stacks made of a higher number of layers are more difficult to compact to a given fibre volume fraction v_f with compaction curves progressively shifting to the left (lower fibre volume fractions), mostly at higher pressures. At the onset of compaction multilayer stacks are easier to compact because of nesting; however, at higher pressures these preforms are usually more difficult to compact to a given v_f, hence fibre volume fractions are lower for a given pressure and the curve shifts to the left. While $v_{f,o}$ usually increases with NOL, $v_{f,r}$ shows the opposite trend at high pressures. Therefore both the rigidity and stiffening index tend to increase with NOL (Fig. 2.27). Other authors who have used lower compaction pressures, which are more indicative of manufacturing using VI for example, find that multilayer preforms are easier to compact to higher fibre volume fractions. Their results are not contradictory as they use different pressure levels. Besides, the effect of the number of layers on the compaction behaviour is often relatively weak and some authors have observed conflicting trends.

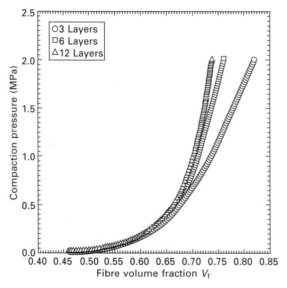

2.27 Textile compaction data for stacks with different numbers of layers for a non-crimp fabric reinforcement.

Preforms made of a higher number of layers show more relaxation in pressure at constant thickness. The pressure decay, defined as the ratio of the pressure observed after 5 mins to the maximum applied pressure, is typically reduced by up to 10% for thicker preforms.

The effects of compacting a reinforcement saturated in a fluid, say resin, are as follows. It should be noted that what is discussed here is quasi-static compaction, i.e. compaction at displacement rates that are sufficiently low to render insignificant the pressure generated by the fluid as it is squeezed out of the reinforcement stack. The phenomenon investigated in the reported experiments is the effect of saturation, and lubrication, on the individual contacts between fibres where forces are transmitted. Typical results for the effect of saturation are given in Fig. 2.28. Experiments show that saturated (v_f, P) compaction curves are generally shifted to the right, that is towards higher fibre volume fractions. The initial fibre volume fraction $v_{f,o}$ and representative fibre volume fraction $v_{f,r}$ are generally higher by a few per cent. As both values of the fibre volume fraction, the rigidity and stiffening behaviour, are generally unaffected; the compaction behaviour remains essentially identical but marginally higher values of fibre volume fraction are observed at given values of the compaction pressure. This can be observed practically for VI as the preform usually compacts to a slightly lower thickness (higher v_f) as the flow front reaches a given position. The effect cannot be observed easily for RTM; however, once enclosed in a mould it is thought that the load on a preform will progressively reduce as a result of time

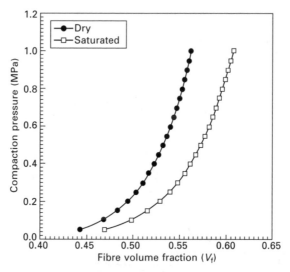

2.28 Effect of fluid saturation in textile compaction behaviour.

relaxation, and that a further reduction will be observed upon impregnation. It is therefore important to use a sufficiently thick preform that will lead to appropriate initial levels of compaction pressure, in order to ensure proper positioning of the preform in the tool throughout the process.

Saturation usually has a strong effect on the relaxation part of the cycle. Saturated stacks reach somewhat higher volume fractions when compacted, and they exhibit significantly less relaxation; after 5 mins the pressure usually remains between 75 and 90% of its initial value. It is believed that this behaviour can be explained by the lubrication of fibre-to-fibre contact points. Individual fibres will slide more easily for a saturated stack and are therefore more likely to attain a higher degree of stability for a given compaction pressure. As a result, fibres are less likely to move under pressure, exhibiting less relaxation.

The above comments are corroborated when one considers the compaction behaviour of a preform subjected to repeated compaction cycles. Figure 2.29 shows typical compaction cycles imposed on a multilayer preform; the maximum pressure level is constant at 1.0 MPa[41]. The figure clearly shows that the compaction and load release curves shift to the right on successive cycles. The shift is more marked on the first repeat and progressively diminishes on successive cycles. Both $v_{f,o}$ and $v_{f,r}$ increase, and this increase is typically more marked than those observed with saturated preforms, notably on the first cycle. The fibre volume fraction observed at higher pressures increases, but less than it does at low fibre volume fractions. Consequently, the rigidity and stiffening index also increase markedly on successive cycles. Such behaviour is observed over a number of cycles; the difference over successive

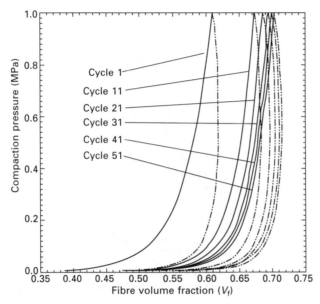

2.29 Pressure curves for successive compaction cycles applied to a multilayer plain weave preform.

cycles diminishes progressively but does not appear to stabilise, at least over 50 cycles. This, along with the fact that an important hysteresis is observed on each cycle, points to the fact that reorganisation of the fibres happens as the textile is compacted and keeps happening on successive cycles. While of little practical importance for processes such as RTM, this provides good insight on the behaviour of textile reinforcements. Furthermore, this behaviour may have consequences for VI where the preform is compacted under vacuum, unloaded upon injection and possibly reloaded during bleeding.

The above trends can be put in perspective by looking at the natural variability observed upon compaction of nominally identical stacks of textile reinforcements. Figure 2.30 shows typical data scatter for 12 identical compaction samples. Here the textile reinforcements were compacted to a lower maximum pressure of 0.1 MPa, which usually leads to more scatter for a given pressure. However, and although the trends discussed above emerge clearly from carefully conducted experiments, the amplitude of the scatter shown in Fig. 2.30 is similar to some of the weaker trends, say for example the effect of the number of layers on compaction curves. Actual production equipment is generally not selected for operation at its limit; in spite of this, variability data should be included when considering necessary forces and control accuracy.

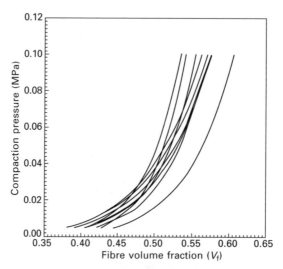

2.30 Compaction data for 12 tests conducted on samples of the same reinforcement, showing the significant data scatter that is to be expected.

2.4.5   Modelling

A number of authors have proposed models for the compaction behaviour of textiles (a thorough review of these is given by Robitaille and Gauvin[39]). Such models usually aim to fulfil one of two objectives: phenomenological models attempt to explain the behaviour from the fundamental principles of mechanics while empirical models provide a simple and usable representation of experimental data. Although both approaches were investigated for compaction, no credible explanation is available for the time relaxation of textile reinforcements and in this case a limited number of empirical models simply aim at quantifying the phenomenon.

The most widely known and used phenomenological compaction model was proposed by Cai and Gutowski[14]. The main concept behind this model is that individual fibres form arches between their contact points; as compaction progresses it is assumed that the number of contact points increases as the length of the arches is essentially proportional to the thickness of the reinforcement being compacted. Consequently, beam bending theory dictates that the arches progressively become more rigid as their number increases, resulting in a marked stiffening behaviour. The authors have proposed different equations (see equation 2.5 for an alternative example), but perhaps the most general is expressed as:

$$P_{comp} = A_s \frac{\left(\dfrac{v_f}{v_{fo}} - 1 \right)}{\left(\dfrac{1}{v_f} - \dfrac{1}{v_{fa}} \right)^4}$$

2.27

where $v_{f,o}$ and $v_{f,a}$ are the initial and maximal fibre volume fractions, A_s is an empirical constant, and v_f and P_{comp} are the fibre volume fraction and the compaction pressure. It should be noted that owing to its nature and phenomenological origin the model uses the fibre volume fraction as the independent variable, which is not ideal; the form of the equation can make the extraction of a fibre volume fraction at a given compaction pressure awkward.

Gutowski's work provides very interesting and valuable insight into the compaction phenomenon; however, it really applies to homogeneous assemblies of parallel fibres. Results of experiments conducted on reinforcements of various architectures clearly show that the textile structure, or the way in which tows or yarns of various sizes are assembled into a planar assembly that is homogeneous and continuous at the macroscopic scale, but heterogeneous and discontinuous at the mesoscopic scale, strongly affect the behaviour. It is possible, for example, to classify compaction curves for random mats, woven textiles and stitched directional reinforcements into relatively distinct groups; this important point is discussed further in the following paragraphs. Furthermore, getting a good fit for the diverse coefficients of the models can be challenging in some cases, and values obtained for coefficients with a clear physical meaning such as $v_{f,o}$ and $v_{f,a}$ often do not match reality when the model is applied to heterogeneous textiles. The model was really developed for assemblies of aligned fibres and the lack of correspondence between the fitted and experimental values reduce its attractiveness to some extent.

The alternative consists of using empirical models. While studying changes in the values of the coefficients of such models with experimental parameters helps in identifying trends, there is no pretence at a quantitative, mechanical description of the phenomena involved in the process. However, this can be compensated by the fact that similar models can be used to describe relaxation behaviour, enabling both phenomena to be included in process simulations and treated in the same way. Empirical models of compaction and relaxation are not satisfactory in all circumstances. Numerous published works have provided remarkable phenomenological insight into compaction and other properties of fibre assemblies, as well as models that can represent real behaviour for actual fibrous structures such as ropes. Finite element analyses similar to those described in section 2.3.5 would appear to be a promising approach here, and this is expected to be an area of significant activity in the near future. At this time most of these models require very significant computational power, but work is ongoing and in the future, more efficient models that can reasonably be used to support actual production of composite parts may appear.

Most empirical models of the compaction revolve around some form of the power law. The authors and numerous others have used the following expression to fit experimental compaction data:

$$v_f = v_{fo} \cdot P_{comp}^B \qquad\qquad 2.28$$

where $v_{f,o}$ is the fibre volume fraction recorded under a pressure of 1 Pa and B is an empirical factor called the stiffening index. Because of the mathematical form of the equation the latter value provides a good indication of the rate of increase of the rigidity as a function of v_f; as such it is practically very useful. Here the pressure is the independent variable, and the equation is easily fitted to data.

Robitaille and Gauvin[39] proposed a similar model for time relaxation:

$$\frac{P}{P_o} = 1 - Ct^{(1/D)} \qquad\qquad 2.29$$

where C is the pressure decay after 1 sec, D is the relaxation index, P_o is the initially applied pressure and P is the compaction pressure observed at time t which is the independent variable.

The above empirical formulae have allowed the formal identification of all the trends mentioned above. The compaction model was also successfully integrated into a number of process simulation models. Tables of coefficients for both equations can be found in the literature for a range of materials[39, 40] and it is very straightforward to fit experimental data to the equations. While these models do not explain the physics of compaction and relaxation and one must identify the coefficients associated to each stack of reinforcements that may be used in production, they are very useful in practice.

A recent study further extends the benefits of the compaction model when considering a number of structurally different reinforcements[42]. The curve shown in Fig. 2.31 is a compaction master curve, which presents the evolution

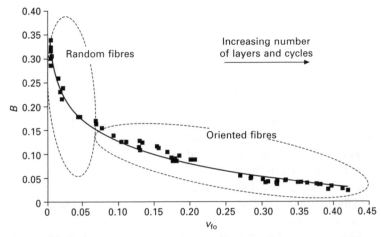

2.31 Compaction master curve, showing a relationship between stiffening index and initial fibre volume fraction values (equation 2.28) for a range of reinforcements.

of the stiffening index B as a function of the initial fibre volume fraction $v_{f,o}$ for an array of reinforcements. The curve shows a fairly clear general relationship between the two parameters. Furthermore, all random reinforcements are found on the left side of the curve, most woven fabrics occupy the centre and textiles featuring highly oriented fibres (e.g. NCFs) are found on the right. Practically this translates into the fairly simple fact that random textiles initially show low fibre volume fractions and that they stiffen relatively slowly, while crimp-free assemblies of unidirectional fibres feature high initial fibre volume fractions and stiffen very rapidly.

References

1. Blanchard P., Cao J., Chen J. and Long A.C. (co-chairs), *NSF workshop on composite sheet forming*, Lowell, MA, Sept 2001. http://www.mech.northwestern.edu/fac/cao/nsfworkshop/index.htm

2. Saville B.P., *Physical Testing of Textiles*, Cambridge, Woodhead Publishing Ltd, 1999.

3. Lomov S.V., Verpoest I., Barburski M. and Laperre J., 'Carbon composites based on multiaxial multiply stitched preforms. Part 2. KES-F characterisation of the deformability of the preforms at low loads', *Composites Part A*, 2003 **34**(4) 359–370.

4. Potter K.D., 'The influence of accurate stretch data for reinforcements on the production of complex structural mouldings', *Composites*, July 1979 161–167.

5. Skelton J., 'Fundamentals of fabric shear', *Textile Research J.*, Dec. 1976 862–869.

6. Wang J., Page R. and Paton R., 'Experimental investigation of the draping properties of reinforcement fabrics', *Composites Sci. Technol.*, 1998 **58** 229–237.

7. Canavan R.A., McGuinness G.B. and O'Braidaigh C.M., 'Experimental intraply shear testing of glass-fabric reinforced thermoplastic melts', *Proc. 4th Int. Conf. on Automated Composites*, Nottingham, Sept. 1995, 127–138.

8. Long A.C., Rudd C.D., Blagdon M. and Johnson M.S., 'Experimental analysis of fabric deformation mechanisms during preform manufacture', *Proc. 11th Int. Conf. on Composite Materials*, Gold Coast, Australia, July 1997, V: 238–248.

9. Prodromou A.G. and Chen J., 'On the relationship between shear angle and wrinkling of textile composite preforms', *Composites Part A*, 1997 **28A** 491–503.

10. Mohammad U., Lekakou C., Dong L. and Bader M.G., 'Shear deformation and micromechanics of woven fabrics', *Composites Part A*, 2000 **31** 299–308.

11. Souter B.J., Effects of fibre architecture on formability of textile preforms, PhD Thesis, University of Nottingham, 2001.

12. Kawabata S., Niwa M. and Kawai H., 'The finite-deformation theory of plain-weave fabrics Part III: The shear-deformation theory', *J. Textile Institute*, 1973 **64**(2) 61–85.

13. McBride T.M. and Chen J., 'Unit-cell geometry in plain-weave fabrics during shear deformations', *Composites Sci. Technol.*, 1997 **57** 345–351.

14. Cai Z. and Gutowski T.G., 'The 3D deformation behavior of a lubricated fiber bundle', *J. Composite Materials*, 1992 **26**(8) 1207–1237.

15. McBride T.M., The large deformation behavior of woven fabric and microstructural evolution in formed textile composites, PhD Thesis, Boston University, 1997.

16. Harrison P., Wiggers J., Long A.C. and Rudd C.D., 'A constitutive model based on meso and micro kinematics for woven and stitched dry fabrics, *Proc. 14th Int. Conf. on Composite Materials*, San Diego, July 2003.

17. Harrison P., Yu W.-R., Wiggers J. and Long A.C., 'Finite element simulation of fabric forming, incorporating predictions of a meso-mechanical energy model', *Proc. 7th Int. European Scientific Association for Material Forming Conf.*, Trondheim, Norway, April 2004, 139–142.

18. Kawabata S., Niwa M. and Kawai H., 'The finite-deformation theory of plain weave fabrics Part I: The biaxial deformation theory', *J. Textile Institute*, 1973 **64**(2) 21–46.

19. Kawabata S., 'Nonlinear mechanics of woven and knitted materials', in Chou T.W. and Ko F.K. (eds) *Textile Structural Composites*, Amsterdam, Elsevier, 1989, pp 67–116.

20. Ferron G., 'Dispositif de traction biaxiale DAX2', Document Tech-metal, Mézières-Lès-Metz, 1992 (in French).

21. Boisse P., Gasser A. and Hivet G., 'Analyses of fabric tensile behaviour. Determination of the biaxatension-strain surfaces and their use in forming simulations', *Composites Part A*, 2001 **32**(10) 1395–1414.

22. Launay J., Lahmar F., Boisse P. and Vacher P., 'Strain measurement in tests on fibre fabric by image correlation method', *Adv. Composite Lett.*, 2002, **11**(1) 7–12.

23. Dumont F., Hivet G., Rotinat R., Launay J., Boisse P. and Vacher P., 'Identification des caractéristiques mécaniques de renforts tissés à partir de mesures de déformations par corrélation d'images', *Mécanique & Industries*, in press (in French).

24. Buet-Gautier K, Analyse et simulation du comportement mécanique des renforts composites tissés, PhD Thesis, University of Orléans, 1998 (in French).

25. Buet-Gautier K. and Boisse P., 'Experimental analysis and models for biaxial mechanical behaviour of composite woven reinforcements', *Experimental Mechanics*, 2001, **41**(3), 260–269.

26. Anandjiwala R.D. and Leaf G.A.V., 'Large-scale extension and recovery of plain woven fabrics', *Textile Research J.*, 1991 **61**(11) 619–634.

27. Realff M.L., Boyce M.C. and Backer S., 'A micromechanical approach to modelling tensile behavior of woven fabrics', in Stokes V.K. (ed) *Use of Plastics and Plastic Composites: Materials And Mechanics Issues*, New York, ASME MD-Vol. 46, 1993, 285–294.

28. Dastoor P.H., Ghosh T.K., Batra S.K. and Hersh S.P., 'Computer-assisted structural design of industrial woven fabrics. Part III: Modelling of fabric uniaxial/biaxial load-deformation', *J. Textile Institute*, 1994 **85**(2) 135–157.

29. Lomov S.V., Gusakov A.V., Huysmans G., Prodromou A. and Verpoest I., 'Textile geometry preprocessor for meso-mechanical models of woven composites', *Composites Sci. Technol.*, 2000 **60** 2083–2095.

30. Grishanov S.A., Lomov S.V., Harwood R.J., Cassidy T. and Farrer C., 'The simulation of the geometry of two-component yarns. Part I: The mechanics of strand compression: simulating yarn cross-section shape', *J. Textile Institute*, 1997 **88**(2) 118–131.

31. Harwood R.J., Grishanov S.A., Lomov S.V. and Cassidy T., 'Modelling of two-component yarns. Part I: The compressibility of yarns', *J. Textile Institute*, 1997 **88**(4) 373–384.

32. Baoxing C. and Chou T.W., 'Compaction of woven-fabric preforms in liquid composite molding processes: single layer deformation', *Composite Sci. Technol.*, 1999 **59** 1519–1526.

33. Marquardt D.W., 'An algorithm for least squares estimation of nonlinear parameters', *J. Soc. Indus. Appl. Math*, 1963 **11**(2) 431–441.

34. Schnur D.S. and Zabaras N., 'An inverse method for determining elastic material properties and a material interface', *Int. J. Numerical Methods Eng.*, 1992 **33** 2039–2057.

35. Hivet G., Modélisation mesoscopique du comportement biaxial et de la mise en forme des renforts de composites tissés, PhD Thesis, University of Orléans, 1998 (in French).

36. Gasser A., Boisse P. and Hanklar S., 'Analysis of the mechanical behaviour of dry fabric reinforcements. 3D simulations versus biaxial tests', *Comp. Material Sci.*, 2000 **17** 7–20.

37. Flanagan D.P. and Belytschko T., 'A uniform strain hexahedron and quadrilateral with orthogonal hourglass control', *Int. J. Numerical Methods Eng.*, 1981 **17** 679–706.

38. Pian T.H. and Chen D., 'On the suppression of zero energy deformation modes', *Int. J. Numerical Methods Eng.*, 1983 **19** 1741–1752.

39. Robitaille F. and Gauvin R., 'Compaction of textile reinforcements for composites manufacturing. I: Review of experimental results', *Polymer Composites*, 1998 **19**(2) 198–216.

40. Robitaille F. and Gauvin R., 'Compaction of textile reinforcements for composites manufacturing. II: Compaction and relaxation of dry and H_2O-saturated woven reinforcements', *Polymer Composites*, 1998 **19**(5) 543–557.

41. Robitaille F. and Gauvin R., 'Compaction of textile reinforcements for composites manufacturing. III: Reorganization of the fiber network', *Polymer Composites*, 1999 **20**(1), 48–61.

42. Correia N., Analysis of the vacuum infusion process, PhD Thesis, University of Nottingham, 2004.

Rheological behaviour of pre-impregnated textile composites

P HARRISON and M CLIFFORD,
University of Nottingham, UK

3.1   Introduction

Virtual design and manufacture of industrial components can provide significant increases in efficiency and reductions of cost in the manufacturing process. This factor provides strong motivation for the composites industry to develop simulation technology for continuous fibre reinforced composites (CFRC) (e.g. Smiley and Pipes 1988; O'Bradaigh and Pipes 1991; de Luca *et al.* 1996, 1998; Pickett *et al.* 1995; McEntee and O'Bradaigh 1998; O'Bradaigh *et al.* 1997; Sutcliffe *et al.* 2002; Lamers *et al.* 2002a,b; Yu *et al.* 2002; Sharma and Sutcliffe 2003). The overriding goal is to make CFRC materials a more attractive option, in a market where composites are often viewed as expensive alternatives to more traditional materials, such as metal and plastic. One of the main challenges in the simulation process is the development of constitutive models that can predict accurately the unique and complex rheological behaviour of viscous CFRC, including textile reinforced thermoplastic composites and thermosetting prepreg composites.

Continuous fibre reinforced composites can be divided into uniaxial CFRC that contain reinforcement in one direction and fabric or textile reinforced composites. The constitutive modelling and associated experimental work on viscous uniaxial and cross-ply CFRC laminates (i.e. laminates consisting of multiple layers of uniaxial CFRC) have received considerable attention over a period of more than a decade. However, until recently, relatively little attention was paid to the characterisation and modelling of viscous textile composites. This situation has changed in recent years; indeed, characterisation and constitutive modelling of viscous textile composites now constitute an active area of research.

The aim of this chapter is to provide the reader with an overview of the current state of rheological modelling of viscous textile composites together with the experimental methods available to characterise their rheology. In section 3.2, a brief discussion of the deformation mechanisms involved in the forming of viscous textile composites is given. Section 3.3 is a review of the constitutive modelling work for viscous textile composites. The ideal fibre reinforced model (IFRM) is discussed in sections 3.3.1 to 3.3.3. Here the term 'ideal' refers to fibre reinforced incompressible materials that are inextensible along the fibre direction (Hull *et al.* 1994). Attention is drawn to the difficulty in formulating constituent-based predictive models for viscous textile composites using this approach. Attempts to deal with this limitation appear in the literature and these are discussed in sections 3.3.4 and 3.3.6. The inextensibility constraint has been relaxed in some models, as discussed in section 3.3.5. Section 3.4 provides a description and discussion of experimental test methods typically employed in characterising the rheology of viscous textile composites. Finally, section 3.5 describes a visual analysis system used in determining the deformation mechanisms acting during the forming of viscous textile composites over complex shaped components. This technique provides a convenient tool for evaluating the predictions of numerical forming codes.

3.2   Deformation mechanisms

Various deformation mechanisms can occur during forming of a sheet of textile composite material (e.g. Cogswell 1992, 1994; Friedrich *et al.* 1997; Murtagh and Mallon 1997; Mallon and O'Bradaigh 2000). These deformations can be classified according to the length scale over which they occur. Figure 3.1 illustrates various macroscopic deformations. When considering a single layer of textile composite sheet, a single mode of deformation known as *intra-ply shear*, dominates; see Fig. 3.1(a). As will be shown, intra-ply shear is by far the most important and widely studied mechanism in continuum models for textile composites. Intra-ply shear, or trellis deformation, can be visualised by thinking of a woven composite material as a pin-jointed net made from inextensible fibres, with the joints occurring at tow crossover points. As the net is forced to deform, the tows rotate about the crossover points, which results in the material shearing. This is the only mode of deformation considered by the IFRM (see section 3.3), and is also the only mode of deformation modelled by kinematic draping codes (see section 4.2).

Given the relatively high tensile modulus of the reinforcement in, for example, glass or carbon fibre-based textile composites *intra-ply extension*, Fig. 3.1(b), may seem an irrelevant deformation mechanism during forming (i.e. intra-ply shear should offer a much lower energy, and therefore preferable, mode of deformation). However, fibres may extend during forming without excessive forces due to uncrimping of the tows (fibre bundles or yarns) in the textile. Thus, a small degree of intra-ply extension may be expected even in textile composites with high modulus reinforcement. Recent attempts to account for fibre extensibility (and at the same time avoid numerical problems associated with the inextensibility constraint) in finite element (FE) simulations

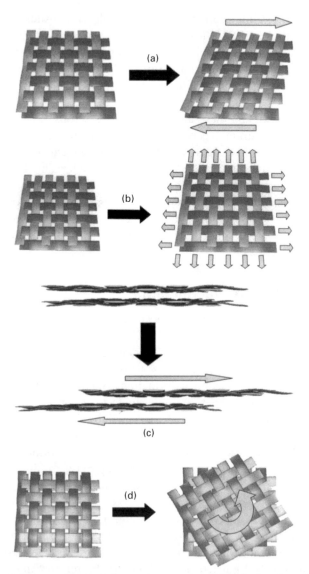

3.1 Macroscopic deformation mechanisms during forming of textile composite material, including (a) intra-ply shear, (b) intra-ply extension, (c) inter-ply slip and (d) inter-ply rotation.

have been made using different approaches (e.g. Lamers *et al*. 2002a; Xue *et al*. 2003; Yu *et al*. 2003) (see section 3.3.6). It is interesting to note that recent interest in self-reinforced polymer–polymer composites (e.g. Prosser *et al*. 2000; Harrison *et al*. 2002d) has made intra-ply extension of even greater relevance, as tows in this type of composite have much lower modulus than for typical woven composites.

For multilayered composites, slip between adjacent layers during forming must also be considered, including *inter-ply slip*, Fig. 3.1(c) and *inter-ply rotation*, Fig. 3.1(d). Inter-ply slip and rotation describe the relative motion of one ply over another as components are formed from multi-ply lay-ups (e.g. Friedrich *et al.* 1997; Murtagh and Mallon 1997; Lamers *et al.* 2003; Phung *et al.* 2003). These modes of deformation are essential if geometries with double curvature are to be formed from multilayer composites without wrinkling, particularly when using plies of initially different orientations.

Recently, homogenisation methods have been used to predict the behaviour of representative volume elements within textiles from consideration of individual tow (fibre bundle) and inter-tow deformations (i.e. mesoscale deformations). Indeed, the potential of textile modelling at the mesoscale has been greatly enhanced recently by increases in computing power and is becoming a growing area of interest for researchers (e.g. Hsiao and Kikuchi 1999; Harrison *et al.* 2004a; Lomov *et al.* 2002; Hagege *et al.* 2003). Examples of mesoscale deformations are shown in Fig. 3.2. Various distinct modes of tow deformation can be identified including *tow-shear*, Fig. 3.2(a), and *tow-squeezing* plus *tow-twisting* (one mode) and *tow-bending* (two modes) about

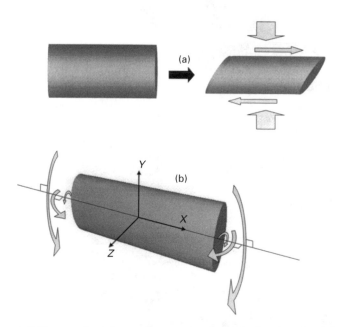

3.2 Mesoscale deformation mechanisms during forming of textile composite material, including (a) tow shear and (b) tow squeezing, twisting and bending. Relative motion between tows can also occur within the ply, including (c) crossover shear and (d) inter-tow shear. Finally, intra-ply slip is another potential deformation mechanism and can be classified as either (e) crossover slip or (f) inter-tow slip.

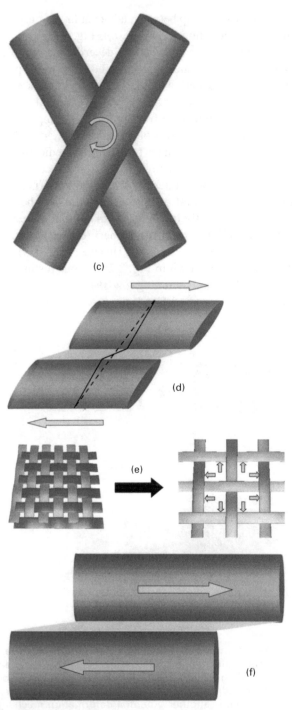

3.2 Continued

the tow axes, Fig. 3.2(b). Mesoscale tow deformations permit macroscale forming of the textile as well as facilitating textile specific deformations such as *mesoscale fibre buckling*. The relative motion of neighbouring tows and their subsequent interaction is also an important factor when modelling textile deformation at the mesoscale. Examples include *crossover shear*, Fig. 3.2(c) (Harrison *et al.* 2002a,b; Zouari *et al.* 2003), which potentially induces large shear strain rates in the matrix fluid between superposed tows, and *inter-tow shear*, Fig. 3.2(d), which arises when the shear rate across tows is less than the average shear rate across the textile composite sheet (see section 3.3.6).

The nature of mesoscale models means that they can include information such as tow architecture, tow and inter-tow kinematics and can also possibly incorporate so-called *intra-ply slip* mechanisms. Two distinct types of intra-ply slip can be identified, *crossover slip*, Fig. 3.2(e) (e.g. Lai and Young 1999), and *inter-tow slip*, Fig. 3.2(f) (Harrison *et al.* 2004b). These mechanisms are a direct result of relative displacements between neighbouring tows and have been observed to play an important role during the deformation of textile composites. Intra-ply slip occurs most readily in loosely woven composites and in some cases can actually improve the formability of the textile (Long *et al.* 2002). However, too much crossover slip can leave resin-rich regions or, in extreme cases, unwanted holes in components as the tows move apart. Crossover slip is important during press forming and hand lay-up, while inter-tow slip is apparent during bias extension tests, Fig. 3.2(f) (Harrison *et al.* 2004b).

Finally, prediction of individual tow properties can be made using knowledge of factors such as matrix rheology and fibre volume fraction, from consideration of micro-scale fibre–fluid interactions within the individual tows. These interactions are illustrated in Fig. 3.3 and include both (a) *intra-tow transverse shear* and (b) *intra-tow axial shear* (see section 3.3.2).

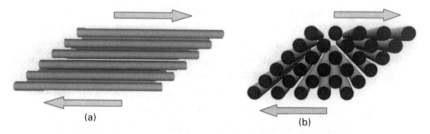

(a)   (b)

3.3 (a) Shearing the composite parallel to the fibre direction gives a measure of the longitudinal viscosity, η_L. (b) Shearing the composite across or transverse to the fibre direction gives a measure of the transverse viscosity, η_T. Micromechanical models based on these fibre organisations have been developed that relate the matrix viscosity and fibre volume fraction to the two viscosities.

Many of the modelling approaches developed to describe the deformations required to form components have as their goal the elimination of fibre buckling (mainly in relation to uniaxial CFRC) and wrinkles. Although the prediction of these is therefore of prime importance, relatively little work has concentrated on the development and subsequent growth of wrinkles (Hull *et al.* 1991, 1994; Friedrich *et al.* 1997; Li and Gutowski 1997; O'Bradaigh *et al.* 1997; Mander *et al.* 1997; Martin *et al.* 1997b; De Luca *et al.* 1998; Yu *et al.* 2003). The prediction of wrinkles is an important area for future work.

3.3   Review of constitutive modelling work

In this section, a review of the constitutive modelling of viscous textile composites is presented. Examination of the literature reveals various modelling approaches. The first and most widely studied is that of continuum mechanics, in particular models based on the IFRM (section 3.3.1). The IFRM has proved successful in many respects, not least in obtaining analytical solutions to specific well-defined flow deformations. However, certain limitations of the theory have become apparent. These include difficulties found in model-parameter evaluation for 'real' textile composites through experimental characterisation and problems involved in implementing these equations in FE codes.

The review of IFRMs includes work on both viscous uniaxial (section 3.3.2) and biaxial (section 3.3.3) CFRC. While at first sight, the case of the uniaxial IFRM is not directly applicable to textile composites, this work is nevertheless considered for two reasons. The first is to illustrate one of the more ambitious yet desirable goals of CFRC constitutive models, that is, the development of fully predictive models based on properties such as fibre volume fraction and resin viscosity (see also sections 3.3.4 and 3.3.6). Success in this area would dramatically reduce the need for extensive experimental characterisation programmes (section 3.4). The second is because at least one approach to textile deformation modelling, the energy summation method (section 3.3.6), is based on the uniaxial rather than biaxial IFRM.

When implementing a material model in an FE code, an important consideration is the effect of the kinematic constraints of the model on the numerical calculations. The inextensibility constraint of the IFRM leads to arbitrary tensile stresses in the reinforcement directions (see section 3.3.1), stresses that can cause potential problems when calculating force equilibrium. This issue has led certain workers to develop alternative constitutive models for sheet forming predictions (section 3.3.5).

3.3.1   The ideal fibre reinforced model for viscous fluids

The continuum theory of the mechanics of CFRC in a solid or elastic state was developed initially by Spencer and referred to as the elastic ideal fibre reinforced model (Spencer 1972, 1984). This work was adapted by Rogers (1989a) to the case of viscous CFRC through use of the 'correspondence principle' of viscoelasticity. The underlying assumptions of the IFRM for viscous composites have been outlined many times previously (e.g. Hull *et al.* 1994; Rogers 1989a, 1990; Spencer 2000; McGuiness and O'Bradiagh 1998; Mallon and O'Bradaigh 2000) and are repeated here for convenience.

As with the elastic IFRM, two important approximations are used in obtaining solutions in the stress analysis, those of incompressibility and fibre-inextensibility. These conditions can be expressed mathematically as follows. Using D, the rate of deformation tensor which can be written in component form as

$$D_{ij} = \frac{1}{2}\left(\frac{\partial v_i}{\partial x_j} + \frac{\partial v_j}{\partial x_i}\right)$$ 3.1

where v_i are the velocity components of a particle at position x_i, the incompressibility condition can be written as

$$D_{ii} = \frac{\partial v_i}{\partial x_i} = 0$$ 3.2

where the repeated suffix summation convention applies. This first familiar constraint is often used in fluid mechanics. The second constraint of fibre inextensibility is more particular to CFRC theory. The reinforcement fibres are considered as continuously distributed throughout the material and for uniaxial reinforcement are defined at any point by a unit vector, $a(x, t)$ or for textile composites with two directions of reinforcement by $a(x, t)$ and $b(x, t)$. Thus, the inextensibility conditions can be expressed in component form as

$$a_i a_j D_{ij} = a_i a_j \frac{\partial v_i}{\partial x_j} = 0, \qquad b_i b_j D_{ij} = b_i b_j \frac{\partial v_i}{\partial x_j} = 0$$ 3.3

The fibres convect with the material and so vary in both space and time, hence the material derivative can be used to find the rate of change of a (or b):

$$\frac{Da_i}{Dt} = \frac{\partial a_i}{\partial t} + v_j \frac{\partial a_i}{\partial x_j} = (\delta_{ij} - a_i a_j) a_k \frac{\partial v_j}{\partial x_k}$$ 3.4

where δ_{ij} is the Kronecker delta. If eqn 3.3 holds then eqn 3.4 reduces to

$$\frac{Da_i}{Dt} = a_k \frac{\partial v_i}{\partial x_k}$$ 3.5

Furthermore, for quasi-static flows, the equilibrium equations

$$\frac{\partial \sigma_{ij}}{\partial t} = 0 \qquad\qquad 3.6$$

also hold, where σ_{ij} is the stress tensor. A consequence of these assumptions is that the stress tensor may be decomposed as:

$$\boldsymbol{\sigma} = -p\boldsymbol{I} + T_A\boldsymbol{A} + \boldsymbol{\tau} \qquad\qquad 3.7$$

for the uniaxial case with one direction of reinforcement or:

$$\boldsymbol{\sigma} = -p\boldsymbol{I} + T_A\boldsymbol{A} + T_B\boldsymbol{B} + \boldsymbol{\tau} \qquad\qquad 3.8$$

for the viscous textile composite with two directions of reinforcement. Here $\boldsymbol{I}$ denotes the unit tensor and $\boldsymbol{A}$ and $\boldsymbol{B}$ are the tensor products:

$$\boldsymbol{A} = \mathbf{a} \otimes \mathbf{a},\, \boldsymbol{B} = \mathbf{b} \otimes \mathbf{b}$$

The first two terms appearing on the right-hand side of eqn 3.7 and the first three on the right-hand side of eqn 3.8 are reaction stresses, $-p\boldsymbol{I}$ is an arbitrary hydrostatic term arising from the incompressibility condition and T is an arbitrary tension in the fibre direction arising from the inextensibility constraint. Subscripts A and B indicate the reinforcement direction. $\boldsymbol{\tau}$ is the extra stress tensor. Determining an appropriate form of $\boldsymbol{\tau}$ is the goal of the constitutive modelling using continuum theory.

In the fibre reinforced viscous fluid model the assumption is generally made that $\boldsymbol{\tau}$ depends on both the rate of deformation tensor, $\boldsymbol{D}$, and the fibre reinforcement directions (one direction for uniaxial composites). According to Spencer (2000) $\boldsymbol{\tau}$ must be form invariant with respect to rigid rotations and so has to be an isotropic function of its arguments. The representation of a tensor function of vectors and tensors is an algebraic problem whose solution is known and can be read off from tables (Spencer 1971; Zheng 1994). The eventual form taken by the extra stress tensor depends on numerous factors, such as whether $\boldsymbol{\tau}$ is linear or non-linear with $\boldsymbol{D}$, the number of reinforcement directions in the constitutive model and the presence of material symmetry (Spencer 2000). Various forms of $\boldsymbol{\tau}$ presented in the literature are discussed in sections 3.3.2 and 3.3.3.

However, before discussing these constitutive models, it is useful to provide a brief review of the work done on uniaxial CFRC. In doing so, an important objective in the work on CFRC becomes clear, namely prediction of composite rheology through knowledge of the matrix rheology, the fibre volume fraction of the composite and, in the case of textile composites, the weave architecture. The motivations for such constituent-based predictive modelling are two-fold. Firstly, once the material behaviours of the constituent components are known, then the rheology of any composite comprised of the matrix phase plus continuous inextensible fibres could be predicted, thus allowing the premanufacture optimisation of a composite to suit a potential application. Secondly, characterising the rheological behaviour of, for example, a thermoplastic matrix polymer at different shear rates and temperatures is relatively easy using modern rheometers, compared with the difficult task of characterising a composite material at different shear rates and temperatures (see Section 3.4). Both factors would lead to significant reductions in time and cost in the manufacturing process. For these reasons, much effort has already been made in pursuing predictive models for uniaxial CFRC. The same motivations also apply for textile composites, though as will be seen, the difficulties in producing such a predictive model for textile reinforced viscous composites are greater than for the case of viscous uniaxial CFRC.

3.3.2 Constitutive behaviour of viscous uniaxial CFRC

The rheological modelling of viscous uniaxial CFRC, composites containing fibre reinforcement in just one direction, is closely related to the modelling of viscous textile composites. Rogers (1989a, 1990) was the first to present a three-dimensional linear form for the extra stress tensor for viscous uniaxial CFRC:

$$\tau = 2\eta_T D + 2(\eta_L - \eta_T)(AD + DA) \qquad 3.9$$

A notable feature of this model is the appearance of two constant model parameters, η_L and η_T, which can be related to two distinct material properties; the 'longitudinal' and 'transverse' viscosities of the composite. These two viscosities result from the dynamic interaction between fibre and matrix phases during shear of the composite. This interaction occurs on a microscopic scale and is shown in the schematic of Fig. 3.3. The diagram shows that the composite's transverse viscosity results from individual fibres 'rolling' past one another, while the longitudinal viscosity results from the fibres sliding past one another along their length. Both η_L and η_T are very sensitive to the fibre volume fraction and can be orders of magnitude higher than the matrix viscosity.

Equation (3.9) has been used successfully to perform theoretical investigations of the flow of uniaxial CFRC. Indeed, analytical solutions for the stresses occurring in certain simple forming operations have been suggested as test cases for numerical codes (Rogers 1990; Rogers and O'Neill 1991) and the model has also been applied in producing theoretical analyses of fibre buckling and wrinkling in shear-flows (Hull et al. 1991, 1994). However, fundamental to performing any accurate numerical forming simulation for 'real' materials, based on such a constitutive model, is knowledge of the relevant model parameters. Evidently, the question of measuring and perhaps even of predicting the model parameters appearing in eqn 3.9 is crucial. Indeed much work has been performed in pursuing these goals.

Various methods have been devised to measure the two viscosities (Advani *et al.* 1997). Dynamic testing has been conducted using both rotational (Cogswell and Groves 1988; Scobbo and Nakajima 1989; Groves 1989; Groves and Stocks 1991; Cogswell 1992) and linear (Scobbo and Nakajima 1989; Wheeler and Jones 1991; Roberts and Jones 1995) oscillatory experiments. Large strain experiments have also been devised and include steady shear linear 'pull-out' experiments (Goshawk and Jones 1996), squeeze flow experiments (Rogers 1989b; Barnes and Cogswell 1989; Balasubramanyam *et al.* 1989; Jones and Roberts 1991; Wheeler and Jones 1995; Shuler and Advani 1996; Goshawk *et al.* 1997), a vee-bending method (Martin *et al.* 1995, 1997b Mander 1998; Dykes *et al.* 1998) and a picture frame test (McGuiness and O'Bradaigh 1998).

Many of these investigations have been concerned not just with measuring these two viscosities, but also with relating η_T and η_L directly to the matrix viscosity, η_m (Cogswell and Groves 1988; Groves 1989; Scobbo and Nakajima 1989; Balasubramanyam *et al.* 1989; Cogswell 1992; Roberts and Jones 1995; Goshawk and Jones 1996; Shuler and Advani 1996). Indeed, the interest in relating η_L and η_T to η_m has led to the development of analytical models, for example, eqn 3.10 and 3.11 (Christensen 1993) and eqn 3.12 (Coffin *et al.* 1995). Analytical predictions are plotted together with experimental data in Fig. 3.4.

$$\eta_{12} = \left\{ \frac{1 + \alpha(v_f/F)}{\sqrt{[1 - \beta(v_f/F)][1 - (v_f/F)]}} \right\} \eta_m \quad \alpha = 0.8730, \beta = 0.8815 \quad 3.10$$

$$\eta_{23} = \left\{ \frac{1 + \alpha(v_f/F)}{\sqrt{[1 - \beta(v_f/F)][1 - (v_f/F)]}} \right\}^3 \eta_m \quad \alpha = -0.1930, \beta = 0.5952 \quad 3.11$$

$$\eta_{12} = \eta_{23} = \left\{ \frac{1}{1 - \sqrt{v_f/F}} \right\} \eta_m \quad 3.12$$

where v_f is the fibre volume fraction and F is the maximum packing fraction, $F = \pi/4$ for a square array or $F = \pi/(2\sqrt{3})$ for hexagonal array of fibre. Predictions using eqns 3.10–3.12 in Fig. 3.4 are calculated using the maximum packing fraction for a square array of fibres. Other analytical models developed for long discontinuous fibres include those presented by Binding (1991) and Pipes (1992).

Materials tested in the experimental investigations are specified in the legends of Fig. 3.4(a) and (b). APC-2 refers to a thermoplastic composite consisting of carbon fibres (≈ 7 μm diameter) in a poly(etheretherketone) (PEEK) matrix while AS-4 refers to a thermoplastic composite consisting of carbon fibres in a poly(arylenesulphide) (PAS) matrix. Other model materials

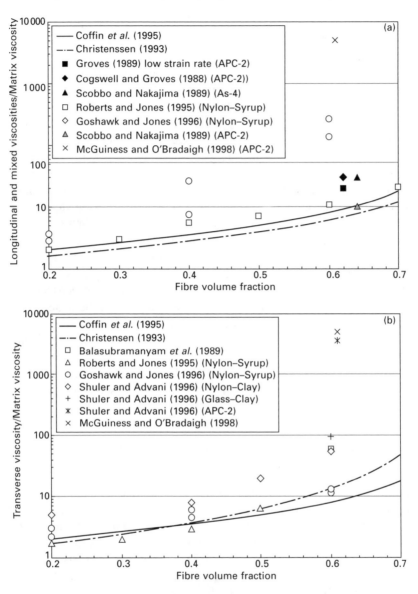

3.4 Data showing ratio between composite and matrix viscosities versus fibre volume fraction from a number of different experimental investigations on uniaxial CFRC. Micromechanical model predictions are also shown. (a) Longitudinal plus mixed mode viscosity measurements and predictions. (b) Transverse viscosity measurements and predictions. The investigation and material are given in the legends.

used in the various investigations include glass ($\approx$ 10–17 µm diameter) or nylon ($\approx$ 210–1000 µm diameter) reinforcing fibres embedded in a matrix of either modelling clay or golden syrup.

Figure 3.4(a) shows longitudinal viscosities plus other apparent viscosities that represent mixed modes of longitudinal and transverse flow induced by rotational rheometers (closed points) while Fig. 3.4(b) shows transverse viscosities. Most experimental data reviewed in Fig. 3.4 revealed a highly non-linear response to shear rate; thus data are subject to the conditions under which they were obtained (e.g. shear strain rate, steady or dynamic data) and are therefore approximate values. Even so, it is clear from Fig. 3.4 that while analytical predictions give a good approximate for a lower limit to the real viscosities of uniaxial composites, these predictions can dramatically underestimate the actual viscosities by an order of magnitude or more. Indeed, viscosities measured on the same thermoplastic composite (APC-2) in different investigations using picture frame, squeeze flow and torsional rheometry testing methods have been reported to vary enormously, by a factor of up to 1000 (McGuiness and O'Bradaigh 1998). Thus, comparison between theoretical predictions and experimental results may well depend on the materials and experimental methods used to collect the data. At present, the reason for the discrepancy between results remains unclear, although this is possibly related to fibre entanglements, non-ideal organisations of fibres, fibre bending rigidity and sample dimensions. A bench-marking exercise, involving several experimental methods and materials, would serve to clarify this ambiguity.

3.3.3   Constitutive models for viscous textile composites based on the IFRM

The initial approach to modelling textile composites, as with uniaxial CFRC, was through the use of the IFRM. However, in this case two fibre directions of reinforcement are incorporated in the assumptions of the continuum theory, as outlined in section 3.3.1. In this section, the biaxial continuum models appearing in the literature are reviewed. Their advantages and disadvantages are discussed before turning attention to methods and models that have been proposed recently as alternatives to the IFRM.

The most general form of the extra stress tensor that is linear with D, for a material with two directions of reinforcement, was presented by Spencer (2000)[1]:

---

[1] An excellent introduction to the continuum mechanical modelling of viscous textile composites.

$$\tau = 2\eta_1 D + 2\eta_2(AD + DA) + 2\eta_3(BD + DB) + 2\eta_4(CD + DC^T)$$

$$+ 2\eta_5(C^T D + DC) \qquad 3.13$$

where $\quad C = a \otimes b, C^T = b \otimes a \qquad 3.14$

and η_1 to η_5 are viscosity parameters of the model. Following Spencer (2000), eqn (3.13) can be simplified by making certain basic assumptions. For example, assuming the two families of fibres are mechanically equivalent, eqn 3.13 reduces to the form

$$\tau = 2\eta_1 D + 2\eta_2(AD + DA + BD + DB) + \eta_5 (trCD)(C + C^T) \quad 3.15$$

Equivalent forms of eqn 3.12 have been presented previously (Rogers 1989a; Johnson 1995). Equation 3.15 simplifies further by considering the case of a plane sheet, a usual practice when analysing the constitutive equations of uniaxial CFRC and textile composites. Equation 3.15 is of particular interest since numerous attempts have been made to fit this model to experimental data, though none of these comparisons have proved particularly successful (McGuiness et al. 1995; Murtagh and Mallon 1995; Canavan et al. 1995; Yu et al. 2000). Indeed, McGuiness et al. (1995) concluded that elastic effects must be included in any successful model, while Yu et al. (2000) suggested that no possible combination of the parameters η_1, η_2, η_5 could tune the shape of the predicted curve to match their own experimental data.

For these reasons, various modifications to the linear biaxial continuum model with constant coefficients have since been proposed. The most general non-linear form of the extra stress tensor for an incompressible anisotropic viscous fluid with two directions of inextensible reinforcement was presented by Spencer (2000):

$$\tau = \psi_1 D + \psi_2 D^2 + \psi_3 (AD + DA) = \psi_4 (BD + DB) + \psi_5(CD + DC^T)$$

$$+ \psi_6(C^T D + DC) + \psi_7(AD^2 + D^2 A) + \psi_8(BD^2 + D^2 B) \qquad 3.16$$

where the response functions $\psi_1, \ldots, \psi_8$ are functions of the invariants

$$trCD, trD^2, trAD^2, trBD^2, trCD^2, \cos 2\phi \qquad 3.17$$

where tr represents the trace operation and 2ϕ is the angle between the two fibre reinforcement families.

Equation 3.16 is too complicated to be of practical use without prior consideration of certain simplifying assumptions. Thus, the special case of a plane sheet with power law behaviour was considered and it was suggested that by using this model, the material parameters could be measured by fitting predictions to picture frame data, where experiments are performed at different strain rates. Indeed, in an earlier investigation, McGuiness and O'Bradaigh (1997) proposed isotropic non-linear viscous and viscoelastic models as well as their equivalent anisotropic counterparts. Two viscosities, η_{12} and η_{33}, with power law behaviours were introduced to represent the resistance to in-plane and through thickness deformations and a Kelvin model (spring and dashpot in parallel) was used to introduce viscoelasticity, as recommended by Canavan *et al.* (1995). Models proposed by McGuiness and O'Bradaigh (1997) are summarised by Mallon and O'Bradaigh (2000). Comparisons between experimental data produced by the picture frame experiment (see section 3.4) and model predictions were presented. The most appropriate choice of constitutive model was found to depend on the particular material under investigation. Thus, McGuiness and O'Bradiagh (1997) were able to demonstrate how picture frame tests can be used to characterise the rheological response of a given textile composite in terms of non-linear rheological models. However, the main drawback with this approach is the large number of experiments required to fit the model parameters.

Another notable success of the continuum approach has been its capacity to predict the asymmetric stress response to shear in opposite directions of certain textile architectures, through consideration of fabric symmetries (Spencer 2000). This unexpected stress response was measured experimentally in picture frame tests (Canavan *et al.* 1995; McGuinness and O'Bradaigh 1997).

The most recent modification to biaxial continuum models incorporating the inextensibility constraint has been to include plastic yield behaviour (Yu *et al.* 2000; Spencer 2002). Yu *et al.* (2000) use a modified version of eqn 3.15 by including an equivalent von Mises yield criteria in the model and compared the model with experimental data. Spencer (2002) described a more general model including both plastic and non-linear behaviour though, as yet, the model remains untested against experimental data.

It should be noted that there is an important difference between these biaxial constitutive models and those applicable to uniaxial CRFC (see section 3.3.2). That is, for uniaxial CFRC the viscosity parameters, for example in eqn 3.12, can be related directly to simple micro-mechanical shearing mechanisms (see Fig. 3.3). However, while the viscosity parameters in eqns 3.13, 3.15 and 3.16 may be related indirectly to these micro-mechanics (Spencer 2000) a simple relationship between biaxial-model viscosity parameters and micro-mechanical mechanisms no longer exists. For this reason, the development of simple micro-mechanical models, which predict the relationship between the matrix viscosity, η_m, and the model parameters appearing in these equations is impossible. Since determination of the parameters in the various biaxial constitutive models requires a considerable amount of experimental work, especially if various temperatures are to be considered, the need for constituent-based predictive models is just as important for modelling textile composites as for modelling uniaxial CFRC. Thus, alternative modelling techniques are now being employed to meet this end (see sections 3.3.4 and 3.3.6).

Finally, as was noted in the introduction, the strong constraints inherent in the IFRM have been found to cause problems when attempting to implement this model in numerical codes. Thus, while the IFRM is useful in providing analytical solutions to specific simple flow problems, relaxation of these constraints is often required when implementing constitutive models in FE software. This point has played a significant role in the development of some of the more recent constitutive models that are suitable for implementation in FE codes.

3.3.4   Predictive models for viscous textile composites based on homogenisation methods

The homogenisation technique can be applied to periodic materials and consequently has proved useful in modelling the deformation stresses (and thermal properties) of textile composites (Hsiao and Kikuchi 1999; Peng and Cao 2002). The method involves modelling the composite using FE analysis on various length scales and using the smaller length scale models to predict the material properties at a larger length scale. Results from the larger-scale predictions can then be fed back into the smaller scale and the process is repeated. Thus, use of the homogenisation method facilitates forming predictions based on fibre angle, volume fraction and properties of the constituent materials in the textile composite (Hsiao and Kikuchi 1999). The method potentially solves one of the main limitations of the biaxial IFRM, i.e. the need for expensive and time-consuming characterisation experiments required to fit model viscosity parameters. A second problem associated with the IFRM, namely numerical problems caused by the fibre inextensibility constraint, was tackled by use of a numerical penalty method. However, the homogenisation method does have its drawbacks, including the difficulty in producing FE meshes of the potentially complicated weave-architectures, the associated large amount of computational power required to run these simulations and the difficulty involved in implementing such a model in commercially available FE codes owing to the lack of an explicitly written constitutive equation.

3.3.5   Explicit constitutive models for viscous textile composites with fibre extensibility

As with the IFRM this family of constitutive models is expressible in explicit matrix form; however, the models differ in that they avoid the strong constraint of fibre inextensibility. Indeed a main aim here has been to develop models that are easy to implement in commercial FE codes. Different approaches have been employed, including the development of entirely new non-orthogonal constitutive models (Xue *et al.* 2003; Yu *et al.* 2002, 2003) as well as direct modification of the IFRM (Lamers *et al.* 2002a,b, 2003). The model of Yu *et al.* (2002) was developed through consideration of stresses induced in a conceptualised pin-jointed net structure consisting of non-orthogonal tows of given tensile modulus. The resulting constitutive equation was formulated by calculating linear elastic stresses induced in the non-orthogonal net structure by a superposition of imposed tensile and simple shear strains. Consequently, the equation is valid for small axial strain deformations along the reinforcement directions. However, the incremental form of the equation means that it can be implemented in numerical codes and used to model large strain behaviour. Use of the non-orthogonal reference frame allows the shear behaviour measured in picture frame tests to be incorporated directly into the model.

In another non-orthogonal model (Xue *et al.* 2003) material tensile properties were determined through tests on textiles with orthogonal tows, i.e. before shear (Boisse *et al.* 2001), while shear properties were determined from picture frame tests (see section 3.4). The stress tensor is expressed in the 2D non-orthogonal reference frame using these properties and then transformed to the global Cartesian frame for use in FE calculations. The resulting model is valid for large strains. Finally, another approach has been simply to relax the strong constraints of fibre inextensibility and incompressibility to produce a modified IFRM (Lamers *et al.* 2002a,b, 2003). At present the main drawback of these explicit models is that, as with the IFRM, a large amount of characterisation experiments are required to fit the material parameters. Indeed, this problem is exacerbated for some of the latest non-orthogonal models since these have been developed for rate-independent mechanics, i.e. shear force is characterised by an elastic rather than viscous material response. A potential solution to this problem is to fit these explicit models to shear force versus shear angle curves predicted using energy summation methods (see section 3.3.6). Further information regarding the implementation of explicit models in FE software can be found in Chapter 4.

3.3.6   Shear force prediction for viscous textile composites based on energy summation

A predictive method of calculating the shear force versus shear angle behaviour of textile composites has been proposed based on energy summation. The main inputs of the model are the fibre volume fraction and viscosity of the viscous matrix (Harrison *et al.* 2002a,b, 2003). The method is based on a combination of simple models that are used to calculate the energy dissipation of various micro- and mesoscale deformations. Deformation of the tows is modelled using uniaxial continuum theory (Rogers 1989a) plus associated transverse and longitudinal viscosity micro-mechanical models (e.g. Christensen 1993; Coffin *et al.* 1995). In addition, mesoscale observations of inter-tow shear within the textile composite have resulted in the development of a simple 'crossover shear' model (see Fig. 3.2d). This model is based on observations of heterogeneous shear strain within the textile (see Fig. 3.5a) (Harrison 2002a; Potter 2002), which results in the relative rotation of superposed tows. Note that a homogeneous shear across the textile would imply that tows within the textile should shear at the same rate as the average shear rate across the textile. If this were the case then the white lines marked on the textile, shown in Fig. 3.5a, would be continuous rather than broken. This would imply zero relative velocity and hence zero shearing of the matrix fluid between the crossovers.

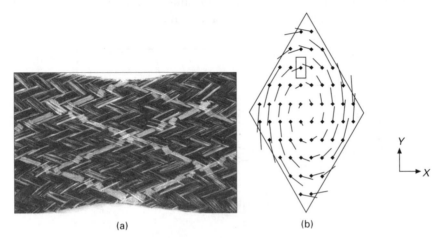

(a)   (b)

3.5 (a) Example of discontinuous shear strain profiles across various textile materials. The broken form of initially continuous lines drawn on the 2 × 2 twill weave glass/polypropylene thermoplastic composite, sheared in a bias extension test at 180 °C, serves to illustrate the in-plane strain profile. (b) Relative velocity field beneath a single tow crossover where velocities are indicated by vector lines. A rectangular element is drawn around one of the points in order to discretise the crossover area for energy calculations. In this example, the tow shear strain rate $\dot{\gamma}_t = \dot{\theta}/2$ and the shear angle $\theta = 30°$.

Incorporation of the observed mesoscale kinematics (i.e. heterogeneous shear) has two important effects. Firstly, modified versions of the homogeneous rate of deformation tensor must be used for both tow and inter-tow regions. Secondly, a relative velocity exists between the two sets of reinforcement tows at the tow crossovers thus energy dissipation in this region must be considered (see Fig. 3.5b).

The advantage in this method is the low computational power required to make shear force versus shear angle predictions, making it a fast, efficient method of producing first estimates of the material behaviour as a function of shear rate, shear angle and temperature. Example results from the energy model are compared against normalised experimental bias extension (see section 3.4.2) measurements in Fig. 3.6 (from Harrison *et al.* 2003). The normalisation procedure is described in section 3.4.3. Experiments were performed on a pre-consolidated glass/polypropylene 2 × 2 twill weave thermoplastic composite at 165 and 180 °C. Valid experimental data for high shear angles are impossible to obtain using the bias extension test. However, picture frame tests (see section 3.4.1) can produce this data and have corroborated the sharp increase in shear force predicted by the model. Inputs into the energy model included matrix shear rheology, fibre volume fraction, observed tow kinematics and weave architecture. Future work is still required in order to predict tow kinematics during forming.

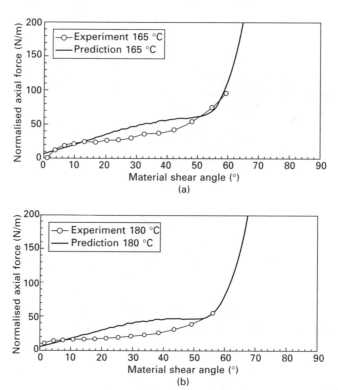

3.6 Bias extension test results together with theoretical predictions. Experiments were performed on a pre-consolidated glass/ polypropylene 2 × 2 twill weave thermoplastic composite at (a) 165 °C and (b) 180 °C.

Since energy calculations produce force versus shear angle curves rather than predicting stresses within the material, this energy summation approach is not strictly speaking a constitutive model. However, results can be used to fit parameters of the constitutive models described in sections 3.3.3 and 3.3.5. In so doing, far less picture frame and bias extension experiments would be required and their intended purpose would be model evaluation rather than characterisation and parameter fitting.

3.4   Characterisation methods

Various experimental methods for characterising the rheology of woven impregnated CFRC have been proposed. Experimental data from such tests can play two important roles; either in fitting the parameters of constitutive models derived from continuum mechanical assumptions, or else in evaluating the accuracy of constitutive models based on a constituent-based predictive approach. The characterisation of textile composites using rotational rheometers was suggested originally by Rogers (1989a), though more specialised test methods have since been developed. Two test methods in particular have seen widespread application, namely the picture frame test (McGuiness *et al.* 1995; Canavan *et al.* 1995; Long *et al.* 1997; McGuiness and O'Bradaigh 1997, 1998; Harrison *et al.* 2002a; Wilks *et al.* 1999; Mallon and O'Bradaigh 2000; Lussier *et al.* 2002; Lussier and Chen 2000; Lebrun *et al.* 2003; Milani *et al.* 2003) and the bias extension test (Yu *et al.* 2002; Murtagh and Mallon 1995; Lebrun and Denault 2000; Lebrun *et al.* 2003; Milani *et al.* 2003; Harrison *et al.* 2003). As industry makes greater use of manufacturing design tools such as finite element simulations, the need to standardise these test procedures is urgent. Work in this direction has been conducted recently (Blanchard *et al.* 2001; Lebrun *et al.* 2003; Harrison *et al.* 2004b; Peng *et al.* 2004).

3.4.1   Picture frame test

The picture frame (or rhombus) test has been used to measure the force generated by shearing technical textiles and textile composites. A sample of fabric is held in a 'picture frame' hinged at each corner (see Fig. 3.7). The frame is loaded into a tensile test machine, and two diagonally opposite corners are extended, imparting pure, uniform shear in the test specimen on a macroscopic scale. There is no uniformly applied standard test procedure, and individual researchers have modified the test method to improve the repeatability and accuracy of test data. One point requiring particular attention is the restraining technique used to hold the sample in the picture frame (McGuiness and O'Bradaigh 1998). Some researchers prefer to constrain the ends of the fabric using pins and a very low clamping force to avoid fibre tensile strain and excessive bending of tows (Harrison *et al.* 2003). Others prefer to prevent fibres from slipping, by tightly clamping the edges of the test specimens under a clamping plate (Milani *et al.* 2003).

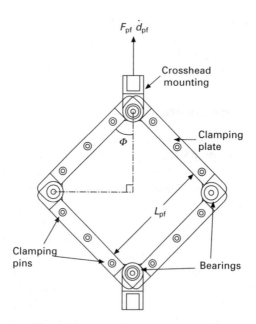

$F_{pf}\ \dot{d}_{pf}$

Crosshead
mounting

Φ

Clamping
plate

L_{pf}

Clamping
pins

Bearings

3.7 Picture frame shear rig. L_{pf} is the side length measured from the centre of the hinges, F_{pf} is the axial force measured by the crosshead mounting, d_{pf} is the rate of displacement of the crosshead mounting and the material shear angle, θ, is defined as $\pi/2 - 2\Phi$.

Recently a method of reducing error due to fibre tensile strain through the use of sample 'tabs' has been suggested (Lebrun *et al.* 2003). The choice of restraining method is made according to the variability of the experimental data, which depends on the type of material under investigation. The axial force required to pull the crosshead of the testing machine is recorded and the shear force can be calculated subsequently using:

$$F_s = \frac{F_{pf}}{2\cos\Phi} \qquad\qquad 3.18$$

where Φ is the frame angle and F_{pf} is the measured axial picture frame force. Test data are often analysed to produce graphs of shear force against shear angle, where the shear angle is defined as:

$$\theta = \pi/2 - 2\Phi \qquad\qquad 3.19$$

Alternatively, axial force can be plotted rather than the shear force to facilitate comparison of the data with bias extension results, which are not necessarily the result of pure shear (see section 3.2). Consideration of the geometry of the test shows that θ can be related directly to the displacement of the crosshead, d_{pf}, by

$$\theta = \frac{\pi}{2} - 2\cos^{-1}\left[\frac{1}{\sqrt{2}} + \frac{d_{pf}}{2L_{pf}}\right] \qquad\qquad 3.20$$

where L_{pf} is the side length of the picture frame, i.e. the distance between centres of the picture frame bearings and d_{pf} is the displacement of the crosshead mounting (see Fig. 3.7).

The picture frame test procedure is simple to perform and results are reasonably repeatable. Since the deformation of the material is essentially homogeneous throughout the deforming sample (edge effects being ignored), the kinematics of the material deformation are readily calculated, facilitating rheological analysis of the results (McGuiness and O'Bradaigh 1997). A major benefit of the test is that the shear angle and current angular shear rate can be related directly to the crosshead displacement and displacement rate. However, one of the main concerns with the test is the boundary condition imposed on the sample. Loose pinning of the sample edges in the side clamps may fail to induce the required kinematics, whereas tight clamping of the sample edges can cause spurious results if the sample is even slightly misaligned. Another concern with the boundary conditions of the picture frame test is related to the influence of the metal rig on the temperature profile across the sample. During the high-temperature testing required for thermoplastic composites, the rig has been shown to cool the material immediately adjacent to the deforming sample, which may influence results significantly (see section 3.4.3).

Normalised results (shear force divided by side length of picture frame, L_{pf}) from picture frame experiments conducted at room temperature at three constant angular shear rates, 0.93, 4.65 and 9.3° s^{-1}, on a 5-harness satin weave carbon–epoxy prepreg are shown in Fig. 3.8. The issues regarding normalisation of shear data from characterisation experiments are discussed in section 3.4.3. Each curve is an average of four tests. The data show a non-linear increase in shear force with increasing shear rate, indicative of a shear-thinning viscous material response.

3.4.2   Bias extension test

The bias extension test involves clamping a rectangular piece of woven material such that the warp and weft directions of the tows are orientated initially at ±45° to the direction of the applied tensile force. The material sample can be characterised by the length/width ratio, $\lambda = L_o/w_o$, where the total length of the material sample, L_o, must be at least twice the width, w_o. The reason for this can be seen when analysing the idealised deformation of a material sample in a bias extension test.

Figure 3.9 shows an idealised bias extension test sample with $\lambda = 2$, in which the material is divided into three regions. If the tows within the sample

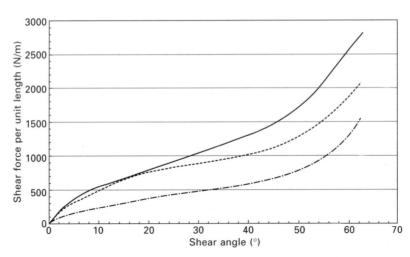

3.8 Picture frame experiments conducted at room temperature at three constant angular shear rates, 0.93, 4.65 and 9.3° s⁻¹ (from bottom to top) on a 5-harness satin weave carbon–epoxy prepreg.

are considered inextensible and no inter-tow or crossover slip occurs within the sample, then one can show that the shear angle in region A is always twice that in region B, while region C remains undeformed. The deformation in region A can be considered equivalent to the deformation produced by the pure shear of a picture frame test. The length of the material sample must be at least twice its width in order for the three different deformation regions to exist. Increasing the length/width ratio, λ, to higher values serves to increase the area of region A.

Like picture frame tests, bias extension tests are simple to perform and provide reasonably repeatable results. The test provides an excellent method of estimating a material's locking angle, i.e. the angle at which intra-ply slip (see section 3.2) and/or out-of-plane bending (wrinkling) become significant deformation mechanisms during bias extension tests. A convenient method of detecting the onset of these alternative deformation mechanisms is to monitor the test kinematics. For example, when $\lambda = 2$ the shear angle in region A of Fig. 3.9 should obey eqn (3.20) as long as deformation mechanisms, such as intra-ply slip, are insignificant compared with trellis shearing (Harrison *et al.* 2002c 2004b).

Unlike the picture frame test, as long as the material sample is tightly clamped, the boundary conditions are much less relevant to the test result. Optimum clamping conditions can be achieved by cutting the clamping areas (shown in Fig. 3.9) along the fibre reinforcement directions. Also, owing to the nature of the sample deformation induced by the bias extension test, the influence of relatively cool material adjacent to the metal clamps during high-temperature testing is of less importance than in picture frame tests.

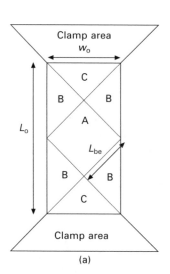
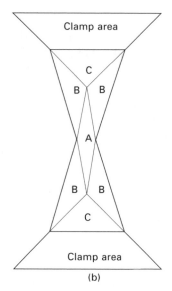

(a)   (b)

3.9 An idealised bias extension test sample with $\lambda = L_o/w_o = 2$ where L_o and w_o are the initial length and width of the bias extension sample respectively. Region A undergoes pure shear. It can be shown that the shear angle in region B is half that in region A. Region C remains undeformed throughout the test and consequently does not contribute to the shear force. For bias extension tests with $\lambda = 2$, L_{be} is the side length of region A and is used to normalise force, displacement and displacement rates.

However, the test does suffer from certain drawbacks. Firstly, the shear angle and angular shear rate in the material must be measured either by time-consuming visual analysis (which can be complicated further if the sample is to be heated in an oven) or by stopping the test at various displacements and taking measurements from the cooled samples. Secondly, the deformation field within the material is not homogeneous, preventing simple rheological analysis of the results. Finally, the test induces a combination of both pure shear and intra-ply slip. In terms of rheological analysis this presents extra difficulty, though conversely, this deformation may be used to advantage as a means to investigate intra-ply slip as a potential deformation mechanism of woven CFRC. Normalised test results from bias extension tests are presented in section 3.4.3.

3.4.3   Normalisation of characterisation test results

Ideally, the determination of material properties in a characterisation test should be independent of the test method and sample geometry. For example, the shear modulus or shear viscosity (depending on modelling approach) and consequently the shear force produced during the testing of a material should be independent of both sample dimensions (picture frame and bias extension tests) and length/width ratio (bias extension test). The differences in sample geometry and test kinematics induced by the picture frame and bias extension tests mean that an appropriate normalisation technique must be used before results from different tests can be compared directly.

Normalisation of shear force data from picture frame tests has been considered recently (Harrison *et al.* 2002c, 2004b; Peng *et al.* 2004). Energy arguments have been used to show that, when the shape of the original test specimen is square, i.e. only a small amount of material is missing from the corners of the sample when loaded in the picture frame, force measurements recorded by the picture frame test should be normalised with respect to a characteristic length. A convenient length to use is the side length of the picture frame, L_{pf}. Thus:

$$\frac{F_1}{L_{pf1}} = \frac{F_2}{L_{pf2}} \qquad\qquad 3.21$$

where the subscripts 1 and 2 refer to two different picture frames of arbitrary size. Energy arguments can also be applied in normalising bias extension test results. Indeed, bias extension results could be normalised by a characteristic length for comparison between results from tests on different sized samples with the same length/width ratio, λ, conducted at the same shear rate. However, a method of normalising bias extension data for comparison with picture frame tests is less obvious because of the different deformations occurring in the samples of the two experiments. In Harrison *et al.* (2003) an energy argument is presented that involved determining the relative contribution to the deformation energy from regions A and B of a sample of arbitrary length/width ratio (see Fig. 3.10).

The argument is based on a number of simple approximations and so a certain amount of disagreement between normalised data from the picture frame and bias extension experiments is expected. Nevertheless, the bias extension normalisation procedure provides a convenient method of using the two different test methods to check the validity of experimentally produced results. By assuming that the textile composite deforms as a 'constrained Newtonian fluid', the energy argument leads to the following relationship (Harrison *et al.* 2003):

$$\frac{F_{pf}}{L_{pf}} = \frac{(\lambda - 1)}{R(2\lambda - 3 + 2X)} \frac{F_{be}}{L_{be}} \qquad\qquad 3.22$$

where F_{be} is the axial force measured by the bias extension test, L_{be} is the characteristic length shown in Fig. 3.10, and

$$X = \frac{1}{4} \left\{ \frac{\cos^2 \theta [1 + 3 \sin^2 (\theta/2)]}{\cos^2 (\theta/2)[1 + 3 \sin^2 (\theta)]} \right\} \qquad\qquad 3.23$$

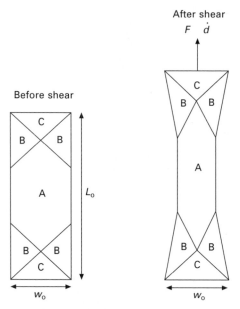

3.10 Bias extension test sample of arbitrary length/width ratio, λ.

and

$$R = \frac{\dot{d}_{be} L_{pf}}{(\lambda - 1)\dot{d}_{pf} L_{be}}$$ 3.24

Error introduced by any non-Newtonian response will become more significant as the ratio R diverges from unity. The above analysis holds only as long as the bias extension test sample follows idealised kinematics. This is a reasonable approximation for low shear angles but begins to fail at higher shear angles because of the occurrence of inter-ply slip in bias extension tests.

In order to plot and compare results of picture frame and bias extension tests it is convenient to define a normalised crosshead displacement that gives an alternative measure of deformation in the material than the shear angle. For the picture frame test this can be defined simply as d_{pf}/L_{pf} whereas for a bias extension test the normalised displacement is defined as $d_{be}/L_{be}(\lambda - 1)$, where d_{be} is the displacement of the crosshead mounting in the bias extension tests. The value of this normalisation technique is illustrated in Fig. 3.11.

Figure 3.11(a) and (b) show comparisons of the normalised force measured using picture frame and bias extension tests on a 2×2 twill weave pre-consolidated glass/polypropylene thermoplastic composite (fibre volume fraction = 0.35, melt temperature approximately 160 °C, thickness = 0.54 mm) at 165 and 180 °C and at a normalised displacement rate of 5.7 min^{-1}.

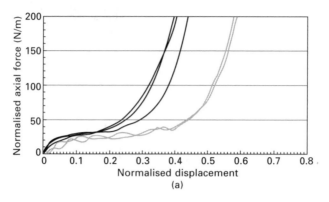

(a)

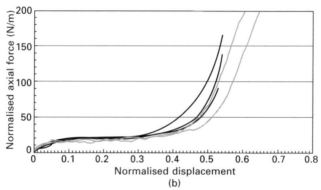

(b)

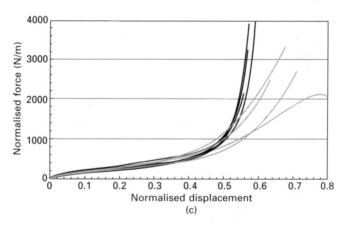

(c)

3.11 Normalised picture frame (black lines) and bias extension (grey lines) results for a pre-consolidated glass/polypropylene 2 × 2 twill weave thermoplastic composite at (a) 165 °C and (b) 180 °C and (c) for a 5-harness satin weave carbon/epoxy prepreg at room temperature.

Likewise, Fig. 3.11(c) shows a comparison of the normalised force measured using picture frame and bias extension tests on a thermosetting 5-harness satin weave carbon/epoxy prepreg (fibre volume fraction = 0.55, thickness = 0.5 mm) at room temperature and at a normalised displacement rate of 0.69 min^{-1}. The force data from each experiment were normalised according to the procedures described above. Results are taken from Harrison *et al.* (2004b).

For the thermoplastic composite the normalised force data show excellent agreement at 180 °C but at 165 °C the results diverge at high normalised-displacements. This was because of the cooling effect of the picture frame, an effect that inhibited melting of the material near the metal clamps. Thus the normalisation technique can be used to compare the results of the two tests to check for possible error due to the experimental method. In Fig. 3.11(c) results from experiments on the thermosetting prepreg composite serve to illustrate the large difference in shear force typically produced by a thermosetting prepreg composite (sheared at room temperature) and a thermoplastic composite sheared above its melt temperature (Fig. 3.11a and b).

3.5   Forming evaluation methods

Once appropriate constitutive model parameters (section 3.3) have been determined or evaluated using characterisation tests (section 3.4), they can be implemented in numerical codes (see Chapter 4). Checking the accuracy of the predictions of these forming simulations against formed components provides the final step in validating the whole virtual forming operation. Grid strain analysis and its application in composite sheet forming have been discussed previously (Martin *et al.* 1997a). Here an optical system is reported, known as CAMSYS Automated Strain Analysis and Measurement Environment (ASAME), that can measure strains and fibre shear angles from formed components both efficiently and accurately. This commercial metal forming analysis system has been adapted to determine the deformation of dry fabrics and composites (Long *et al.* 1997, 2002). The procedure is as follows:

1. A square grid with a (known) regular spacing is marked on a sheet of the material to be formed. For 0/90° woven composites, the grid should coincide with the tow directions.
2. A component is formed using the prepared material. Any process can be analysed in theory, as long as the manufacturing route does not obscure the gridlines.
3. Photographs are taken of the formed component. At least two photographs have to be taken of each region of interest. The photographs must also include a target cube, supplied by the manufacturer of the system. The cube allows the processing software to compute the 3D coordinates of each gridline.

4. The photographs are combined and analysed by computer software to compute the coordinates of gridlines and nodes.
5. By computing the angle formed at the intersection of gridlines, intra-ply shear can be calculated.
6. By comparing the gridline spacing with the marked grid, fibre slip or fibre extension can be calculated.

As an example, the system is used to analyse two hemispherical components formed using two different materials by the vacuum forming process. The first material was a double layer of woven thermoplastic commingled glass/ polypropylene composite material – a balanced 2 × 2 twill woven fabric of density 1.5 g/cm^3 consisting of a yarn of 60% by mass commingled E-glass and polypropylene, denoted Material A. The second material was a single layer of woven (5-harness satin weave) preimpregnated thermosetting carbon/ epoxy material, of density 1.301 g/cm^3, consisting of a tow of 55% by volume carbon fibre with epoxy matrix, denoted Material B. A female mould was used for each component, with a radius of 60 mm.

3.5.1   Shear deformation

Using the optical system, the shear angle distribution for each hemisphere was measured. The results are displayed in Fig. 3.12 for Materials A and B. The distributions are reasonably symmetric between quadrants. Maximum shear occurs along the bias axis of each quadrant. However, Material B (prepreg) exhibits slightly higher levels of shear than Material A (thermoplastic) in the highly deformed regions. The maximum shear angle for the prepreg was 55°, compared with 46° for the thermoplastic. This difference is presumably due to the different rheological behaviours of the two composite materials as well as uncertainties associated with the hand lay-up procedure.

The information captured using this technique can be presented in numerous ways for comparison with numerical simulation results. For example, Fig. 3.13 shows the fibre pattern predicted by a kinematic drape model (KDM). This model does not take any material properties into account and is based on a mapping procedure that calculates the draping of a pin jointed net across the surface of the component. The method incorporates the fibre inextensibility constraint.

In order to visualise graphically the difference in shear angle between the formed components and the kinematic drape predictions, shear angle is plotted along a section of maximum shear in Fig. 3.14. For the hemisphere formed using Material A, two quadrants were analysed. Variation in the two sets of data illustrates the degree of asymmetry in the formed component. For each component, shear angle increases along the section in a manner similar to that predicted by the KDM, until a maximum shear angle is reached at the flange region.

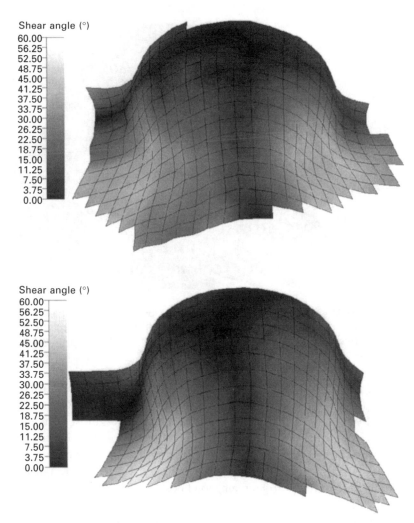

3.12 Shear angle distribution for hemisphere formed from Material A (commingled glass/polypropylene, top) and Material B (carbon/epoxy prepreg, bottom).

As another example, a global picture of the difference in shear angle across the components may be obtained by displaying the frequency with which any particular shear angle occurs. A graph showing such data is presented in Fig. 3.15, along with data for the KDM. Lower shear angles were obtained more frequently for Material B than for Material A, although the range of angles was higher for Material B. The distribution predicted by a kinematic drape simulation covers a wider range of angles than achieved with either material. Lower shear angles (<5°) were predicted more frequently than

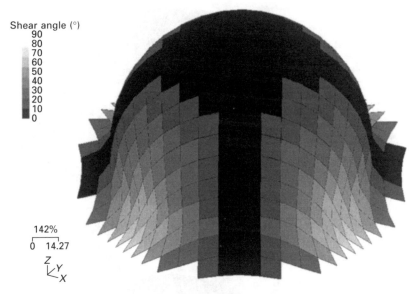

3.13 Shear angle distribution predicted by kinematic drape model. The kinematic drape model is based on a geometric mapping of a pin-jointed fibre network, and ignores material forming rheology when calculating the fibre pattern.

occurred in the experimental measurements, probably due to the difficulty in laying up the material symmetrically (i.e. in practice each quadrant was pre-sheared by a small amount). The predicted distribution is most similar to that obtained using Material B (prepreg). This frequency distribution method of visualising the data can also be used to compare the shear distributions required to form different components, making it a useful design tool in deciding which lay-up or geometry to choose.

3.5.2   Secondary deformation mechanisms

When a textile composite is formed to a 3D shape such as a hemisphere, a number of deformation mechanisms occur (see section 3.2). The primary mechanism is intra-ply shear. However, un-crimping of tows and crossover slip can also contribute to the material deformation. The optical system described here is capable of measuring not just the shear angle between reinforcement directions at any point but also the distance between nodes of the deformed grid (see section 3.5.1). Using this information extension and contraction along the lengths of the reinforcement directions can also be measured. Figure 3.16 shows the extension ratio (final grid spacing divided by initial spacing) experienced by the two materials during forming.

The extension ratio is positive in the region close to the pole of the

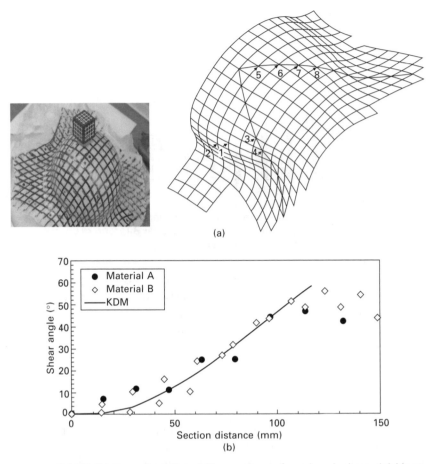

3.14 Validation of predicted fibre patterns for a hemisphere. (a) Lines along sections of maximum shear, with insert showing formed component and measuring cube. (b) Comparison between measured and predicted shear angles along a section of the hemispheres.

hemisphere and negative on the part close to the flange. Positive extension ratio is due to the lay-up of the material over the tool. The material is pushed down into the tool leading to the un-crimping of tows in the region close to the pole. However, yarns close to the flange are pushed together, producing a negative extension ratio. Extension ratios for Material A were in the range 0.78 to 1.24, while those for Material B were 0.87 to 1.10. Although Material A has a large tow width and inter-tow spacing, Material B has a small tow width and little inter-tow spacing. The Material A textile structure facilitates relative movement between tows, leading to higher extension ratios. Similar trends are observed in each direction for the each material. Higher changes in grid spacing were observed for Material A, which would explain the lower shear angles observed for this material. As before, information captured

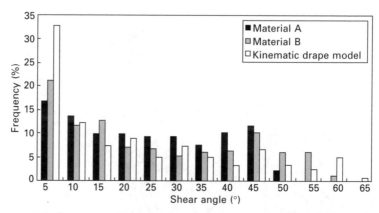

3.15 Frequency distribution comparing shear angles obtained for
Material A (commingled fabric) and Material B (prepreg)
hemispheres with kinematic drape model predictions.

using this optical technique can be presented in numerous ways for comparison
with numerical simulation results. Indeed, observations of material extensions
have prompted various workers to perform numerical simulations incorporating
extensible fibre reinforcement behaviour (Lamers *et al.* 2002a, 2003; Sharma
and Sutcliffe 2003).

3.6   Summary

The rheological modelling and characterisation of viscous textile composites
have progressed significantly since 1995. Through incorporation of non-
linear material behaviour in continuum models (section 3.3.3), various workers
have shown how these materials can be characterised using specialised
experimental techniques (section 3.4). However, the number of parameters
appearing in these models has meant an extensive number of tests must be
performed for each characterisation, which is a long and costly process. A
response to this problem has been the use of homogenisation and energy
summation techniques, facilitated by recent dramatic increases in computational
speed. These methods predict the rheological behaviour of viscous textile
composites using knowledge of the composites structure and behaviour of
the constituent materials. Methods of evaluating predictions of computer
simulations of complex material deformations that implement these rheological
models are becoming more sophisticated and rigorous (section 3.5).

3.7   Acknowledgements

We thank the following organisations for their support: EPSRC, BAE
SYSTEMS, BP Amoco, ESI Software, Ford Motor Company, QinetiQ, Saint-

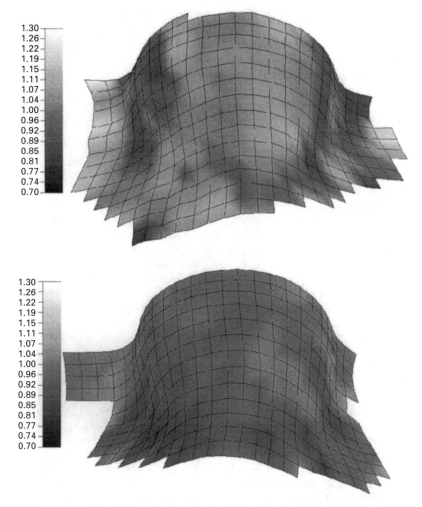

3.16 Extension ratio distribution for a hemisphere formed and consolidated using Material A (thermoplastic, top) and Material B (prepreg, bottom).

Gobain Vetrotex, MSC Software Ltd, and the Universities of Cambridge and Leeds. The authors also thank both Prof. A.J. Spencer and Dr W.R. Yu for helpful discussions on this work. We also acknowledge R. Garcia Gil and M. Sherburn for their contributions.

3.8   References

Advani S.G., Creasy T.S. and Shuler S.F. (1997), *Composite Sheet Forming: Rheology of Long Fiber-reinforced Composites in Sheet Forming*, 323–369, 8, ed. Bhattacharyya D., S.G. Pipes Elsevier, Amsterdam.

Balasubramanyam R., Jones R.S. and Wheeler A.B. (1989), Modelling transverse flows of reinforced thermoplastic materials. *Composites*, 20, 1, 33–37.

Barnes J.A. and Cogswell F.N. (1989), Transverse flow processes in continuous fibre-reinforced thermoplastic composites. *Composites*, 20, 1, 38–42.

Binding D.M. (1991), Capillary and contraction flow of long-(glass) fibre filled polypropylene *Composites Manufacturing*, 2, 3/4, 243–252.

Blanchard P., Cao J., Chen J. and Long A.C. (2001), National Science Foundation Workshop on Composite Sheet Forming, http://www.mech.northwestern.edu/fac/cao/nsfworkshop/index.htm, Lowell, MA, 5–7 September.

Boisse P., Gasser A. and Hivet G. (2001), Analyses of fabric tensile behaviour: determination of the biaxial tension-strain surfaces and their use in forming simulations. *Composites: Part A*, 32, 1395–1414.

Canavan R.A., McGuiness G.B. and O'Bradaigh C.M. (1995), Experimental intraply shear testing of glass-fabric reinforced thermoplastic melts, *4th International Conference on Automated Composites (ICAC '95)*, Nottingham, September, 127–139.

Christensen R.M. (1993), Effective viscous flow properties for fibre suspensions under concentrated conditions. *Journal of Rheology*, 37, 1, 103–121.

Coffin D.W., Pipes R.B. and Simacek P. (1995), First-order approximations for the effective shearing viscosities of continuous-fiber suspensions. *Journal of Composite Materials*, 29, 9, 1196–1180.

Cogswell F.N. (1992), *Thermoplastic Aromatic Polymer Composites*, Butterworth-Heinemann Ltd, Oxford.

Cogswell F.N. (1994), Continuous fibre systems, in *Flow and Rheology in Composite Manufacturing*, 127–144, Ch. 10, Ed. Advani S.G., Pipes R.B., Elsevier, Amsterdam.

Cogswell F.N. and Groves D.J. (1988), The melt rheology of continuous fibre reinforced structural composite materials. *Xth International Congress on Rheology*, Sydney, 275–277.

De Luca P., Lefebure P., Pickett A.K., Voldermayer A.M. and Werner W. (1996), The numerical simulation of press forming of continuous fiber reinforced thermoplastics. *28th International SAMPE Technical Conference*, 4–7, November Seattle, Washington.

De Luca P., Lefebure P. and Pickett A.K. (1998), Numerical and experimental investigation of some press forming parameters of two fibre reinforced thermoplastics: APC2-AS4 and PEI-CETEX. *Composites Part A*, 29A, 101–110.

Dykes R.J., Martin T.A. and Bhattacharyya, D. (1998), Determination of longitudinal and transverse shear behaviour of continuous fibre-reinforced composites from vee-bending. *Composites Part A*, 29A, 39–49.

Friedrich K., Hou M. and Krebs J. (1997), *Composite Sheet Forming: Thermoforming of Continuous Fibre/Thermoplastic Composite Sheets*, 91–162, Ch. 4, Ed. Bhattacharyya D., Pipes S.G., Elsevier, Amsterdam.

Goshawk J.A. and Jones R.S. (1996), Structure reorganisation during the rheological charaterisation of continuous fibre-reinforced composites in plane shear. *Composites Part A*, 27A, 279–286.

Goshawk J.A., Navez V.P. and Jones R.S. (1997), Squeezing flow of continuous fibre-reinforced composites. *Journal of Non-Newtonian Fluid Mechanics*, 73, 327–342.

Groves D.J. (1989), A characterisation of shear flow in continuous fibre thermoplastic laminates. *Composites*, 20, 1, 28–32.

Groves D.J. and Stocks D.M. (1991), Rheology of thermoplastic composite carbon fibre composite in the elastic and viscoelastic states. *Composites Manufacturing*, 2, 3/4, 179–184.

Hagege B., Boisse P. and Billoet J.L. (2003), Specific simulation tools for the shaping process of knitted reinforcements. *6th International ESAFORM Conference on Materials Forming*, 28–30 April, Salerno, 871–874.

Harrison P., Clifford M.J., Long A.C. and Rudd C.D. (2002a), Constitutive modelling of impregnated continuous fibre reinforced composites: a micro-mechanical approach. *Plastics, Rubber and Composites*, 2, 31, 76–86.

Harrison P., Clifford M.J., Long A.C. and Rudd C.D. (2002b), A micro-mechanical constitutive model for continuous fibre reinforced impregnated composites. *5th International ESAFORM Conference on Materials Forming*, 15–17 April, Krakow, 275–278.

Harrison P., Clifford M.J. and Long A.C. (2002c), Shear characterisation of woven textile composites, *10th European Conference on Composite Materials*, 3–7 June, Brugge.

Harrison P., Long AC., Clifford M.J., Garcia Gil R., Ward I.M. and Hine P.J. (2002d), Investigation of thermoformability and molecular structure of Curv^{TM}: *Automotive Composites and Plastics*, 3–4 December, Basildon, ESSEX, UK.

Harrison P., Clifford M.J., Long A.C. and Rudd C.D. (2003), Constitutive modelling of shear deformation for impregnated textile composites. *6th International ESAFORM Conference on Materials Forming*, 28–30 April, Salerno, 847–850.

Harrison P., Clifford M.J., Long A.C. and Rudd C.D. (2004a), A constituent based predictive approach to modelling the rheology of viscous textile composites. *Composites: Part A* 38, 7–8, 915–931.

Harrison P., Clifford M.J. and Long A.C. (2004b), Shear characterisation of woven textile composites: a comparison between picture frame and bias extension experiments. *Composites Science and Technology*, 64, 10–11, 1453–1465.

Hsiao S.W. and Kikuchi N. (1999), Numerical analysis and optimal design of composite thermoforming process. *Computing Methods in Applied Mechanical Engineering*, 177, 1–34.

Hull B.D., Rogers T.G. and Spencer A.J.M. (1991), Theory of fibre buckling and wrinkling in shear flows of fibre-reinforced composites. *Composites Manufacturing*, 2, 3/4.

Hull B.D., Rogers T.G. and Spencer A.J.M. (1994), Theoretical analysis of forming flows of continuous fibre resin systems. *Flow and Rheology in Composite Manufacturing*, 203–256, Ch. 10, Ed. Advani S.G., Pipes R.B., Elsevier, Amsterdam.

Johnson, A.F. (1995), Rheological model for the forming of fabric reinforced thermoplastic sheets. *Composites Manufacturing*, 6, 153–160.

Jones R.S. and Roberts R.W. (1991), Ply reorientation in compression. *Composites Manufacturing*, 2, 3/4, 259–266.

Lai C.L. and Young W.B. (1999), Modelling fibre slippage during the preforming process. *Polymer Composites*, 20, 4, 594–603.

Lamers E.A.D., Akkerman R. and Wijskamp S. (2002a), Fibre orientation modelling for rubber press forming of thermoplastic laminates. *5th International ESAFORM Conference on Material Forming*, Krakow, 323–326, 14–17, April.

Lamers E.A.D., Akkerman R. and Wijskamp S. (2002b), Drape modelling of multi-layer fabric reinforced thermoplastic laminates. *Proc. 6th International Conference on Textile Composites (TexComp 6)*, Philadelphia, USA, September.

Lamers E.A.D., Wijskamp S. and Akkerman R. (2003), Drape modelling of multi-layered composites. *6th International ESAFORM Conference on Material Forming*, Salerno, 823–826, 28–30, April.

Lebrun G. and Denault J. (2000), Influence of the temperature and loading rate on the intraply shear properties of a polypropylene fabric. *15th American Association for composites*, 25–27 September, College Station, Texas, 659–667.

Lebrun G., Bureau M.N. and Denault J. (2003), Evaluation of bias-extension and picture-frame test methods for the measurement of intraply shear properties of PP/glass commingled fabrics. *Composite Structures*, 61, 341–352.

Li H. and Gutowski T. (1997), *Composite Sheet Forming: The Forming of Thermoset Composites*, 441–472, Ch. 11, Ed. Bhattacharyya D., Pipes S.G., Elsevier, Amsterdam.

Lomov S.V., Belov E.B. and Verpoest I. (2002), Behaviour of textile reinforcement in composite forming: mathematical modelling of deformability, internal geometry and permeability of the perform. *5th International ESAFORM Conference on Material Forming*, Krakow, Poland, 271–274, 14–17, April.

Long A.C. Rudd C.D., Blagdon M. and Johnson M.S. (1997), Experimental analysis of fabric deformation mechanisms during preform manufacture. *Proc. 11th International Conference on Composite Materials (ICCM-11)*, Gold Coast, Australia, 5, July, 238–248.

Long A.C., Gil R.G., Clifford M.J., Harrison P., Sharma S.B. and Sutcliffe M.P.F. (2002), Experimental analysis of fabric deformation during forming of textile composites. *5th International ESAFORM Conference on Material Forming*, Krakow, Poland, 279–282, 14–17, April.

Lussier D. and Chen J. (2000), Material characterisation of woven fabrics for thermoforming of composites. *15th American Association for Composites*, 25–27 September, College Station, Texas, 301–310.

Lussier D.S., Chen J. and Sherwood J.A. (2002), Viscosity based models for shear of glass/thermoplastic fabrics. *5th International ESAFORM Conference on Material Forming*, Krakow, Poland, 283–286, 14–17, April.

Mallon P.J. and O'Bradaigh C.M. (2000), in *Comprehensive Composite Materials, Volume 2: Polymer Matrix Composites*, Eds Talreja R., Manson J.A.E., Series eds Kelly A., Zweben C., Elsevier, Amsterdam, 874–913.

Mander S.J. (1998), Roll forming of fibre reinforced thermoplastic composites. PhD Thesis, University of Auckland, New Zealand.

Mander S.J., Panton S.M., Dykes R.J. and Battacharyya D. (1997), *Composite Sheet Forming: Roll Forming of Sheet Materials*, 473–515, Ch. 12, Ed. Bhattacharyya D., Pipes S.G. Elsevier, Amsterdam.

Martin T.A., Bhattacharyya D. and Collins I.F. (1995), Bending of fibre-reinforced thermoplastic sheets. *Composites Manufacturing*, 6, 177–187.

Martin T.A., Christie G.R. and Bhattacharyya D. (1997a), *Composite Sheet Forming: Grid Strain Analysis and its Application in Composite Sheet Forming*, 217–245, Ch. 6, Ed. Bhattacharyya D., Pipes S.G. Elsevier, Amsterdam.

Martin T.A., Mander S.J., Dykes R.J. and Bhattacharyya D. (1997b), *Composite Sheet Forming: Bending of Continuous Fibre-Reinforced Thermoplastic sheets*, 371–401, Ch. 9, Ed. Bhattacharyya D., Pipes S.G. Elsevier, Amsterdam.

McEntee S.P. and O'Bradaigh C.M. (1998), Large deformation finite element modelling of single-curvature composite sheet forming with tool contact. *Composites Part A*, 29A, 207–213.

McGuiness G.B. and O'Bradaigh C.M. (1997), Development of rheological models for forming flows and picture frame testing of fabric reinforced thermoplastic sheets. *Journal of Non-Newtonian Fluid Mechanics*, 73, 1–28.

McGuiness G.B. and O'Bradaigh C.M. (1998), Characterisation of thermoplastic composite melts in rhombus-shear: the picture frame experiment. *Composites Part A*, 29A, 115–132.

McGuiness G.B., Canavan R.A., Nestor T.A. and O'Bradaigh M.O. (1995), A picture frame intra-ply sheet forming of composite materials. *Proceedings of the American Society of Mechanical Engineers (ASME) Materials Division,* ASME, 1107–1118.

Milani A.S., Nemes J.A., Pham X.T. and Lebrun G. (2003), The effect of fibre misalignment on parameter determination using picture frame test. *Proc. 14th International Conference on Composite Materials (ICCM-14),* San Diego, US, July.

Murtagh A.M. and Mallon P.J. (1995), Shear characterisation of unidirectional and fabric reinforced thermoplastic composites for pressforming applications. *Proc. 10th International Conference on Composite Materials (ICCM-10),* Whistler, Canada, 373–380, August.

Murtagh A.M. and Mallon P.J. (1997), *Composite Sheet Forming: Characterisation of Shearing and Frictional Behaviour during Sheet Forming,* 163–216, 5, Eds Bhattacharyya D., Pipes S.G., Elsevier, Amsterdam.

O'Bradaigh C.M. and Pipes R.B. (1991), Finite element analysis of composite sheet-forming process. *Composites Manufacturing,* 2, 3/4, 161–170.

O'Bradaigh C.M., McGuiness G.B. and McEntee S.P. (1997), *Composite Sheet Forming: Implicit Finite Element Modelling of Composites Sheet Forming Processes,* 247–322, Ch. 7, Eds Bhattacharyya D., Pipes S.G., Elsevier, Amsterdam.

Peng X.Q. and Cao J. (2002), A dual homogenization and finite element approach for material characterization of textile composites. *Composites: Part B,* 33, 45–56.

Peng X.Q., Cao J., Chen J., Xue P., Lussier D.S. and Liu L. (2004), Experimental and numerical analysis on normalization of picture frame tests for composite materials. *Composites Science and Technology,* 64, 11–21.

Phung T., Paton R. and Mouritz A.P. (2003), Characterisation of interply shearing resistance of carbon–epoxy unidirectional tape and fabric prepregs. *6th International ESAFORM Conference on Materials Forming,* 28–30 April, Salerno, Italy, 867–870.

Pickett A.K., Queckborner T., De Luca P. and Haug E. (1995), An explicit finite element solution for the forming prediction of continuous fibre-reinforced thermoplastic sheets. *Composites Manufacturing,* 6, 237–243.

Pipes R.B. (1992), Anisotropic viscosities of an orientated fibre composite with a power law matrix. *Journal of Composite Materials,* 26, 10.

Potter K. (2002), Bias extension measurements on cross-plied unidirectional prepreg. *Composites Part A,* 33, 63–73.

Prosser W., Hine P.J. and Ward I.M. (2000), Investigation into thermoformability of hot compacted polypropylene sheet. *Plastics, Rubber and Composites,* 29, 8, 401–410.

Roberts R.W. and Jones R.S. (1995), Rheological characterisation of continuous fibre composites in oscillatory flow. *Composites Manufacturing,* 6, 161–167.

Rogers T.G. (1989a), Rheological characterisation of anisotropic materials. *Composites,* 20(1), 21–27.

Rogers T.G. (1989b), Squeezing flow of fibre reinforced viscous fluids. *Journal of Engineering Mathematics,* 23, 81–89.

Rogers T.G. (1990), Shear characterisation and inelastic torsion of fibre reinforced materials', in *Inelastic Deformation of Composite Materials,* ed. Dvorak G.J., Springer-Verlag, New York.

Rogers T.G. and O'Neill J.M. (1991), Theoretical analysis of forming flows of fibre-reinforced composites. *Composites Manufacturing,* 2, 3/4, 153–160.

Scobbo J.J. and Nakajima N. (1989), Dynamic mechanical analysis of molten thermoplastic/continuous graphite fiber composites in simple shear deformation. *21st International*

Society for the Advancement of Materials and Process Engineering (SAMPE-21) Technical Conference, 25–28, September, Atlantic City, New Jersey.

Sharma S.B. and Sutcliffe M.P.F. (2003), A simplified finite element approach to draping of woven fabric. *6th International ESAFORM Conference on Materials Forming*, 28–30 April, Salerno, 887–890.

Shuler S.F. and Advani S.G. (1996), Transverse squeeze flow of concentrated aligned fibers in viscous fluids. *Journal of Non-Newtonian Fluid Mechanics*, 65, 47–74.

Smiley A.J. and Pipes R.B. (1988), Analyses of the diaphragm forming of continuous fiber reinforced thermoplastics. *Journal of Thermoplastic Composite Materials*, 1, 298–321.

Spencer A.J.M. (1971), Theory of invariants, in *Continuum Physics*, Ed. Eringen A.C., Academic Press, New York, 240–353.

Spencer A.J.M. (1972), *Deformation of Fibre-reinforced Materials*, Clarendon Press, Oxford.

Spencer A.J.M. (1984), *Continuum Theory for Strongly Anisotropic Solids,* CISM Courses and Lectures No. 282, Spinger-Verlag, New York.

Spencer A.J.M. (2000), Theory of fabric-reinforced viscous fluids. *Composites Part A*, 31, 1311–1321.

Sutcliffe M.P.F., Sharma S.B., Long A.C., Clifford M.J., Gil R.G., Harrison P. and Rudd C.D. (2002), A comparison of simulation approaches for forming of textile composites. *5th International ESAFORM Conference on Materials Forming*, 15–17 April, Krakow, 311–314.

Wheeler A.B. and Jones R.S. (1991), A characterisation of anisotropic shear flow in continuous fibre composite materials. *Composites Manufacturing*, 3/4, 192–196.

Wheeler A.B. and Jones R.S. (1995), Numerical simulation of fibre reorientation in the consolidation of a continuous fibre composite material. *Composites Manufacturing*, 6, 263–268.

Wilks C.E., Rudd C.D., Long A.C. and Johnson C.F. (1999), Rate dependency during processing of glass/thermoplastic composites. *Proc. 12th International Conference on Composite Materials (ICCM-12)*, Paris.

Xue P., Peng X. and Cao J. (2003), A non-orthogonal constitutive model for characterising woven composites. *Composites Part A*, 34, 183–193.

Yu W.R., Pourboghrat F., Chung K., Zampaloni M. and Kang T.J. (2002), Non-orthogonal constitutive equation for woven fabric reinforced thermoplastic composites. *Composites Part A*, 33, 1095–1105.

Yu W.R., Zampaloni M., Pourboghrat F., Chung K. and Kang T.J. (2003), Sheet hydroforming of woven FRT composites: non-orthogonal constitutive equation considering shear stiffness and undulation of woven structure. *Composite Structures*, 61, 353–362.

Yu X., Zhang L. and Mai Y.W. (2000), Modelling and finite element treatment of intra-ply shearing of woven fabric. *International Manufacturing Conference in China (IMCC 2000)*, 16–17 August, Hong Kong.

Zheng Q.S. (1994), Theory of representations for tensor functions. *Applied Mechanical Review*, 47, 554–587.

Zouari B., Dumont F., Daniel J.L. and Boisse P. (2003), Analyses of woven fabric shearing by optical method and implementation in a finite element program. *6th International ESAFORM Conference on Materials Forming*, 28–30 April, Salerno, Italy, 875–878.

4

Forming textile composites

W - R Y U, Seoul National University, Korea and
A C L O N G, University of Nottingham, UK

4.1 Introduction

This chapter describes modelling approaches for forming of textile composite sheets into three-dimensional components. Forming processes considered here involve both dry fabrics (for liquid moulding) and pre-impregnated composites. A number of simulation techniques exist within the research community, and several of these have been implemented within commercial software packages. The aim of these models is to predict the fibre orientations within the formed component, from which additional information can be deduced, e.g. ply templates for net-shape forming, component thicknesses (for prepreg) or fibre volume fractions (for dry fabric), and composite mechanical properties. Virtually all the work in this field has concentrated on fabrics with two orthogonal fibre directions (i.e. bidirectional fabrics and their composites), and hence this description applies exclusively to such materials. Two modelling approaches are described in detail in this chapter. The first is based on a simple mapping procedure, and the second on a mechanical approach utilising finite element (FE) analysis.

4.2 Mapping approaches

4.2.1 Kinematic models

A number of researchers have developed kinematic simulations of draping or forming for textile reinforcements[1–4]. Several of these models have been implemented within commercially available software packages (e.g. MSC.Patran Laminate Modeler[5], Vistagy FiberSIM[6]). In this approach, the fabric is modelled as a pin-jointed net (or 'fishnet'), which is mapped onto the surface of the forming tool by assuming that tow segments are able to shear at the joints (tow crossovers). A unique draped pattern can be obtained by specifying two intersecting tow paths, referred to as generators, on the surface of the forming tool. The remaining tows are positioned using a mapping approach, which involves solving geometric equations to determine the intersection of the surface with possible crossover points for the tow segments. Several strategies are available for specifying the generator paths, with either geodesic or projected paths used typically. Correct specification of the generators is critical, as these will determine the positions of all remaining tows. An example of this type of analysis is shown in Fig. 4.1,

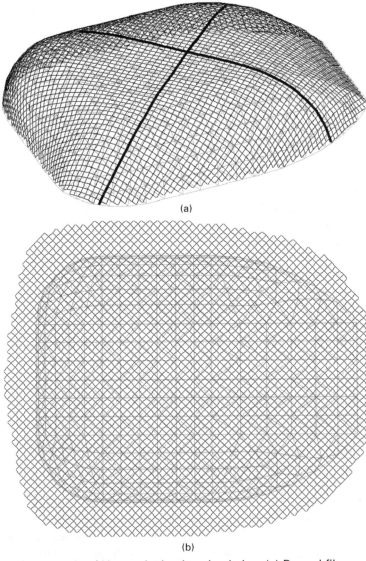

(a)

(b)

4.1 An example of kinematic draping simulation. (a) Draped fibre pattern based on geodesic generator paths (marked in bold). (b) Predicted net-shape for ± 45° ply.

which illustrates the predicted fibre pattern obtained using a kinematic model with geodesic generator paths. Also shown is the predicted net-shape, which can be used as a ply cutting template.

The kinematic approach provides a very fast solution, with typical run times of less than 10 seconds for complex components. Commercial packages have established interfaces with computer-aided design (CAD) and mesh generation software, and interfaces with ply cutting and placement (laser guide) systems also exist. However, as stated above the solution is dependent on accurate specification of the generator paths. In practice these may be affected by material and process boundary conditions. One other issue is that purely kinematic models are unable to differentiate between materials other than in the specification of the locking angle, which is used to indicate possible areas of wrinkling. Consequently an identical fibre pattern is obtained, regardless of variations in material forming characteristics or processing technique. Nevertheless the ease of use of commercially available kinematic models has meant that they are used widely by composites manufacturers, notably in the aerospace and high-performance automotive (e.g. Formula 1) sectors.

4.2.2   Iterative draping simulation

Generally the geometric/kinematic approach to drape simulation works well for symmetric shapes draped with balanced materials. However for non-symmetric geometries, correct placement of the two generator tow paths may be problematic, and it may not be possible to identify their positions intuitively. In addition, as has been shown in Chapter 2, materials such as non-crimp fabrics (NCFs) exhibit a preferential direction for deformation. This means they are more difficult to shear in one direction, with shear forces for a given shear angle typically up to five times higher for shearing parallel to the stitch than for shearing perpendicular to the stitch. A typical example of this is shown in Fig. 4.2, which shows the shear resistance for a ±45° glass fabric retained using a tricot stitch.

During preform manufacture with NCFs, the resulting fibre pattern may be different from that obtained using a balanced fabric. The use of a mechanical forming simulation would allow consideration of material directionality, as described in section 4.3.3, although this requires a significant level of expertise and increased computation time. The alternative approach presented here is based on the use of a geometric mapping algorithm within an iterative scheme (an approach first suggested by Bergsma[3]). The energy required to produce each mapping is calculated, with the mapping resulting in the lowest energy assumed to represent the actual behaviour of the fabric. In reality the deformation energy is the sum of several components, although here only intra-ply (in-plane) shear resistance is considered, as this is thought to be

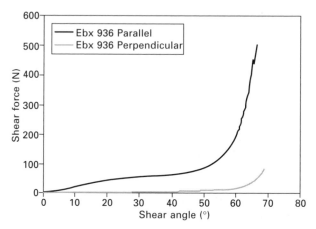

4.2 Shear force versus shear angle curves for a ±45° glass NCF with
a tricot stitch, sheared both parallel and perpendicular to the
stitching direction.

most indicative of the effect of fabric construction on deformation.

The fabric shear energy (work done during shearing, U_s) can be calculated
simply from the area under the torque-shear angle curve:

$$U_s(\theta) = \int_0^\theta T(\gamma)\,\mathrm{d}\gamma \qquad 4.1$$

where $T(\gamma)$ is the torque required to achieve a shear angle γ. This expression
can be evaluated by fitting an empirical relationship to the shear force versus
shear angle curve, measured for example using a picture frame shear test
(see Chapter 2). For NCFs, two curves may be specified to represent shearing
parallel and perpendicular to the stitching thread. The total energy is calculated
within the drape simulation by summing the contribution at each node (tow
crossover).

A simple way to determine the mapping resulting with the minimum
energy is to use an iterative scheme based on the two generator tow paths.
This approach involves finding the two intersecting paths that result in the
lowest total energy. A Hooke and Jeeves minimisation method is used[7],
where the generator path is defined one step at a time from a user-defined
starting point. Each successive set of nodes is positioned by iterating the
generator path angle to achieve the minimum increase in shear energy. For
reasons of computational efficiency, this technique is preferred to a global
minimisation algorithm in which the entire generator path is modified at
each stage of the process. However as nodes in contact with the tool are
subject to additional constraints due to friction, the present approach is likely
to be reasonably accurate.

The above algorithm has been implemented within the 'DrapeIt' software package at the University of Nottingham (as described in detail by Souter[8]). An automotive component is used here to evaluate and to validate the model. The geometry is based on a transmission tunnel for a high-performance vehicle, produced by resin transfer moulding (RTM). Preforms were produced by hand lay-up over a male former. An arrangement of several discrete forming pads was developed to hold the fabric onto the component surface during lay-up. These assisted in the lay-up process and resulted in improved repeatability during preform manufacture. Figure 4.3 compares predicted fibre patterns for this geometry obtained using a variety of techniques. Figure 4.3(a) was obtained using a purely kinematic algorithm with geodesic generator paths. Figure 4.3(b) was generated using the energy minimisation algorithm with shear data for a plain weave fabric. It is clear that the energy minimisation algorithm has reduced the shear deformation over the surface significantly compared with the kinematic model; if the shear strain energies for each simulation are analysed, a reduction of 35% is recorded. Figure 4.3(c) was generated using shear data for a ±45° carbon NCF, having similar shear compliance data to that given in Fig. 4.2. In this case fabric deformation was increased in quadrants that were sheared in the preferential direction.

Figure 4.4 compares predicted and measured shear angles along the length of the component for the NCF. Fabric layers were marked with an orthogonal grid, and shear angles were determined by measuring the relative position of grid points using digital vernier callipers. The results agree over the majority of the length, although the model over-estimates the shear deformation at the rear of the tunnel. For experimentally produced preforms, wrinkles were present in this location. Darts (triangular cuts) were used to alleviate wrinkling, reducing the overall shear deformation in the region. These discontinuities were not represented in the forming simulation.

From the above it may be concluded that software packages based on a kinematic mapping approach are easy to use, and that they can provide a fast and accurate solution for relatively simple geometries. Where either the component or material behaviour is non-symmetric, an iterative energy-based mapping may provide greater accuracy. Nevertheless this approach is still a simplification when compared with the real process. Specifically, it does not consider all the boundary conditions or material properties that are relevant to real forming processes. For example an automated process may include a pinching frame or blank-holder to support the material and to suppress wrinkling. Thermoplastic composites usually require a non-isothermal forming process, where material properties will change as the material is cooled during forming. Such material behaviour and boundary conditions may require the use of a mechanical modelling approach, as described in the following section.

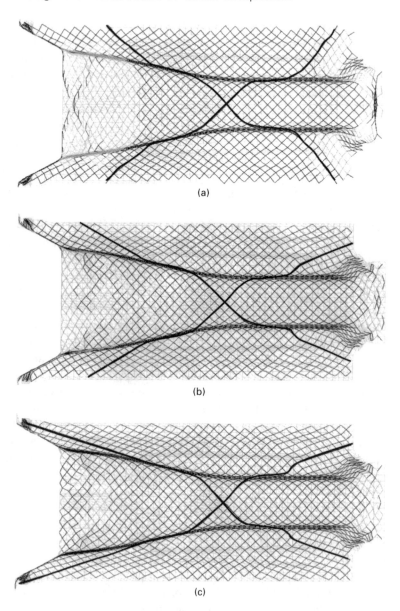

4.3 Predicted fibre patterns for an automotive transmission tunnel.
(a) Geometric mapping with geodesic generator paths. (b) Iterative
analysis with shear data for a plain weave. (c) Iterative analysis with
shear data for a ±45° tricot stitched NCF.

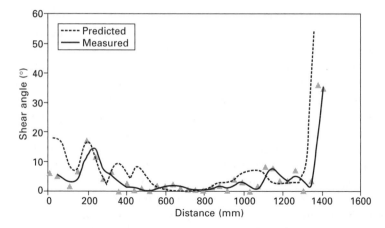

4.4 Comparison of predicted and measured shear angles along the length of the transmission tunnel draped with a ±45° tricot stitched NCF. Measurements were taken along the centre line from left to right with reference to Fig. 4.3(c).

4.3   Constitutive modelling approach

4.3.1   Introduction

In developing composites manufacturing processes, process engineers need to analyse the full forming process, considering factors such as tool configuration and processing conditions. This requires detailed analysis of the material during the forming process, considering material behaviour and process variables. For this purpose the FE method is considered useful, despite long computation times and requirements for an in-depth knowledge of FE analysis and constitutive models. This is because it can predict the final formed shape including stress and strain distributions, and the effects on these from the tool configuration and its interaction with the material. For FE modelling and analysis, constitutive equations are required to describe the mechanical behaviour of the material during forming. This is dependent on the reinforcement structure and (for prepreg) the matrix rheology. Mechanical behaviour of dry fabrics and prepreg were described in detail in Chapters 2 and 3 respectively. In this chapter, constitutive modelling for continuous fibre reinforced composites will be discussed, along with its application to forming simulation.

Fabric-based composites not only improve mechanical and physical properties but also enable rapid, automated production of composite structures. Numerous manufacturing processes have been proposed over the years to shape composite prepreg or preforms, ranging from compression moulding to diaphragm forming and hydroforming. Owing to their high success with metals, various attempts have also been made to apply sheet forming techniques to composites[9, 10]. One difficulty here is the limited draping capability of fabrics, which are prone to wrinkling. To design an appropriate sheet forming process for fabric-based composites, significant research on constitutive modelling for woven structures has been performed. This can be classified into two general categories: continuum mechanics and non-continuum mechanics-based models. The non-continuum approach can be used for modelling of textile preforms without resin. This enables draping simulations before resin injection during RTM, with (for example) particle dynamics representing the intersection point between warp and weft threads[11].

For continuous fibre reinforced composite forming analysis, the continuum-based approach is preferred because of its ability to model complex deformation behaviour. In general, for composite modelling the material can be treated as either homogeneous or non-homogeneous. Based on non-homogeneous concepts, truss and shell (membrane) elements can be used to model the textile structure and resin matrix, respectively[12, 13]. In this meso-cell approach, the microstructural interaction between the matrix and the reinforcement is accounted for by using common nodes to define the truss and shell (or membrane) elements. Therefore, the meso-cell approach can account for the evolution of the microstructure during forming. However, since this approach takes a large amount of computation and pre-processing time, the homogeneous continuum approach is usually preferred.

Many models assuming homogeneous material properties have been developed, utilising well-defined mathematical theories (such as orthotropic constitutive equations), thereby reducing the computational cost (e.g. Dong et al.[14]). These studies assume the preservation of the initial orthotropy of the fabric during forming, which is common practice in sheet metal forming analysis. This assumption is not accurate for the analysis of fabric-based composites, since the change in fibre angle is significant and hence the principal material directions change as the material is formed. Several efforts have been made to develop numerical models capable of capturing this fibre angle evolution. This has resulted in non-orthogonal continuum-based constitutive equations for unidirectional composites that with fibre inextensibility and incompressibility assumptions[15] based on the *ideal fibre reinforced model* (IFRM)[16, 17]. This approach has been applied to diaphragm forming analysis for unidirectional thermoplastic composites[18], and has recently been extended to thermoplastic composites with woven fabric reinforcements[19].

A great deal of research has been devoted to the use of homogenisation methods to account for the evolution of microstructural parameters and also to predict the macroscopic deformed shape of composites. Homogenisation methods involve modelling a unit cell in 3D and then solving a set of governing equations to obtain the homogenised material properties of the composite based on a geometric description of the unit cell[20]. In spite of the computational disadvantages, including long computational times, the homogenisation method has been used to model a multitude of domains such as the micro-level of the fibre and the meso-level of the fabric[21]. Recently, a new method has been developed that reduces computational cost and accounts for microstructural effects on the forming behaviour of thermoplastic composites. The method utilises numerical techniques to extract non-linear elastic material properties based on the homogenisation method under a continuum mechanics framework[22]. An extensive review of constitutive modelling and its application to forming analysis is provided by Lim and Ramakrishna[23].

In order to simulate forming behaviour, constitutive models are usually implemented within FE software. Since commercial FE software such as ABAQUS[24], MARC[25], PAM-FORM[26] and LS-DYNA[27] are able to model non-linear behaviour, including material and geometric (contact) non-linearity, it is beneficial to use such software. Here, two solution schemes are available: the implicit and explicit FE methods. Although analyses for some simple forming problems can utilise the implicit method, the increase in complexity of forming problems brings with it a number of difficulties, particularly variable contact conditions. For this reason the explicit approach is preferable for forming simulation of textile composites[28, 29]. Implicit and explicit FE methods are described in detail in Chapter 8 in the context of composite mechanical behaviour.

4.3.2   Non-orthogonal constitutive model

It is the shearing mechanism that enables bidirectional fabrics to conform onto the tool during sheet forming, because fibres in tows are very stiff and can be assumed to be inextensible (apart from a small initial extension in woven fabrics due to fibre straightening or 'uncrimping'). Because of shearing, material axes represented by the warp and weft directions become non-orthogonal. Before discussing a method to treat the non-orthogonal nature of the material axes, the limitations of an orthotropic assumption of material axes are discussed to emphasise the necessity for a non-orthogonal constitutive equation.

As an example, a square sheet of woven composite was subjected to a shear boundary force on its right-hand edge while the left-hand edge was constrained not to move in the x-direction but was free to move in the y-direction. To prevent rigid body motion of the sheet, the lower left corner was completely fixed as shown in Fig. 4.5(a). A total of 10×10 rectangular plane stress elements were used with an orthotropic, elastic constitutive equation. A small shear modulus was used with large Young's moduli in both the x and y material directions. The predicted shape in Fig. 4.5(b) shows a significant change in length of the left edge. This length change comes from the assumption that the material axes remain orthogonal during deformation

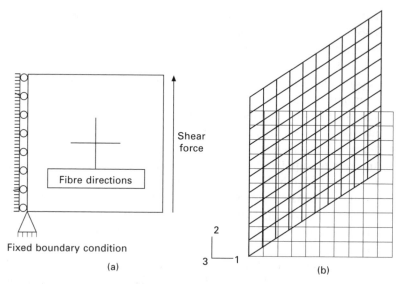

Fibre directions

Shear force

Fixed boundary condition

(a)

2

3 ——1

(b)

4.5 Simple shear simulation[30]. (a) Geometry and boundary condtions. (b) Deformed shape using orthotropic constitutive equation (E_{11} = 19 GPa, E_{22} = 19 GPa, v = 0.1, G = 0.01 MPa).

(they can rotate, but remain orthogonal). This issue is discussed in more detail by Yu *et al.*[30]

Given the obvious problems associated with the orthotropic material law, several approaches have been proposed to predict deformation behaviour of bidirectional textiles. Structural elements are often used to represent the tows within the fabric structure, for example one-dimensional beam elements can be used to represent tow sections between crossovers. This approach can simulate the non-orthogonality of two fibre directions; however, interaction properties (e.g. at tow crossovers or resistance to deformation offered by the matrix for prepregs) are usually determined empirically through repeated numerical simulations, with material properties adjusted to provide correlation with experiments[12, 13].

An alternative approach relies on continuum mechanics, where the fabric structure is modelled using one element type[30, 31]. One advantage of this approach is that the constitutive equation developed is not dependent on the element type; also out-of-plane deformation can be simulated. Here, a non-orthogonal constitutive equation is introduced and its applications to forming simulation are discussed. This was developed by Yu *et al.*[30], using a type of homogenisation method to extract macro-material properties by deriving an analytical form for the stress–strain relationship. The constitutive equation consists of two parts: contributions from fibre directional properties and shear properties.

Fibre directional properties

To model the stress and strain relationship dependent on the fibre directional properties, it is important for a constitutive equation to include fibre directions, tow spacing and tow parameters (e.g. fibre diameter). As shown in Fig. 4.6, a structural net comprising several unit cells, one of which is a pair of warp and weft tows with pinned joints, was considered. In modelling of the contribution of fibre directional properties to the constitutive equation, the warp and weft tows are assumed to rotate freely at crossover points. Here, the final form of the equation is presented to explain the difference between the current constitutive form and other forms such as isotropic and orthotropic elastic equations:

$$
\begin{bmatrix} \Delta\sigma_{xx} \\ \Delta\sigma_{yy} \\ \Delta\sigma_{xy} \end{bmatrix}
$$

$$
= \begin{bmatrix}
\dfrac{\tilde{E}^{\alpha}}{bc} + \Gamma\left(\dfrac{a}{\bar{h}}\right)\left(\dfrac{a^2}{c}\right) & \Gamma\left(\dfrac{a}{\bar{h}}\right)\left(\dfrac{b^2}{c}\right) & \Gamma\left(\dfrac{a}{\bar{h}}\right)\left(\dfrac{ab}{c}\right) \\[2ex]
\Gamma\left(\dfrac{b}{\bar{a}}\right)\left(\dfrac{a^2}{c}\right) & \Gamma\left(\dfrac{b}{\bar{a}}\right)\left(\dfrac{b^2}{c}\right) + \dfrac{\tilde{E}^{\beta}}{\bar{a}c} & \Gamma\left(\dfrac{b}{\bar{a}}\right)\left(\dfrac{ab}{c}\right) \\[2ex]
\Gamma\left(\dfrac{b}{\bar{h}}\right)\left(\dfrac{a^2}{c}\right) & \Gamma\left(\dfrac{b}{\bar{h}}\right)\left(\dfrac{b^2}{c}\right) & \Gamma\left(\dfrac{b}{\bar{h}}\right)\left(\dfrac{ab}{c}\right)
\end{bmatrix}
\begin{bmatrix} \Delta\varepsilon_{xx} \\ \Delta\varepsilon_{yy} \\ 2\Delta\varepsilon_{xy} \end{bmatrix}
$$

4.2

where $a (= |\vec{b}| \cos \Phi)$ and $b (= |\vec{b}| \sin \Phi)$ are the projections of the weft tow on the local x- (warp direction) and y-axes respectively, while $\bar{a}\ (= |\vec{a}|)$ and $|\vec{h}|\ (= (b/a)\,\bar{a})$ are the lengths of the warp within a unit cell and the extended height of the weft tow, respectively, as shown in Fig. 4.6 (c = thickness). The

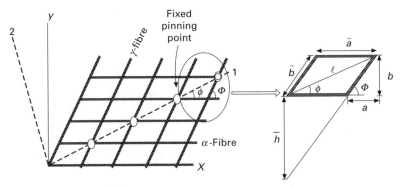

4.6 A structural net (left) and unit cell (right) of a fabric reinforced composite.

two vector quantities, $\vec{a}$ and $\vec{b}$, express the magnitudes and directions of the warp and the weft in a unit cell and define the main internal variables for the constitutive equation. The $\bar{E}^\alpha$ and Γ terms in eqn 4.2 are introduced to represent the stiffness of the warp and the weft tows, and are defined as follows:

$$\bar{E}^\alpha = E^\alpha A^\alpha, \; \Gamma = A^\gamma E^\gamma/(a^2 + b^2)^{3/2} \qquad\qquad 4.3$$

where A^α is the cross-sectional area of the warp tow, which is calculated from the number of filaments in the tow and the diameters of the filaments, and E^α is the elastic modulus of the fibres, while A^α and E^γ are the corresponding properties for the weft tow.

Shear properties

Since eqn 4.2 considers only fibre directional properties, a model to incorporate shear stiffness into the constitutive equation is also required. Here, two methods are discussed, one of which utilises a fictitious spring, while another incorporates measured shear properties directly. The former is simple and effective to represent shear stiffness, but it has a drawback in determining the spring constants, which may require repeated numerical simulation, adjusting the spring constants to match the simulation results to the experiment. This idea for treating the inter-tow friction and subsequent locking represents the effect of tow compaction, as the originally orthogonal tows are sheared and hence move closer together (see Fig. 4.7). In eqn 4.2, an additional stiffness term, $\bar{E}^\beta$, can be added to represent the shear stiffness. A full description of this approach is given by Yu et al.[30]; a similar approach has been developed recently by Sharma and Sutcliffe[32].

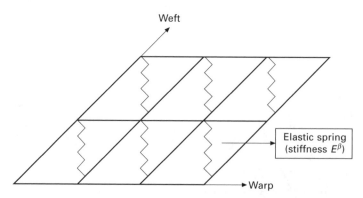

4.7 Scheme for modelling of shear stiffness by a fictitious spring.

An alternative method has been developed, which is rigorous enough to incorporate the material shear force curve directly into the constitutive equation.

The material shear behaviour can be determined either experimentally or analytically as described in Chapters 2 and 3. First, the stresses acting on the non-orthogonal frame are identified during picture frame shear using a covariant basis and the definition of the traction vector. A short description for the shear stiffness model is given in the following section and readers can refer to Yu *et al.*[33, 34] for a detailed derivation.

Consider a square picture frame hinged at each corner as illustrated in Fig. 4.8. The resultant force, $F_R(\gamma)$, is recorded, and this can be converted to the shear force acting on the sides of the frame as follows:

$$F_s = \frac{F_R(\gamma)}{2\cos(\pi/4 - \gamma/2)} \qquad 4.4$$

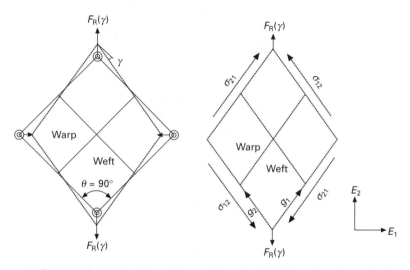

4.8 Schematic of the picture frame shear test (left) and contra-variant stress (σ^{12}, σ^{21}) components acting along normalised covariant base vectors (g_1, g_2).

On the other hand, during picture frame shear, the actual stresses acting on the specimen can be expressed based on the normalised covariant base vectors (g_1 and g_2). Note that bold characters represent vector or tensor quantities. The covariant bases (representing two fibre directions), g_1 and g_2, are unit vectors that may enable two diagonals of the contra-variant metric tensor to be obtained as follows:

$$g_1 = g_1^1 E_1 + g_1^2 E_2, \ \sqrt{(g_1^1)^2 + (g_1^2)^2} = 1 \qquad 4.5$$

$$g_2 = g_2^1 E_1 + g_2^2 E_2, \ \sqrt{(g_2^1)^2 + (g_2^2)^2} = 1 \qquad 4.6$$

$$g^{11} = g^1 \cdot g^1 = \frac{1}{\sin^2\theta} = g^{22} = g^2 \cdot g^2 \qquad 4.7$$

where E_1 and E_2 are inertial bases as shown in Fig. 4.8. Then, the stresses acting on the specimen may be expressed in contra-variant component form from which the tangent stiffness $(d\sigma/d\gamma)$ due to shear angle increment can be obtained. The stress increment $(d\sigma)$ due to shear angle increment $(d\gamma)$ is then given by:

$$\begin{bmatrix} \Delta\sigma_{xx} \\ \Delta\sigma_{yy} \\ \Delta\sigma_{xy} \end{bmatrix} = \begin{bmatrix} 0 & 0 & 2G_1 g_1^1 g_2^1 + G_2(g_1^1 g_2^2 - g_1^2 g_2^1) \\ 0 & 0 & 2G_1 g_1^2 g_2^2 + G_2(g_1^1 g_2^2 - g_1^2 g_2^1) \\ 0 & 0 & G_1(g_1^1 g_2^2 + g_2^1 g_2^2) \end{bmatrix} \begin{bmatrix} 0 \\ 0 \\ \Delta\gamma \end{bmatrix} \qquad 4.8$$

where

$$G_1 = \frac{1}{lh}\left\{ \frac{dF_s}{d\gamma}\sqrt{g^{11}} + F_s\sqrt{g^{11}(g^{11}-1)} \right\}, \quad G_2 = \left(\frac{F_s}{lh}\right)\sqrt{g^{11}} \qquad 4.9$$

in which lh is cross-sectional area of the specimen.

Note that eqn 4.8 is expressed based on a coordinate system that is different from the coordinate system used in the derivation of eqn 4.2. The coordinate system used in eqn 4.2 was constructed based on the warp direction as local x direction, while eqn 4.8 was derived using a coordinate system that bisects the two fibre directions. Therefore, eqn 4.8 needs to be transformed into the coordinate system of eqn 4.1, with the results added to provide the overall constitutive equation.

4.3.3   Application of non-orthogonal constitutive model to forming simulation

To explore the validity of the non-orthogonal equation and its application to composites manufacturing processes, the deformation behaviour of both dry fabric and thermoplastic prepreg are analysed and discussed below.

Stamping simulation for woven dry fabric

This simulation was performed to investigate the forming behaviour of a plain woven fabric using the non-orthogonal constitutive equation. Predicted results were compared with experiments with respect to the boundary profile and fibre angle distribution of the draped preform. An FE model for the draping simulation was constructed as shown in Fig. 4.9 to represent the forming apparatus described by Mohammed et al.[35] The radius of the upper mould in this model was 100 mm while that of the lower mould was 97 mm. As the upper mould moved downwards, the fabric was draped over the lower mould, which consisted of a hemispherical dome surrounded by a flat base. The distance from the centre of the lower mould to the edge of the flat base

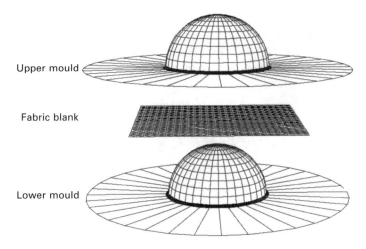

4.9 Finite element model for hemispherical stamping simulation of woven fabric.

was 260 mm. A 320 × 320 mm square of plain woven fabric was used in the experimental study[35]. For the simulation, 1250 triangular shell elements were used to model a quarter section of the woven fabric due to symmetry. The material properties for the fabric blank were calculated based on the structure and material properties of the woven fabric used in the experiment. The friction coefficient between the moulds and the fabric blank was obtained from published data[14]. The material and process variables used in the analysis are listed in Table 4.1. Note that this simulation was performed utilising a spring to represent shear resistance and its stiffness value was tuned to produce good agreement with experimental results.

Table 4.1 Material and structural properties of woven fabric used in draping simulation

	E-glass plain weave
Sheet material	
α fibre stiffness $(\tilde{E}^{\alpha})$	90 kN
β fibre stiffness $(\tilde{E}^{\gamma})$	90 kN
Hardening parameter $(\tilde{E}^{\beta})$	10 N
Structural parameters	
Thickness	0.58 mm
Yarn space	6.25 mm
Friction coefficient	
Upper mould–sheet	0.5
Lower mould–sheet	0.2

Figure 4.10 shows the edge profile of the deformed woven preform. The predicted and measured profiles agree very closely, although there is a little disagreement in the weft direction. This may be due to the assumption that there are no material and structural differences between warp and weft directions in the current simulation.

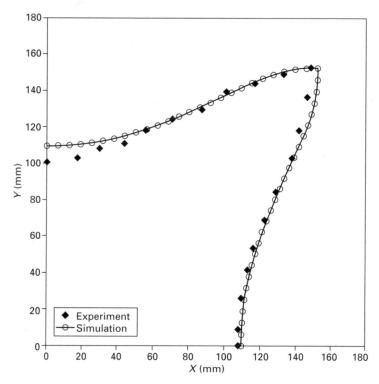

4.10 Comparison between boundary profiles of draped preform from experiment[35] and analysis[30].

Predicted local fibre orientations are represented in Fig. 4.11. Fibre orientation contours were obtained experimentally by measuring the orientation of the deformed gridlines inscribed on the fabric prior to draping[35]. When considering the fibre angle contour for the simulated draped fabric, a region representing 90° inter-fibre angles is observed around the apex of the lower mould, which corresponds to the zero-shear zone. At the equator, where the hemispherical dome changes to the flat surface, the fabric deformation is maximum and an inter-fibre angle of 35.2° was predicted, compared with an experimentally measured fibre angle of 34°. From these comparisons, it can be concluded that the non-orthogonal constitutive equation can simulate the forming behaviour for this material accurately.

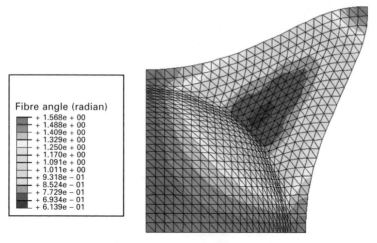

Fibre angle (radian)
+ 1.568e + 00
+ 1.488e + 00
+ 1.409e + 00
+ 1.329e + 00
+ 1.250e + 00
+ 1.170e + 00
+ 1.091e + 00
+ 1.011e + 00
+ 9.318e − 01
+ 8.524e − 01
+ 7.729e − 01
+ 6.934e − 01
+ 6.139e − 01

4.11 Predicted fibre angle distribution (radians) for woven fabric hemisphere[30].

Stamping simulation for woven thermoplastic prepreg

The non-orthogonal equation (4.2) can also be applied to thermoplastic prepreg or thermoset prepreg. Here it is assumed that the resin matrix and its interaction with the fibres can be characterised using a picture frame shear test. The deformation behavior of the prepreg is then simulated using eqns 4.2 and 4.8 with the experimental shear data for the prepreg. Rate and temperature effects during forming are ignored in this approach, although Harrison *et al.*[36] have recently implemented a system to account for such effects using a 'look-up table' of material properties, where shear stiffness data are specified as a function of rate and temperature. The alternative is to use a viscous constitutive law directly (see Chapter 3 and de Luca *et al.*[28]), although this is relatively complicated and usually requires model parameters to be determined empirically by comparison between predicted results and specimen tests.

For experimental forming two types of specimens were prepared using a commingled woven fabric: a dry woven fabric with tows comprising commingled glass and polypropylene filaments (as received from the company) and a pre-consolidated plate of the same material (0.5 mm thickness). The consolidated plate was prepared using a 'hot/cold' process in a double compression moulding press with two sets of platens: one hot (205 °C) and the other cold (66 °C). The fabric samples were placed between two polished steel plates, which were then placed into the hot press for approximately 5 mins under 0.344 MPa (50 psi) pressure. Once the matrix had melted, the plates were shuttled to the cold press. Here a larger pressure 1.032 MPa (150 psi) was applied for another 5 mins. For shear property measurement, four layers of commingled woven fabric were consolidated into a laminate, which was subsequently tested using picture frame shear apparatus[37]. Tests were conducted at the appropriate forming temperature (up to 190 °C) at a constant velocity of 2 mm/s. Typical experimental data for this material are shown in Fig. 4.12.

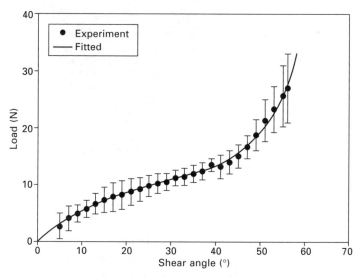

4.12 Averaged experimental load *vs.* shear angle curve during picture-frame shear. Error bar represents the standard deviation of the data from five specimens[37].

The thermoforming press set-up is illustrated in Fig. 4.13. Instead of using a matched female die, the lower chamber was left open to the atmosphere, without any pressure or mould surface used to force the material to take the shape of the punch. To allow the woven thermoplastic blank to draw-in, a 4 mm clearance was maintained between the upper and lower blank-holders. Utilising this set-up specimens were formed and the results were compared against the simulated shapes. The material was drawn to a depth of 0.045 72 m (1.8 inch) using a punch rate of 2 mm/s.

Specimens were formed using both the dry fabric and pre-consolidated materials. The formed shapes showed little difference, despite the fact that the pre-consolidated sheet contained an even distribution of matrix prior to forming whereas the fabric consisted of separate reinforcement and matrix filaments. Fibre angle variations measured along an arc at 45° to the warp direction are shown in Fig. 4.14, in which inter-fibre angle is plotted against arc length from the centre of the specimen.

A circular mesh was constructed using shell elements to simulate stamp forming with the material properties given in Table 4.2. For this simulation using the non-orthogonal constitutive equation, an explicit solver was used

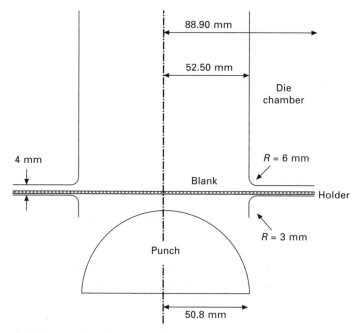

4.13 Thermoforming set-up for woven thermoplastic composites.

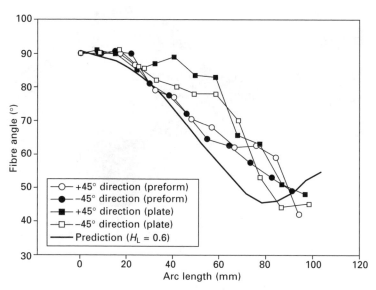

4.14 Inter-fibre angle variations along arc length in ±45° direction for dry commingled fabric (preform) and pre-consolidated (plate) samples[33].

Table 4.2 Structural and material properties of commingled glass–polypropylene plain woven material

Description	Property
Structural parameter	
Elastic modulus of fibres	73 GPa
Yarn spacing	5.25 mm
Plate thickness (1 layer)	0.5 mm
Yarn parameter	
Filaments in warp yarn	1600
Filaments in weft yarn	1600
Filament diameter	18.5 μm

to avoid any divergence problems inherent in the implicit solver due to material non-linearity and contact between the punch and the material. Figure 4.15 shows the stamped shape and fibre angle distribution, which is compared with experimental data in Fig. 4.14. Good agreement was obtained between experimental and simulation results using the non-orthogonal constitutive equation, with the only significant discrepancies around the equator of the hemispherical dome.

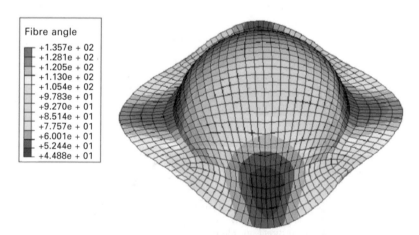

Fibre angle

+1.357e + 02
+1.281e + 02
+1.205e + 02
+1.130e + 02
+1.054e + 02
+9.783e + 01
+9.270e + 01
+8.514e + 01
+7.757e + 01
+6.001e + 01
+5.244e + 01
+4.488e + 01

4.15 Formed shape and fibre angle distribution (degrees) predicted using the non-orthogonal constitutive equation and shear material properties[33].

Forming simulation of NCF preform

Studies performed previously at the University of Nottingham[8, 38] have shown that fabric construction can have a significant effect on the shear properties of reinforcements and therefore on their deformed patterns, even for simple shapes such as hemispheres. This is particularly evident for NCFs, in which reinforcement tows are held together using a warp-knit stitching thread. The stitches in an NCF offer resistance to shearing, so that shearing behaviour in the direction parallel to the stitch can be quite different from that perpendicular to the stitch. This issue was discussed in section 4.2.2, where an iterative mapping approach was used to simulate the forming of NCFs. However, as already noted, this approach does not consider all of the boundary conditions relevant to the forming process. This section models the asymmetric shear behaviour of NCFs using the non-orthogonal constitutive equation, including the shear stiffness model in eqn 4.8. It will focus on whether a continuum mechanics model can simulate the skew symmetric shape of NCF preforms when formed using a hemispherical set-up.

For this example, a ± 45° NCF with a tricot stitch was chosen, and its structural parameters are listed in Table 4.3. The shear behaviour of the NCF was measured by picture frame shear testing[8], and results plotted in Fig. 4.2 show a large difference in shear response according to the shear direction. Non-linear shear force data with respect to shear angle were incorporated into the constitutive equation through eqn 4.8, for which the test results were approximated using non-linear regression. It is not unusual to have some fluctuations in the measured data from picture frame shear testing, particularly at low shear angles. Therefore, the measured data were divided at an arbitrary small shear angle (5°), below which an exponential function was used and above which a polynomial function was used as shown in eqn 4.10. The coefficients for these equations are listed in Table 4.4.

$$\text{if } \gamma < 5°, F_s = k_1\gamma + k_2(1 - e^{-k_3\gamma})$$

$$\text{if } \gamma \geq 5°, F_s = a_0 + a_1(\gamma - 5) + a_2(\gamma - 5)^2 + a_3(\gamma - 5)^3$$

$$+ a_4(\gamma - 5)^4 + a_5(\gamma - 5)^5 + a_6(\gamma - 5)^6 \qquad 4.10$$

Table 4.3 Structural and material properties of NCF (COTECH EBX 936)

Description	Property
Structural parameter	
Elastic modulus of fibres	65 GPa
Yarn spacing	0.82 mm
Thickness (assumed)	0.5 mm
Yarn parameter	
Filaments in warp yarn	600
Filaments in weft yarn	600
Filament diameter	16 μm

Next, the shear force data was utilised in the constitutive model to simulate the deformation behaviour of the NCF reinforcement. As a first simulation, the picture frame shear test, which was used to determine the non-linear

Table 4.4 Coefficients for the non-linear curve fitting for NCF shear stiffness parallel and perpendicular to the stitching direction

Coefficients	Parallel	Perpendicular
k_1	1.3639	0.1063
k_2	1.1470	0.9604
k_3	20.9684	6.6138
a_0	7.9670	1.4621
a_1	1.3639	1.0634×10^{-1}
a_2	0.4098	2.8111×10^{-2}
a_3	-4.2656×10^{-2}	-4.1599×10^{-3}
a_4	1.7480×10^{-3}	2.0179×10^{-4}
a_5	-3.2532×10^{-5}	-4.0126×10^{-6}
a_6	2.2958×10^{-7}	2.8625×10^{-8}

shear force curve, was simulated using the same dimensions as the experimental specimen[8]. Membrane finite elements were used to model the picture frame specimen, thus ignoring transverse shear and bending deformations. In this simulation, an implicit solver was utilised. Since the NCF does not show any thickness change during shearing until a large shear angle is reached[8], a constant thickness condition was imposed in this virtual test of picture-frame shear. In Fig. 4.16 predicted load profiles are compared with experimental data, demonstrating good agreement and thus confirming that the shear stiffness model in eqn 4.8 is suitable for simulating NCF shear deformation.

To show the validity of the combined non-orthogonal constitutive equation and shear stiffness model in forming simulation for NCFs, a hemispherical forming process was modelled. The tool for the process consisted of a die, a blank-holder and a punch. The dimension of the punch (radius = 60 mm) was determined based on the forming station at the University of Nottingham[8], while other dimensions were selected to maintain a 2 mm clearance between the punch and the die. Then, an FE model was constructed using triangular elements for the NCF blank and rigid elements for the forming tools, similar to that shown in Fig. 4.9.

An initial study was undertaken to choose the most appropriate element type to represent the fabric. This involved comparison of results obtained using membrane and shell elements. Since the shell element in FE analysis considers a through thickness stress gradient to describe bending behaviour of the material, it involves bending and transverse shear strains in addition to in-plane (membrane) strains. In contrast membrane elements consider only in-plane deformation with no stress gradient through thickness. Hence the use of shell elements requires increased computational time, with analyses here taking three times as long for shell elements than for membrane elements. As the predicted fibre patterns obtained using the two element types were almost identical, membrane elements were used in the present study. However,

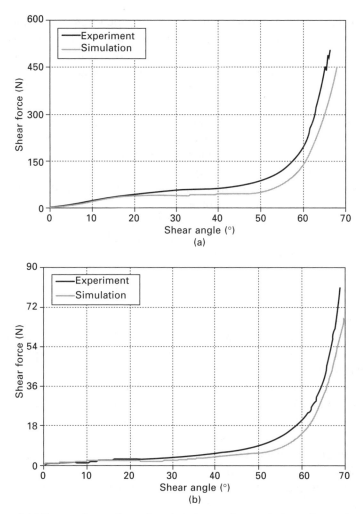

4.16 Comparison of predicted and experimental shear force-shear angle curves. (a) Parallel shear direction and (b) perpendicular shear direction to the stitch in NCF.

shell elements should be considered if the bending behaviour of NCFs is characterised and recognised as an important factor to determine forming behaviour. Furthermore, the shell element is indispensable if spring-forward after forming is considered.

The effect of the blank-holding force (BHF) on the forming behaviour of the NCF was investigated by a parametric study. Here the friction coefficient between the forming tools and the NCF was assumed to be 0.1. BHF values of 0, 1000 and 5000 N (0, 17 421 and 87 105 Pa in pressure) were considered. An increase in BHF is expected to develop tension in the fabric, so that the formed shape becomes more symmetric as experienced during experiments at Nottingham[8, 38]. This may be possible because high tension applied through the blank-holder suppresses the difference between the shear behaviour according to the positive (parallel to stitch) and the negative (perpendicular to stitch) shear directions. Figure 4.17(a) shows an asymmetric fibre pattern and the possibility of wrinkling when forming is simulated with zero BHF (with only a small blank-holder mass, 650 g, included in the simulation). Figure 4.17(b) shows that the formed shape becomes more symmetric as the BHF increases. This information is useful to design the forming process for a symmetric product, such as a safety helmet, from NCF. In Fig. 4.18, predicted shear angles are compared with experimental measurements[8] along the line of maximum shear on the hemisphere surface, i.e. the bias direction. For the experiments a minimal BHF was used to stop the material from folding during forming. Here the difference in shear angle according to the shear direction was significant, with shear angles in quadrants sheared perpendicular to the stitching thread up to $10°$ higher than those sheared parallel to the stitch. Predicted shear angles for zero BHF matched experimental values very closely. An increase in BHF is predicted to reduce the shear angle difference between quadrants sheared parallel and perpendicular to the stitch. This is particularly evident at the equator of the hemisphere, where shear deformation is most severe. This matches the trend observed experimentally for other components[8].

A number of simulations were performed to assess the effect of initial blank shape on the formed pattern. In the metal forming field, many research studies were dedicated to the initial blank shape design, with the aim of optimising deformation for a specific product (e.g. Chung and Richmond[39, 40]). The basic idea behind ideal forming starts with the definition of a path that produces a desired homogeneous deformation with minimum plastic work. It is also assumed that formability of local material elements is optimum when they deform according to a minimum work path. An ideal process is then defined as one having local deformations optimally distributed in the final shape, from which the ideal configuration for the initial stage can be determined. Little work has been reported on ideal forming for composite materials. Tucker and coworkers[41, 42] analysed random fibre mats by utilising a forming limit diagram. They determined a potential describing formability that included stretch of the mat and the forming limit, and minimised the potential to seek an ideal initial blank shape where all elements resided within the forming limit. Experimental observations related to the importance of blank shape design were reported by O'Bradaigh[9], who formed several specimens modified from a square blank by progressively removing material from the blank and found that blank shape affected wrinkling behaviour.

To investigate the effect of blank shape on the formed fibre pattern, a forming simulation was performed for a circular blank with a radius of 150

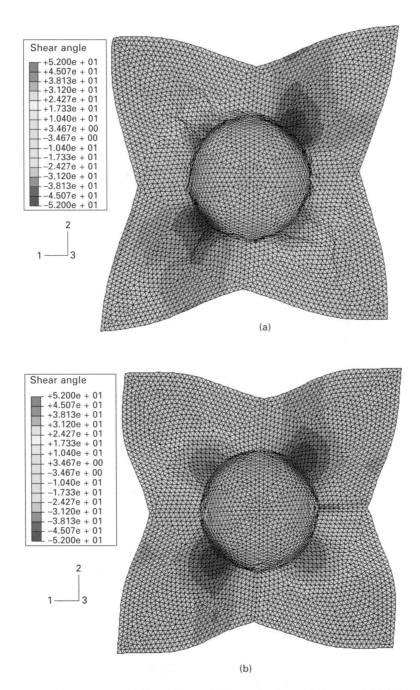

4.17 Effect of blank-holder force (BHF) on the formed shapes: (a) BHF = 0 N; (b) BHF = 5000 N. Shear angles are in degrees.

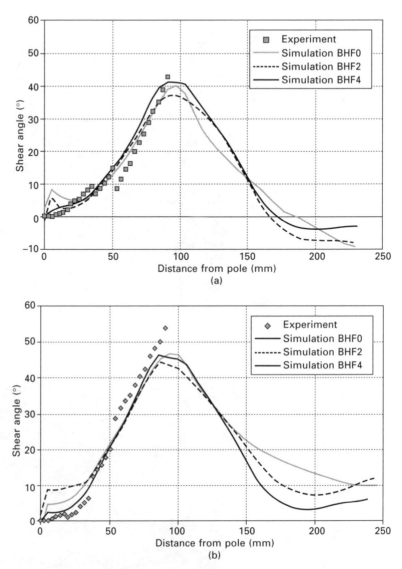

4.18 Comparison between predicted and experimentally measured shear angles for hemisphere formed using an NCF. In the legend BHF0 = 0 N, BHF2 = 1000 N and BHF4 = 5000 N. (a) Quadrant where shear is parallel to the stitching direction. (b) Quadrant where shear is perpendicular to the stitching direction.

mm. The difference between the square and the circular blanks exists in the excess material that lies outside of the blank-holder. The entire edge of the circular blank is initially inside the blank-holder, while the square blank has excess in the bias direction. As the fibres have a high modulus, the edges in the fibre directions of both the square and circular blanks will be drawn in during forming. Then, they will be kept under the blank-holder throughout the forming process. However, the edges in the bias directions in the square blank, which are extensible via in-plane shearing, will add resistance to forming. This resistance may make shearing in the bias direction difficult, resulting in less shear deformation for the square blank than for the circular blank. As a result the square blank shows a more symmetric formed shape than the circular blank, as shown in Fig. 4.19. This difference is highlighted in Fig. 4.20, which compares the predicted shear angle distributions for both blank shapes. This feature may be advantageous for producing symmetric formed shapes from NCF. More generally this study suggests that blank shape is an important parameter in the design of an optimum forming process for NCF.

4.4   Concluding remarks and future direction

This chapter has presented modelling techniques for forming of bidirectional fabrics and their composites. Kinematic drape models are now relatively mature, and are used widely particularly for hand- lay-up of prepreg. Here the material is modelled as a pin-jointed net, with a geometric approach based on in-plane shear used to map the fabric onto the component surface. This offers a rapid solution, and integration within established CAD/CAM (computer-aided design/manufacture) environments. However the approach is limited in that it cannot represent effects of manufacturing process conditions (e.g. rate, temperature, blank-holder force) or material variables (shear stiffness, tool/ply friction). Some of these limitations are alleviated by modifying the mapping algorithm, using an iterative method where total deformation energy is minimised to select the appropriate mapping. The first step here is to incorporate shear compliance data within the analysis, so that non-symmetric geometries and materials with preferential deformation directions can be modelled. Such an approach was demonstrated in this chapter, and has been shown to provide a more accurate solution than kinematic models for non-crimp fabrics. However there are still limitations here. While current research activities at Nottingham aim to introduce additional energy dissipation terms (e.g. ply-tool friction), a mapping approach is unlikely to represent the physical process accurately, particularly for forming of thermoset or thermoplastic prepregs, where rate and temperature effects may be important. Mapping approaches also only offer a crude approach to predicting wrinkling, via specification of a 'locking angle'. In practice material wrinkling is a mechanical problem, and occurs in regions subjected to compressive forces beyond the local buckling limit.

The alternative to a mapping approach is to use a mechanical model, where deformation is simulated over a number of time steps. Given an

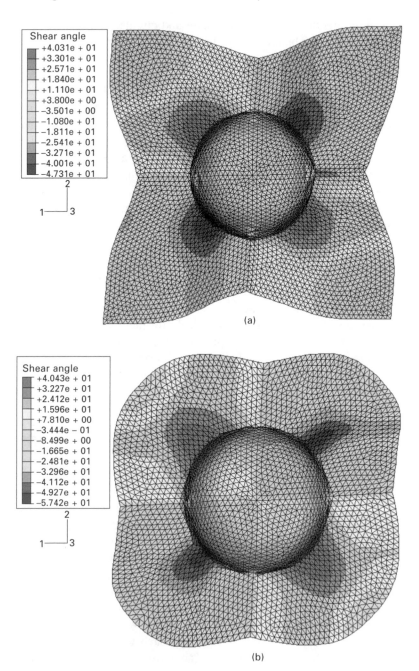

4.19 Effect of blank shape on the formed shape for an NCF hemisphere. (a) Square blank shape (side length = 300 mm). (b) Circular blank shape (diameter = 300 mm). Shear angles are in degrees.

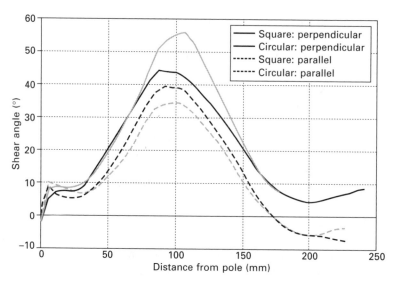

4.20 Predicted shear angle distributions for NCF hemispheres with different blank shapes.

appropriate constitutive model, this approach is able to represent accurate material behaviour and can include realistic boundary conditions. This chapter has focused on one particular model, a non-orthogonal constitutive equation, and its validity and application to forming simulation. Through several simulation examples, forming analyses were performed focusing on three factors: the constitutive equation for the reinforcement, the interface behaviour between the tool and reinforcement, and mechanical processing parameters such as blank-holding force. The model presented here was validated using these examples, and one can conclude that it is reasonably accurate. However, there are numerous published studies on constitutive modelling for forming simulation of continuous fibre reinforced composites. To provide process engineers in industry with an appropriate simulation tool, and to facilitate its utilisation for process design, it is important to recognise both the validity and limitations of a particular constitutive equation. Benchmark simulations for a specified problem would be a good way to compare the many constitutive models available, and an international benchmarking exercise has been initiated recently to address this issue[43].

Since composite sheet forming frequently involves multilayer laminates formed in a single operation, work in this field should be expanded to multilayer analyses by modelling the interface between layers. Two approaches are possible here – either the layers can be modelled separately[28] or a layered element can be used[44] (similar to layered shells used extensively for analysis of laminates). In either case, friction or viscous modelling for the interface is required. Here relatively little data are available, particularly for dry fabrics.

Such data are also required to model automated stamping processes where a blank-holder is used to support the material. Understanding of this type of process is essential if textile-based composites are to progress into high-volume applications.

In this chapter a simulation example was provided to emphasise the importance of blank shape on the deformed fibre pattern for a specific product; this highlights optimisation of blank shape and ideal forming analysis as an important research topic. More generally there is a need for an integrated approach to optimisation in the manufacture of textile composite components. Sheet forming affects several other stages in the manufacturing/performance cycle, for example resin flow in RTM[45, 46] and component mechanical properties[47, 48]. At present each of these stages is analysed in isolation, so that the overall solution may not be optimum. Materials optimisation is also important here, where ideally one should be able to model manufacturing and performance for any combination of polymer and textile without carrying out a long series of characterisation tests. This requires predictive modelling of constitutive behaviour using the approaches discussed in Chapters 2 and 3, and such analyses are now being utilised directly within forming analyses for dry fabric[49] and prepreg[36]. Ultimately such work should result in a general and robust predictive modelling approach for the design and manufacture of textile composite components.

4.5   Acknowledgements

Dr Yu would like to express his gratitude to Dr Farhang Pouboghrat at Michigan State University (USA) and Dr Kwansoo Chung at Seoul National University (Korea), who enthusiastically supported research work on development of the non-orthogonal constitutive equation.

4.6   References

1. Smiley A.J. and Pipes R.B., 'Analysis of the diaphragm forming of continuous fiber reinforced thermoplastics', *J. Thermoplastic Composite Materials*, 1988 **1** 298–321.
2. Van West B.P., Pipes R.B., Keefe M. and Advani S.G., 'The draping and consolidation of comingled fabrics', *Composites Manufacturing*, 1991 **2** 10–22.
3. Bergsma O.K., 'Computer simulation of 3D forming processes of fabric reinforced plastics', *Proc. 9th Int. Conf. on Composite Materials (ICCM-9)*, Madrid, 1993, pp. 560–567.
4. Long A.C. and Rudd C.D., 'A simulation of reinforcement deformation during the production of preforms for liquid moulding processes', *Proc. IMechE Part B*, 1994 **208** 269–278.
5. MSC. Patran Laminate Modeler, www.mscsoftware.com
6. FiberSIM 4.0, www.vistagy.com
7. Bunday B.D., *Basic Optimisation Methods*, London, Edward Arnold Ltd, 1984.

8. Souter B.J., Effects of fibre architecture on formability of textile preforms, PhD Thesis, University of Nottingham, 2001.

9. O'Bradaigh C.M., 'Sheet forming of composite materials', in Advani S.G. (ed) *Flow and Rheology in Polymer Composites Manufacturing*, Amsterdam, Elsevier, 1994, 517–569.

10. Krebs J., Friedrich K. and Bhattacharyya D., 'A direct comparison of matched-die versus diaphragm forming', *Composites Part A*, 1998 **29** 183–188.

11. Breen D.E., House D.H. and Wozny M.J., 'A particle-based model for simulating the draping behavior of woven cloth', *Textile Research J.*, 1994 **64**(11) 663–685.

12. Cherouat A. and Billoet J.L., 'Mechanical and numerical modeling of composite manufacturing processes deep-drawing and laying-up of thin pre-impregnated woven fabrics', *J. Materials Processing Technolo.*, 2001 **118**(1–3) 460–471.

13. Sidhu R.M.J.S., Averill R.C., Riaz M. and Pourboghrat F., 'Finite element analysis of textile composite preform stamping', *Composites Structures*, 2001 **52** 483–497.

14. Dong L., Lekakou C. and Bader M.G., 'Solid-mechanics finite element simulations of the draping of fabrics: a sensitivity analysis', *Composites Part A*, 2000 **31** 639–652.

15. O'Bradaigh C.M. and Pipes R.B. 'A finite element formulation for highly anisotropic incompressible elastic solids', *Int. J. Numer. Meth. Eng.*, 1992 **33** 1573–1596.

16. Spencer A.J.M., *Deformations of Fibre Reinforced Materials*, Oxford, Clarendon Press, 1972.

17. Rogers T.G., 'Rheological characterization of anisotropic materials', *Composites*, 1989 **20**(1) 21–27.

18. O'Bradaigh C.M., McGuinness G.B. and Pipes R.B., 'Numerical analysis of stresses and deformations in composite materials sheet forming:central indentation of a circular sheet', *Composites Manufacturing*, 1993 **4**(2) 67–83.

19. Spencer A.J.M., 'Theory of fabric-reinforced viscous fluids', *Composites Part A*, 2000 **31** 1311–1321.

20. Hsiao S.W. and Kikuchi N., 'Numerical analysis and optimal design of composite thermoforming process', *Comput. Methods Appl. Mech. Engrg.*, 1999 **177** 1–34.

21. Takano N., Uetsuji Y., Kahiwagi Y. and Zako M., 'Hierarchical modeling of textile composite materials and structures by the homogenization method', *Modeling Simul. Mater. Sci. Eng.*, 1999 **7** 207–231.

22. Peng X.Q. and Cao J., 'A dual homogenization and finite element approach for material characterization of textile composites', *Composites Part B*, 2002 **33** 45–56.

23. Lim T.C. and Ramakrishna S., 'Modeling of composite sheet forming: a review', *Composites Part A*, 2002 **33** 515–537.

24. ABAQUS FEA software, www.abaqus.com

25. MSC Marc, www.mscsoftware.com

26. ESI PAM-FORM, www.esi-group.com

27. Livermore Software Technology Corporation LS-DYNA, www.dyna3d.com

28. de Luca P., Lefebure P. and Pickett A.K., 'Numerical and experimental investigation of some press forming parameters of two fibre reinforced thermoplastics: APC2-AS4 and PEI-CETEX', *Composites Part A*, 1998 **29** 101–110.

29. Pickett A.K., Queckborner T., de Luca P. and Haug E., 'An explicit finite element solution for the forming prediction of continous fibre-reinforced thermoplastic sheets', *Composites Manufacturing*, 1995 **6**(3–4) 237–243.

30. Yu W.R., Pourboghrat F., Chung K., Zampaloni M. and Kang T.J., 'Non-orthogonal constitutive equation for woven fabric reinforced thermoplastic composites', *Composites Part A*, 2002 **33** 1095–1105.

31. Xue P., Peng X. and Cao J., 'A non-orthogonal constitutive model for characterizing woven composites', *Composites Part A*, 2003 **34** 183–193.

32. Sharma S.B. and Sutcliffe M.P.F., 'A simplified finite element approach to draping of woven fabric', *Proc. 6th Int. ESAFORM Conf. on Material Forming*, Salerno, Italy, Apr 2003, pp. 887–890.

33. Yu W.R., Zampaloni M., Pourboghrat F., Liu L., Chen J., Chung K. and Kang T.J., 'Sheet forming analysis of woven FRT composites using picture-frame shear test and non-orthogonal constitutive equation', *Int. J. Materials and Product Technology*, 2004 **21**(1–3) 71–88.

34. Yu W.R., Harrison P. and Long A.C., 'Forming simulation of non-crimp fabric using non-orthogonal constitutive equation', *Composites Part A*, 2005 **36** 1079–1093.

35. Mohammed U., Lekakou C. and Bader M.G., 'Experimental studies and analysis of the draping of woven fabrics', *Composites Part A*, 2000 **31** 1409–1420.

36. Harrison P., Yu W.R., Long A.C. and Clifford M.J., 'Simulation of viscous textile composite sheet forming based on a unit cell energy model', *Proc. 11th European Conf. on Composite Materials*, Rhodes, Greece, June 2004.

37. Lussier D. and Chen J., 'Material characterization of woven fabrics for thermoforming of composites', *Proc. American Society for Composites 15th Technical Conference*, College Station, Texas, Sept 2000, pp. 301–310.

38. Long A.C., Souter B.J. and Robitaille F., 'Mechanical modelling of in-plane shear and draping for woven and non-crimp reinforcements', *J. Thermoplastic Composite Materials*, 2001 **14** 316–326.

39. Chung K. and Richmond O., 'Ideal forming–I. Homogeneous deformation with minimum plastic work', *Int. J. Mechanical Sci.*, 1992 **34**(7) 575–591.

40. Chung K. and Richmond O., 'Ideal forming-II. Sheet forming with optimum deformation', *Int. J. Mechanical Sci.*, 1992 **34**(8) 617–633.

41. Dessenberger R.B. and Tucker C.L., 'Ideal forming analysis for random fibre preforms', *Transactions of the ASME*, 2003 **125** 146–153.

42. Tucker C.L., 'Forming of advanced composites', in Gutowski T.G. (ed) *Advanced Composite Manufacturing*, New York, John Wiley & Sons, 1997, pp. 297–372.

43. Blanchard P., Cao J., Chen J. and Long A.C. (co-chairs), *NSF workshop on composite sheet forming*, Lowell, M.A., Sept 2001, http://www.mech.northwestern.edu/fac/cao/nsfworkshop/index.htm

44. Lamers E.A.D., Shape distortions in fabric reinforced composite products due to processing induced fibre reorientation, PhD Thesis, University of Twente, 2004.

45. Long A.C., Preform design for liquid moulding processes, PhD Thesis, University of Nottingham, 1994.

46. Bickerton S., Simacek P., Guglielmi S.E. and Advani S.G., 'Investigation of draping and its effect on the mold filling process during manufacturing of a compound curved composite part', *Composites Part A*, 1997 **28A** 801–816.

47. Crookston J., Long A.C. and Jones I.A., 'Modelling effects of reinforcement deformation during manufacturing on elastic properties of textile composites', *Plastics Rubber and Composites*, 2002 **31**(2) 58–65.

48. van den Broucke B., Tuner F., Lomov S.V., Verpoest I., de Luca P. and Dufort L., 'Micro-macro structural analysis of textile composite parts: case study', *Proc. 25th Int. SAMPE Europe Conf.*, Paris, April 2004, pp. 194–199.

49. Boisse P., Daniel J.L. and Soulat D., 'Meso-macro approach for the simulation of textile composite forming', *Proc. 7th Int. Conf. on Textile Composites*, Yamagata, Japan, Sept 2004.

5

Manufacturing with thermosets

J D O M I N Y, Carbon Concepts Limited, UK and
C R U D D, University of Nottingham, UK

5.1   Introduction

This chapter describes manufacturing processes for textile composites based on thermoset resin systems. A wide range of processes are available here, summarised in Table 5.1 in terms of their ability to be used with different reinforcement or textile forms. In the present chapter processing issues and material considerations associated with two families of processing routes will be addressed: pre-impregnated composites and liquid moulding.

Table 5.1 Comparison of processing routes for textile-based thermoset composites, illustrating capability with respect to different reinforcement forms

	Open mould (wet lay-up)	Prepreg	Liquid moulding
True unidirectional	No	Yes	No
Weaving	Yes	Yes	Yes
Weft knitting	Yes	Unlikely	Yes
Warp knitting	Yes	Yes	Yes
Braiding	Yes	Unlikely	Yes
Non-wovens	Yes	Unlikely	Yes

5.2   Pre-impregnated composites

In many composites processing methods, the matrix is added to the fibres at the time of manufacture. Resin transfer moulding and conventional hand lay-up are examples of this. An alternative approach is for the material supplier to bring the resin and fibre together, so that the user has only to lay the material onto a tool and then process it. These materials are called pre-impregnates, almost universally referred to as prepregs.

Like any other thermosetting composite material, the curing process requires heat and pressure. The trick with prepregs is to allow them a reasonable shelf-life while ensuring that they process at an acceptable temperature. To achieve this the prepreg will typically be stored in a freezer and then cured at an elevated temperature. A further complication is that the prepreg must have a useful 'out-life', which is the maximum time during which the prepreg can be uprocessed without curing. The earliest prepregs produced in the 1960s necessitated quite high cure temperatures and had very short out-lives. Now it is possible for them to be stored in a freezer, have an out-life of possibly several months and a cure temperature of between 50 and 200 °C. Not surprisingly, the higher temperature cure materials have the longer out-lives.

Prepregs tend to be epoxy resins reinforced with carbon, glass or aramid fibres. Other matrix materials are occasionally used, in for example, high-temperature polymer composites.

5.2.1   Material systems

The earliest prepregs were developed for the aerospace industry. They were based on carbon fibre reinforcements and required very high pressures and temperatures to cure them. This drove the users to either autoclaving or hot pressing to achieve the necessary conditions (possibly 3 bar and 200 °C). The cost of these processes became very high. Not only was the equipment, such as the autoclave, expensive, but so also were the materials and the tooling and consumables needed to withstand the aggressive processing conditions.

For around two decades, prepregs were limited to niche applications in large aerospace structures, generally in non-critical parts. For many and complex reasons, the anticipated move of large civil aircraft aerospace to carbon prepregs in primary structures never really happened, and the materials remained inaccessible to industry, generally because of their high costs and difficult processing. Military aircraft did adopt prepreg materials, to the point that now most of the structure is carbon fibre. Light aviation also adopted composites with enthusiasm. Aircraft such as the Slingsby Firefly were essentially entirely composites. However, these have only recently started to move towards prepregs.

In the early 1980s, two factors led to important developments in prepreg technology. The first was that some areas of motorsport, particularly Formula 1, adopted prepreg composites. While this did not significantly change the technology, it did raise the profile. The other was the widespread adoption of carbon fibre composites for racing yachts, particularly in the America's Cup. Along with other applications, the yachts introduced a requirement for prepreg materials on parts that could not be autoclaved. It also meant that the curing temperatures had to be lower as the new customers, often with very large parts (e.g. yacht hulls) would not be able to pay very high 'aerospace' prices for tooling capable of operating at around 200 °C.

This spawned a new technology, quite separate from its aerospace ancestry. In the UK, companies such as the Advanced Composites Group and SP Systems developed a new family of prepregs which offered:

- high quality from a vacuum bag process;
- achievable processing temperatures (from about 50 °C);
- acceptable costs.

It was these developments that opened up prepregs to 'middle industry'.

5.2.2   Advantages of prepreg materials

A prepreg material essentially combines the properties of the reinforcing fibre, which for textile composites may be in the form of woven or non-crimp fabric, with a high-performance epoxy resin matrix. The main advantages are:

- optimum fibre volume fraction;
- minimum variability;
- easy to cut accurately, either by hand methods or machine;
- controllable tack to allow plies to stay in place during fabrication;
- plies hold their shape and can be placed accurately;
- simple vacuum bag processes are available using tooling and processing techniques that are familiar to the industry.

The main disadvantages are that prepregs tend to be more expensive than separate resin and fibre systems, and need to be frozen for long-term storage.

5.2.3   Processing of prepregs

Almost any polymer matrix composite requires some degree of heat and pressure to complete the cure. The first prepregs were consolidated using a matched die moulding process to achieve the pressure, and an oven to achieve the temperature. Today, many of the prepregs used in some parts of aerospace and related industries still require autoclaves for consolidation and cure at relatively high temperatures of around 200 °C. The autoclave is necessary to achieve the minimum possible level of voids within the structure, giving the highest quality. The high-temperature cure is traditionally necessary to give useful working temperatures in the finished composite of between 120 and 150 °C.

When prepregs were developed that could be cured under a vacuum at modest temperatures, they became useful in a much wider range of applications. Prepreg technology became available to many industries where specialist equipment such as the autoclave was not available and temperatures of 200 °C could not be achieved. Some of these prepregs were originally designed as tooling materials where the low weight and high stiffness of composites, particularly those based on carbon fibre, were advantageous, but the parts were often too large for high-temperature autoclave cure. Very large structures, including the hulls and masts of America's Cup yachts, could be produced from prepregs, using vacuum bags and large simple ovens (effectively heated sheds) to cure the composite.

Modern prepreg systems offer extreme flexibility and can be processed under any conditions that can achieve the equivalent pressure to one atmosphere and the cure temperature (typically 50 °C upwards). Far from being restricted to autoclaves, modern processing techniques include:

- vacuum bagging;
- pressing;
- internal pressure bags;
- diaphragm moulding.

A typical, simple prepreg process is the SPRINT™ process illustrated in Fig. 5.1. For conventional prepreg, the only difference is that the resin films are integrated into the fabric rather than the discrete layers of resin shown in the diagram. The mould surface is treated with a proprietary release coating. The prepreg is then stacked on top. Once the lay-up is complete, the prepreg is covered with a release sheet. Some material systems may need this to allow resin to bleed through into the breather felt. The breather layer is necessary to allow even application of the vacuum. If it is not present, the vacuum will pull the bag tight to the surface and some areas will 'lock-off' and not see the full vacuum. Glass tow vents are placed at intervals around the periphery of the laminate to allow air, but not resin, to exit. Finally, the whole stack is covered with the vacuum bag, which is sealed to the mould, generally using a mastic tape. Ports are required to vent the bag to the vacuum pump, and to fit a pressure gauge to monitor the process. For parts that require the very highest quality or some specialist prepreg systems

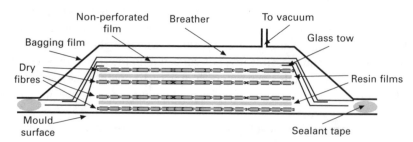

5.1 Process stack for the SPRINT™ process. Conventional prepreg is similar, other than that the resin is integrated into the fabric rather than the discrete layers central to the SPRINT™ concept. Figure reproduced by kind permission of SP Systems Ltd.

which require a very high process pressure, the mould may be placed in an autoclave, which will apply pressures of several atmospheres. While the autoclave does offer very high quality, it is a very expensive piece of equipment with correspondingly high overheads, and its benefit for the bulk of composite applications is questionable. The industry is generally developing processes and materials that avoid the need for autoclaves. Once the vacuum is established, the part will be cured in an oven. Most prepregs will require a cure temperature of between 50 and 200 °C.

5.2.4   The cost equation

The initial cost of purchasing prepreg materials can be around double that of separate resin and fibre systems. However, since prepregs are generally easier to use and better controlled than 'wet' systems, the cost of the finished part may not be very different. Table 5.2 compares the use of a typical, low-temperature prepreg with more conventional wet systems.

For small parts, prepregs are often most suitable as the quantity of material is low and precise handling becomes an important issue. For very large structures, processes such as resin transfer moulding (RTM) and vacuum infusion (VI) (see section 5.3) may generally be more appropriate, owing to the significance of materials in the cost equation. In between these extremes, the manufacturer will have to consider the true cost of all the elements contributing to cost – not just that of the material. A final consideration is the manufacturer's normal preference. For instance a company that normally uses prepreg would incur considerable hidden cost in retraining its staff to use RTM or VI.

5.2.5   Hybrid systems

In recent years there has been a new family of materials developed, which are essentially a combination of conventional prepregs and the dry infusion process, which uses alternating layers of dry fabric and resin film. Materials such as SPRINT^{TM} from SP Systems look very similar to conventional prepreg, but apply the resin to one side of the fabric, rather than distributing it through the thickness. The advantage of this is that the fabric itself provides pathways to allow all of the air to be extracted under vacuum, before the resin is heated to the point that it starts to flow. This produces a high-quality, low-voidage composite without the need for high processing pressures and autoclaves. Figure 5.1 illustrates the process 'stack', and a complete boat hull bagged up and in process is shown in Fig. 5.2.

Table 5.2 Comparison of advantages and disadvantages associated with prepreg with wet lay-up

	Prepreg	Wet lay-up
Material cost	High. Often difficult to obtain in small quantities.	Medium to low.
Tooling	Relatively high cure temperatures mean that some thought must be given to the tooling materials. Even at 50 °C some woods, paints and plastics start to become unstable.	Room temperature cure allows the use of most low-cost materials.
Consumables	The lack of free resin often allows consumables such as breather fabrics, to be reused.	Resin will tend to prevent reuse of consumables.
Cutting	Prepreg fabrics are very stable due to the presence of the resin. It can be cut with anything from a knife to an NC machine. Very accurate cutting is possible.	Dry fabrics tend to be unstable and unravel at the edges. This makes cutting to size very difficult, resulting in a trimming problem after cure. Stabilised fabrics are available but are expensive.
Control of fabric	Prepreg fabrics are stable, easy to lay with precision and drape well.	Wet fabrics can be difficult to lay down evenly and are almost impossible to position precisely. Stabilised fabrics can be difficult to drape.
Surface finish	Excellent with suitable tooling.	Can be difficult to reliably achieve a good surface finish.
Component quality	The controlled resin implicit in the use of prepregs results in very consistent fibre volume fractions and laminate thickness.	Although it is possible to produce very high-quality parts by wet lay-up and vacuum bagging, the repeatability is very dependent upon the experience of the operator.

5.2.6   Conclusion

There is little doubt that, from structural considerations, prepregs offer a material system that will produce a consistent part of very high quality. The negative side of the equation is cost, both in terms of the initial cost of the material and its storage and cure requirements. Only an analysis of the parameters of a particular part within the context of a particular manufacturer will indicate the cost effectiveness of the material.

5.2 A large boat hull being manufactured using the SPRINT™ process. Figure reproduced by kind permission of SP Systems Ltd.

5.3   Liquid moulding of textile composites

Of all the potential conversion routes for textile-based intermediates, liquid moulding is arguably the most flexible, combining the small batch flexibility with the low emissions and higher quality levels of prepreg, without the high on-costs associated with those materials. Liquid moulding describes a family of closed mould processes whereby a dry reinforcement is impregnated with a liquid thermosetting resin. Table 5.3 summarises the most widely used processes and terminology.

Apart from some of the technical reasons for using liquid moulding, it is also of strategic importance that liquid moulding is used almost exclusively for producing engineering sub-assemblies or complete structures – rather than semi-finished goods – with the cost benefits associated with eliminating this intermediate stage. A materials cost save of 40% might be expected, compared with prepregs.

A large part of the flexibility lies in the facility to change textile architectures to suit the application (for processing or performance reasons) without the need to re-engineer the upstream process (as would be the case with prepregs, compounds, etc.). This also offers the only sustainable route (for thermosets) for non-mainstream fabrics – 3D weaves, braids, etc.

Table 5.3 Glossary of terminology related to liquid moulding

RTM	Resin transfer moulding – positive pressure, matched mould variant with polymer resins.
VARI	Vacuum assisted resin injection – adaptation of RTM using vacuum sink to enhance pressure gradient.
SRIM	Structural reaction injection moulding – adaptation of RTM using reactive processing equipment and techniques, usually for urethanes.
VI, VARTM, RIFT, SCRIMP	Vacuum impregnation processes using one hard tool plus vacuum membrane. Usually involves flow-enhancing medium. Some variants are patent protected.
RFI	Resin film infusion – (usually) through thickness impregnation by stacking films and fabrics, increasing materials available in kit form.

The mainstream liquid moulding applications are RTM products in the automotive industry and elsewhere in industrial goods. The principal resins are unsaturated polyester (for automotive body structures) and epoxies or bismaleimides for aerostructures. Several monographs were produced in the mid-1990s which provide a general introduction to the field[1–3]. The last decade of the 20th century was marked by a moderate increase in the RTM market and a much wider acceptance of the technology for low-volume applications. This can be attributed to a combination of legislative changes concerning styrene emissions and cost reduction exercises in the transportation industries. The technology has matured considerably during this period with the emergence of materials, computer-aided design (CAD) systems, tooling solutions all more or less tailored to the process. The developments might be largely summarised as the refinement of polyester RTM and the evolution of a parallel technology, equally robust for high-performance aerospace resins. Less well known, perhaps, have been attempts made over the same period to adapt the technology to thermoplastics resins – notably polyamides and polyesters, an intriguing possibility as this market segment continues its buoyant growth.

The purpose of this section is to summarise the basic science and technology of RTM and to review the important developments which have taken place post-1995 in the related fields of resin film infusion (RFI), structural reaction injection moulding (SRIM) and vacuum infusion. More fundamental issues related to process modelling are described in Chapter 7.

5.3.1   Resin transfer moulding (RTM)

Key features
Matched moulds
Low pressure processing
Structural and cosmetic parts
Low-volume applications
Darcy's law:

$$Q = \frac{KA}{\mu} \frac{\partial p}{\partial x}$$

where Q = volumetric flow; K permeability, A cross-sectional area, p pressure and x distance (parallel to flow direction)

The basic process (Fig. 5.3) involves a positive pressure resin supply and a pair of matched moulds. The fabric charge (preform) is loaded, the mould closed and resin is introduced at superatmospheric pressures (2–20 bar is typical). Darcy's law is usually assumed to apply and the time taken to impregnate the preform therefore depends on the pressure gradient, the resin viscosity and the hydraulic conductivity or *permeability* of the fibre bed. Impregnation times range from a few minutes to several hours and the resin needs to be formulated taking this quantity into account. The resin may be supplied via a simple pressure pot or, more usually, some form of positive displacement pump.

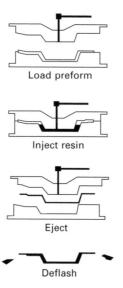

Load preform

Inject resin

Eject

Deflash

5.3 RTM process schematic.

The process can be configured in many ways, depending upon the application and the industrial environment. However, a typical manufacturing cell would involve low stiffness tooling (glass fibre reinforced plastic, GRP, or nickel shells) backed by a fabricated stiffening structure with peripheral clamping. A pneumatic, reciprocating pump would deliver polyester resin, dosed with a low-temperature initiator and passed through a static mixer en route to the mould. The mould may be heated by resistance elements or fluid-carrying pipes; either may be embedded within the mould structure. Porting is essential to admit the resin and to vent the air. These may be inside or outside the main cavity; the latter is often preferable for cosmetic parts to avoid witness marks. Single injection ports are the norm, although it is possible to use multi-porting (e.g. for large area mouldings), but special care is obviously required to avoid air entrapment where flow fronts coalesce.

Unsaturated polyester and epoxy resins account for the majority of applications, the major constraint being viscosity – 0.5 Pa s represents a practical upper limit. Higher viscosity systems such as aerospace epoxies or bismaleimides may require reactive diluents or substantial preheating in order to bring them into this range. Such systems (and other hot-setting formulations) are generally handled as premixed, one pot systems while low temperature curing resins (especially unsaturated polyesters) are dosed with peroxide initiators on-line.

Reinforcements for RTM include a wide range of mats, fabrics and non-woven structures. In-plane permeability in the dominant property for most applications and virtually all intermediates will fall within the range that can be processed. Notable exceptions are forms based largely on monofilaments (owing to the hydraulic radius effect) and many chopped strand mats (owing to the tendency for fibre washing which arises from highly soluble binders). It should be noted also that permeability is a strong function of wetted area (and therefore fibre volume fraction), as well as fibre orientation.

Various proprietary products are used to create high-permeability regions as processing aids. Perhaps the simplest approach is the elimination of alternating tows in, e.g., a woven or multiaxial warp knit fabric, with obvious consequences for mechanical performance. Alternatives include polyester monofilament netting, as used widely during infusion processing (see below) or so-called flow enhancement fabrics, notably Injectex® which includes a proportion of warp-bound tows to create relatively large inter-tow conduits for resin flow.

5.3.2   Vacuum infusion (VI)

Key features
Control of styrene vapour
Thickness, V_f control
Elimination of prepregging stage
High consumables cost

In this process (Fig. 5.4) the upper half of the mould pair is replaced with a vacuum membrane. This reduces the tooling cost significantly and provides an increasingly important fabrication route for large structures where the throughput requirement is small. The key advances over open-mould alternatives are control of volatiles, thickness and fibre volume fraction (V_f) and resin consumption. Although the basic technology is more than 50 years old, recent industrial interest has been strong and driven almost exclusively by legislation on styrene emissions. In this context, the benefits of vacuum consolidation such as higher and more consistent volume fractions are significant, but secondary considerations.

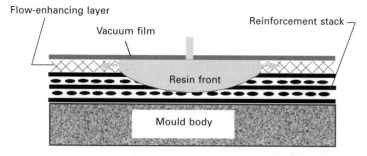

5.4 Vacuum infusion process schematic.

The usual infusion set-up involves a GRP mould operated at room temperature. The fabric is laid-up by hand, followed by the ancillary materials – peel ply, breather felt, infusion mesh and vacuum membrane. Sealing relies on mastic tape, as with conventional vacuum bag moulding, and the facility to pull an effective vacuum. While the impregnation process is relatively fast for most set-ups (owing largely to the infusion mesh), the overall manufacturing time may be quite similar to those for wet laminated products. This is because the mould preparation – reinforcement, infusion mesh placement and bagging – are all critical, manual operations. Nevertheless, infusion continues to make inroads into the wet laminating sector, largely driven by legislative pressures.

Resins for vacuum infusion are generally low-viscosity variants on standard (i.e. laminating resins) epoxies and polyesters. The flow requirements are no more demanding than for conventional RTM, although the process is usually done at room temperature and so there is not usually a temperature-related viscosity reduction. One of the consequences of moulding at sub-atmospheric pressures is the danger of degassing. Infusion temperatures are generally sufficiently low to inhibit styrene boiling but the risk of pulling dissolved gases out of solution remains. This is the subject of a useful discussion by Brouwer et al.[4], who recommend degassing the resin before infusion in the presence of a nucleating agent (e.g. Scotchbrite[TM]) or sparging (bubbling through an air stream at reduced pressure).

Given the limited pressure gradient (since the maximum pressure difference between source and sink is usually 1 atmosphere) most operators rely upon an infusion mesh to create a high permeability zone at the surface of the laminate. Typically, this consists of a knitted monofilament which combines high porosity with low transverse compliance to provide the important flow gallery which makes for reasonable infusion times. Multiple infusion mesh layers are also beneficial – these may be either distributed or contiguous, although obviously this adds to consumables costs since the mesh is not generally reusable and creates a sacrificial resin film.

High-permeability zones, such as those created by the infusion mesh, mean that the flow within a typical laminate with any significant thickness will be three-dimensional. Flow leads at the surface and lags at the underside – the so-called lead-lag effect. This has several consequences:

- The infusion needs to continue until the lagging face is thoroughly impregnated.
- The infusion mesh may need to be cut back from the vented edge of the part to allow the flow front to recover.
- The magnitude of the effect will vary with the reinforcement style and the laminate thickness.

One further consequence of vacuum-driven processes of this sort is the indirect control of laminate thickness. During impregnation, the local thickness depends upon the pressure difference across the vacuum membrane and the transverse compliance of the reinforcement. Both factors will vary with time and position because, firstly, the pressure inside the bag depends upon its distance from the injection point and secondly because the transverse compliance varies with saturation and (probably) with time also. Practically, this means a gradual reduction in laminate thickness from the inlet to the flow front. A small 'ditch' may also be observed at the flow front because of the increased compliance of the saturated reinforcement. The evolution of the thickness post-filling is determined largely by the process control strategy. Many operators will bleed resin from the laminate to increase the fibre fraction and reduce the void fraction. Andersson[5], for example, notes the progressive consolidation during this phase.

5.3.3   Structural reaction injection moulding (SRIM)

Key features
Polyurethane processing
High capital costs
High throughputs

SRIM is essentially the marriage of reaction injection moulding and RTM (see Fig. 5.5). Because the injection involves monomers, the viscosity is low enough to enable extremely short filling times – a few seconds is typical. Although, in principle, many polymers can be processed this way, the important commercial applications involve polyurethanes and polyureas. The automotive industry is more or less the exlusive market for SRIM products – semi-structural items such as fascias, bumper beams and interior trim panels are all current. Although the technology applies equally to ordered textile structures, most of the reinforcement that is processed this way involves various forms of non-woven. This reflects partly the performance criteria for current applications; also the need for rapid impregnation and fairly high washing resistance at fairly low fibre volume fraction.

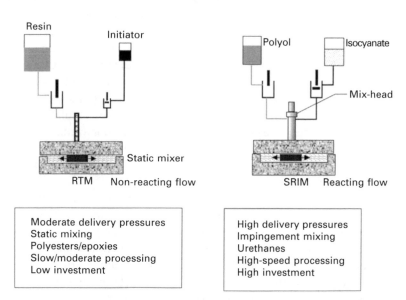

5.5 Comparison of RTM with SRIM.

SRIM also offers some possibilities for processing of thermoplastics. The earliest system for polyamide 6 was developed in the early 1980s and is usually termed NYRIM. This involves the anionic polymerisation of γ-caprolactam, and was developed originally by Monsanto[6–8] and later sold to

DSM. The main market for these materials is industrial cast polyamides (although some operators also incorporate milled glass reinforcement). A more recent formulation for polyamide 12 was developed by EMS Chemie[9], using a liquid initiator for laurolactam polymerisation (although the same system can be used with carolactam to produce PA6). Studies at EPFL[10] have examined a continuous version of the process for producing semi-finished goods via reactive pultrusion, where this somewhat demanding process can be operated in a steady state. Liquid moulding processes involving thermoplastic polymers are described in more detail in section 6.4.

5.3.4   Resin film infusion (RFI)

Key features
B-staged resins
Incremental vacuum bagging technique
Low throughputs

The introduction of several commercial 'kit' systems has renewed industrial interest in RFI, particularly in the aerospace and marine industries. RFI was initially patented in 1986 by Boeing[11] but has not received widespread technical interest outside the aerospace industry, where some work has been performed on composite wing stiffeners. Several semi-finished products have been introduced that package fabric and resin film together in a variety of ways.

The conceptual process involves interleaving fabric and B-staged films in the appropriate ratio before bagging and autoclaving in a similar fashion to prepreg processing. While offering ostensibly lower materials costs to conventional prepregs, progress was hindered by poor availability of suitable resins and limited understanding of the practicalities. Few systematic studies have been published although there has been some research effort in both the USA[12] and Australia.

In response to industrial demands for materials cost reduction and in an effort to preserve existing markets, several materials suppliers now offer materials in kit form for RFI-type processes. Such products are sometimes termed semi-pregs, since the fabrics are either partially impregnated or are tacked to a B-staged resin film. An increasing number of product variants includes surfacing films, syntactic cores, etc., and some of these are illustrated in Fig. 5.6. Here there is some overlap with the 'hybrid systems' described in section 5.2.5.

Whichever approach is used to combine the fabric and resin, several common and distinguishing features can be identified:

- The presence of integral breather channels within the dry portions of the fabric.

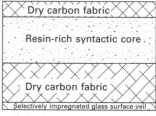

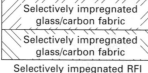

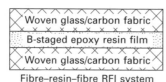

5.6 Schematic representation of RFI kit systems.

- In some cases, superior draping characteristics to conventional prepregs due to the incomplete impregnation.
- Relatively short lay-up times, since the laminate is generally designed to be laid up as a single ply.
- The relatively short flow paths mean that minimum viscosity resins are not required and so higher molecular weight products can be used than are typically necessary in RTM.

As with conventional prepregs, the materials are designed to be processed using vacuum bag and oven or autoclave. Heat and pressure are applied and the resin flows the short distance, either in the plane or through the thickness of the fabric, before crosslinking fully.

Applications include low-volume manufacturing of automotive body panels, wind turbine blades and high-performance marine structures. Body panel systems incorporate typically a syntactic core containing glass microspheres. This reduces panel mass and enables, e.g., carbon fabrics to be used cost effectively. Cosmetic body panels may also include a satin weave cloth to mask fibre strike-through or a sandable surfacing film. Frost et al.[13] have made an initial comparison of several commercial systems.

Most of the published research focuses on process simulation, usually evolved from RTM simulation codes. Loos[14] presents results for a composite stiffener based on their 3D RFI process model, simulating flow front position, impregnation time, temperature distributions, resin viscosity, degree of cure, preform deformations, fibre volume fractions and final cured shape. Other studies include those of Blest *et al.*[15], Antonucci *et al.*[16] and Sevostianov *et al.*[17] which deal specifically with cavitations and resulting void formations in RFI.

5.4   References

1. Rudd C.D., Long A.C., Kendall K.N. and Mangin C.E., *Liquid Moulding Technologies*, Woodhead Publishing Ltd, 1997.
2. Kruckenburg T. and Paton R. (ed.) *Resin Transfer Moulding for Aerospace Structures*, Kluwer Academic Publishers, 1998.
3. Potter K.D., *Resin Transfer Moulding*, Chapman and Hall, 1997.
4. Brouwer W.D., van Herpt E.C.F.C. and Laborus M.,Vacuum injection moulding for large structural applications, *Composites Part A*, 2003 **34** 551–558.
5. Magnus Andersson H., Visualisation of composites manufacturing, Doctoral Thesis, Luleå University of Technology, 2003.
6. Riggs C.D., NYRIM catalyst and prepolymer for polyamide block copolymers, *Proc. National Technical Conference,* Society of Plastics Engineers, 1983, pp. 102–103.
7. Recktenwald D.W., Advances in formulating and processing polyamide block copolymers, *Proc. SPI/SPE Plastics Show & Conference – East*, 1984, Society of Plastics Engineers, pp. 151–158.
8. Gabbert J.D. and Hedrick R.M., 'Advances in systems utilizing polyamide block copolymers for reaction injection moulding', *Kunststoffe – German Plastics*, 1985 **75**(7) 14–16.
9. Mairtin P.O. and O'Bradaigh C.M., Processing and mechanical properties evaluation of a commingled carbon-fibre/PA-12 composite, *Proc. Int. Conf. on Automated Composites*, Bristol, Sept 1999, pp. 55–62. Institute of Materials.
10. Luisier A., Bourban P.E. and Manson J-A.E., Reaction injection pultrusion of PA12 composites: process and modelling, *Composites Part A*, 2003 **34** 583–595.
11. US Patent 4,622,091, 1986.
12. Dow M.B. and Dexter H.B., *Development of stitched, braided and woven composite structures in the ACT Program and at Langley Research Center (1985 to 1997)*, summary and bibliography, NASA report NASA/TP-97-206234, Nov. 1997.
13. Frost M., Solanki D. and Mills A., Resin film infusion process of carbon fibre composite automotive body panels, *SAMPE Journal*, 2003 **39**(4) 44–49.
14. Loos A.C., Rattazzi D. and Batra R.C., A three-dimensional model of the resin film infusion process, *Journal of Composite Materials*, 2002 **36** 1255–1273.
15. Blest D.C., McKee S., Zulkifle A.K. and Marshall P., Curing simulation by autoclave resin infusion, *Composites Science and Technology*, 1999 **59** 2297–2313.
16. Antonucci A., Giordano M., Inserra S. and Nicolais L., A case study of resin film infusion technology: the manufacturing of a composite stringer, *Proc. 10th European Conference on Composite Materials (ECCM-10)*, June 2002, Brugge, Belgium.
17. Sevostianov I., Verijenko V.E. and von Klemperer C.J., Mathematical model of cavitation during resin film infusion process, *Composite Structures*, 2000 **48** 197–203.

Composites manufacturing – thermoplastics

M D WAKEMAN and J-A E MÅNSON,
École Polytechnique de Fédérale Lausanne (EPFL), Switzerland

6.1 Introduction

Thermoplastic-based composites differ from thermoset-based composites due to the nature of the matrix because the often time-consuming chemical reaction (crosslinking) is not required during the processing of thermoplastics. Harmful chemical emissions are generally reduced during processing. The considerable shelf-life of thermoplastics is also an advantage because sub-ambient storage is not required. Thermoplastic composites offer increased recyclability and can be post-formed or reprocessed by the application of heat and pressure. A large range of tough matrix materials is also available.

The manufacture of components from textile thermoplastic composites requires a heating process, either directly before the final moulding process, where an oven plus a cool tool is used (non-isothermal processing), or in a hot mould (isothermal processing). This is needed to raise the polymer matrix temperature sufficiently above the melt temperature (T_m) to reduce viscosity for impregnation of the textile structure during the final conversion process. The application of pressure to give intimate contact and hence heal adjacent yarns and plies, reducing void content, completes consolidation followed by cooling, under pressure, to solidify and crystallise the matrix to finish the cycle (Fig. 6.1)[1]. Heat transfer thus forms the principal boundary condition governing process cycle times, with a corresponding potential for lower conversion costs. These basic steps define the many processing techniques that can be used to transform different material forms into the final product, where considerable flexibility exists to heat and shape the textile composites. One limitation for manufacturing techniques that results from the continuous, well-ordered and close-packed fibre architectures is that these materials do not flow in the same way as a fibre suspension to fill a tool, but must instead be deformed by draping mechanisms (see Chapters 3 and 4). Where flow does occur, this is local to the fibre bundle or, in certain cases, across the stationary fibre bed.

However, this limitation notwithstanding, a wide variety of both materials

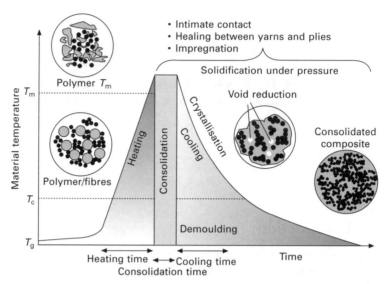

6.1 Thermal cycle for thermoplastic composite processing (adapted from Månson et al.[1]). T_m = melt temperature, T_c = crystallisation temperature, T_g = glass transition temperature.

and processing techniques have been developed for both niche applications and components produced at high volumes. Several commercially available aligned fibre forms are available that are suitable for conversion by a variety of processes and this market sector has an estimated growth rate of 20% per annum.

This chapter starts with an overview of the impregnation and consolidation processes for thermoplastics that have driven both materials development and the choice of final conversion process. This is followed by a summary of thermoplastic composite pre-impregnation manufacturing routes, and a review of final conversion processes for textile thermoplastic composites. The chapter concludes by reviewing forthcoming processing techniques for hybrid textile composite structures.

6.2   Consolidation of thermoplastic composites

6.2.1   Impregnation

The impregnation of fibre beds with thermoplastics is inherently more difficult than with equivalent processes for most thermosets owing to the high viscosity of thermoplastics at melt temperature. For example, polypropylene (PP) at 195 °C has a viscosity[2] of 410 Pa s, which is over 500 times the upper limit of 0.8 Pa s for unsaturated polyester resin. Since flow can be approximated by Darcy's law, the time for impregnation of a fibre bed by a constant viscosity material is given by eqn 1[3], where the impregnation time (or pressure) can be minimised by reducing the mass transfer distance, x, or the viscosity, μ:

$$t_{imp} = \frac{x^2 \mu (1 - V_f)}{2K(p_a + p_c - p_g)} \qquad\qquad 6.1$$

where:

x = impregnation distance
K = fibre bed permeability in the Z-direction
μ = matrix viscosity
p_a = the applied pressure
p_c = the capillary pressure
p_g = the gas pressure
V_f = the reinforcing fibre volume fraction

Intimate mingling techniques reduce the impregnation (mass transfer) distance of the matrix into the fibre bed during processing, reducing consolidation pressures. There is often an associated increase in product cost with increasing homogeneity[2]. Reduced impregnation times improve throughput and minimise thermal degradation of the polymer. Simple increases in pressure, during direct-melt impregnation for example, are likely to be counter-productive since this will also reduce the transverse permeability of the fibre bundles. Notably for lower viscosity systems, capillary pressure can play a role in the spontaneous wetting of fibres as low external pressure may be applied.

The starting distribution of fibres, matrix and porosity (defining the preconsolidation level) dominate the structure of aligned fibre materials, which fall into two principal classes (Fig. 6.2)[1]. The first class consists of

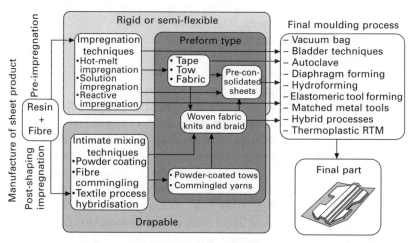

6.2 Material routes, preform types and final moulding processes for aligned fibre thermoplastic composites (adapted from Månson *et al.*[1]).

unconsolidated 'dry' mixtures of polymer and reinforcing fibre, for example as commingled yarn, which is flexible and hence suitable for further processing as a hybrid yarn into textile structures. In addition to the wide range of fibre architectures (woven and non-crimp fabrics, braids and knits), this has the advantage that textiles can be draped to a mould surface while cold. This offers ease of handling for isothermal processes. The disadvantage is that an often-considerable material bulk must be accommodated in the moulding process and that full impregnation and consolidation must occur in the final conversion process. The second class consists of partially or fully preconsolidated sheet, for example via film stacking and melt impregnation, which is subsequently semi- or fully rigid. This aids handling for non-isothermal processes where heating occurs in a first step followed by often automated transfer to the cool moulding tool, but inhibits the use of isothermal conversion processes because the material cannot be conformed to the tool in the cold material state.

Various techniques have been adapted for producing intimate fibre–resin combinations, including hot-melt impregnation, solution impregnation, reactive impregnation, powder coating, fibre commingling and textile process hybridisation. These are discussed in section 6.3.

6.2.2   Consolidation of thermoplastic textile composites

The consolidation of thermoplastic matrix composites consists of three major steps: intimate contact, autohesion and fibre impregnation[1, 4]. Intimate contact is required between the surfaces of successive plies, which are initially rough with gaps consequently existing between the plies. The application of heat and pressure causes viscous deformation of the contact points and the degree of intimate contact is increased. The second step occurs after contact of the two surfaces with an autohesion bonding process occurring at the interface where segments of macromolecules diffuse and interpenetrate across the interface. The third step depends upon the initial impregnation state of the composite. Here impregnation of the fibre bed is either fully accomplished, or completed during the final stages of compaction. The deformation and flow mechanisms involved during the forming stage include resin percolation, squeeze flow or transverse flow, intra-ply shear and inter-ply shear.

The flow mechanism of resin percolation consists of a movement of matrix through the fibres and as such is not a deformation of the fibre arrangement. Without sufficient resin percolation, proper consolidation of the part will not occur, resulting in residual porosity. Squeeze flow or transverse fibre flow is a mechanism whereby parallel fibres displace transverse to the fibre direction due to the normal pressure applied across the laminate surface.

The consolidation stages taking place during final conversion depend on the initial degree of impregnation and consolidation when at process temperature. The consolidation process is illustrated here through the example of commingled materials[3, 5–8] which is shown schematically in Fig. 6.3. The polymer and reinforcing fibres are generally not distributed homogeneously within a yarn. Impregnation initiates after heating, with molten matrix pooled at the edges of the reinforcing fibre bundles. Increased compaction pressure moves the bundles closer together and bridging of bundles occurs with the matrix. Individual matrix pools then coalesce around the dry fibre bundles. Additional compaction pressure results in flow of the matrix to impregnate the fibre bundles and at full impregnation further increases in pressure may result in in-plane matrix flow on the meso-scale[9]. The time taken to impregnate the reinforcing fibre bundles depends upon the structure of the fabric, yarn dimensions and shape, number of fibres per yarn, fibre diameters and the quality of commingling[2].

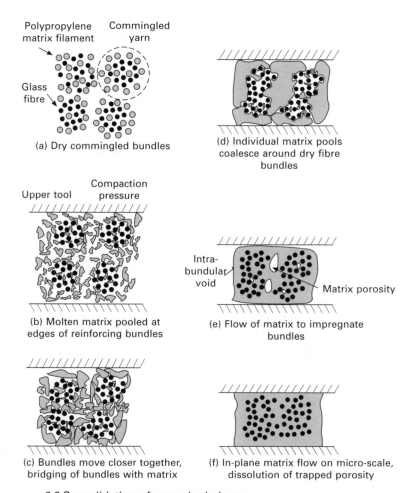

(a) Dry commingled bundles

(b) Molten matrix pooled at edges of reinforcing bundles

(c) Bundles move closer together, bridging of bundles with matrix

(d) Individual matrix pools coalesce around dry fibre bundles

(e) Flow of matrix to impregnate bundles

(f) In-plane matrix flow on micro-scale, dissolution of trapped porosity

6.3 Consolidation of commingled yarns.

6.2.3   Effect of fabric deformation on consolidation

The formation of complex shapes needs the material to have sufficient compliance at the processing temperature to deform to the imposed geometry. When the shaping operation occurs at room temperature (e.g. non-consolidated fabrics), this relies on the draping characteristics of the dry textile. When non-isothermal processes are used, the matrix is in the melt phase and lubricated fabric deformation occurs, with forming and consolidation often occurring in rapid sequence.

The basic deformation mechanisms that permit the material to conform are intra-ply shear and inter-ply shear (see Chapter 3)[4]. The degree of conformability is controlled by the reinforcement architecture that dictates the component complexity. For biaxial reinforcements, the principal deformation is intra-ply shear, whereby the yarns are subject to in-plane rotation about their crossover points, enabling a woven textile to conform without wrinkling to a complex surface. Where multiple plies are present, inter-ply slip also occurs whereby the individual ply layers slip across one another when forming a curved shape.

A consequence of intra-ply shear is varying local superficial fabric density. In the simplest case, where the part thickness is not directly constrained (such as in a diaphragm forming or vacuum consolidation process), the laminate thickness will increase locally in the sheared region. The effect on fibre volume fraction (V_f) depends on the matrix distribution but will remain more or less unchanged. Where rigid tools are used, the cavity thickness is defined and a pressure gradient will result given the necessary force for compaction to the nominal thickness. Regions with greater shear will be subject to higher compaction pressures. In a rapid process, matrix solidification greatly inhibits in-plane flow. Thus the effects of the consolidation pressure variations are frozen in, which implies a variation in void fraction over the laminate. Aligned fibre composites with fibre contents of greater than 50% by volume can only adapt to gap tolerances of less than 5% of the original sheet thickness[10].

The issue of non-uniform consolidation can be approached in three main ways. The first uses drape modelling tools[5, 11] to estimate the arising distribution of superficial densities or consolidated thicknesses. These data can be used to define the cavity thickness in a way that takes into account the fabric deformation. Thus the part would be produced with varying thickness but uniform V_f and at a uniform consolidation pressure. Although this approach is practical from a tool-making perspective, the non-uniform cavity is then dedicated to a specific fabric and its orientation with respect to the axes of the part. Part quality becomes highly susceptible to small deviations in fabric alignment and changes to the fabric specification would require the mould cavity to be reworked.

A more flexible alternative involves the manufacture of a deliberately compliant mould[12]. Here one die is rigid, as before, while the other is cast from an elastomer, offering a useful solution for intermediate manufacturing volumes. Practical difficulties remain since the laminate thickness is not controlled directly, which has implications for subsequent assembly operations, and the elastomeric tool has a limited life and significantly lower thermal conductivity than the metal die, which makes for non-uniform cooling and solidification.

Arguably the most effective solution for high-volume press-based manufacturing is to use a multilayer material of fabric skins with a core structure. During forming, the fabric skins will shear in the usual manner followed by local squeeze flow of the core, which can be either a glass mat thermoplastic (GMT) type material or a polymer film[13, 14]. Use of such a core enables variations in cavity pressure or V_f arising from the fabric deformation to be reduced, but with a lower overall V_f compared with aligned fibre materials. However, owing to its general proximity to the neutral axis, this does not significantly affect flexural stiffness.

6.2.4   Void content reduction

Composite porosity should be reduced to minimise quality variations and maximise mechanical properties. The tendency of preconsolidated sheet materials to deconsolidate, or 'loft' during preheating, or the bulky initial structure of dry textile structures heated to process temperature, results in trapped gas in the consolidating laminate. Much of the gas is expelled from the tool before the matrix viscosity reaches a limiting level. Porosity then reduces with increasing time at pressure to a final residual level. The decreasing temperature and increased pressure of the trapped gas results in volumetric reduction of the void volume, with polymer flowing into the original void cavity while the viscosity remains below a limiting threshold.

Voids can also form from crystallisation shrinkage due to the relation of pressure and temperature on specific volume during cooling. The volumetric shrinkage that occurs between the onset and end of crystallisation is shown in Fig. 6.4[14, 15] for PP-based materials. Data points are superimposed on the pressure–volume–temperature (PVT) curves based upon a non-isothermal prediction that mapped through thickness temperatures[16]. This indicates that if pressure were released before the drop in specific volume (as crystallisation occurs), then any shrinkage from local crystallisation would not be accommodated by a laminate thickness change but by local void growth. Where one area of a tool, for example at a region of decreased thickness, cools first, the tool displacement to accommodate local shrinkage in thicker part regions will be limited by the thinner previously crystallised zone, with the shrinkage resulting in mid-plane voidage.

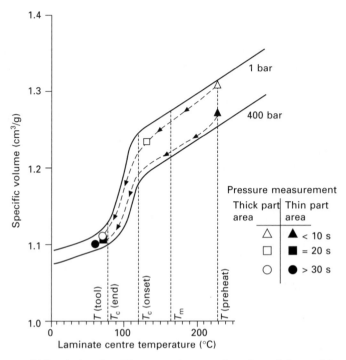

6.4 PVT relation for PP composite as a function of time with staggered cooling times for different laminate thickness (adapted from Janeschitz Kriegl *et al.*[15]).

The final phase of void reduction consists of the compressed gas diffusing into the surrounding polymer. Thermodynamic data suggest that individual molecules dissolve among the polymer chains[17]. For the case of PP, a given matrix element at 200 °C and 300 bar is capable of dissolving 15 times its own volume of air. The diffusion coefficient of air in PP at 200 °C results in a mean free path of 0.4 mm after 60 s for a diffusing molecule, indicating that the dissolved gas cannot escape out of the laminate during the short time of the high-temperature period. The diffusion mechanism is activated thermally, with the rate of dissolution decreasing as the laminate centre cools. The majority of trapped gas is hence dissolved into the matrix before solidification of the melt, with reduced dissolution below the melting temperature. Thus the impregnation state, any local thickness variations and the cooling rate will influence the final void content level and distribution in the part.

6.2.5   Internal stresses and spring-back phenomena

Processing conditions have a direct influence on the build-up of internal stresses in composites, and hence on the final component quality[18–20]. The morphology and mechanical properties of semi-crystalline thermoplastic composites are influenced by the cooling conditions applied during the solidification stage of the processing cycle. Increased cooling rates usually reduce the degree of crystallinity, give smaller crystallite sizes, increase the interlaminar fracture toughness, and reduce the transverse elastic modulus[21, 22]. For example, rapid cooling of a composite surface at the end of a processing cycle induces temperature gradients through the part thickness, leading to internal stress build-up from heterogeneous and anisotropic shrinkage of the material. Non-isothermal solidification of semi-crystalline thermoplastic composites also gives rise to internal stresses due to crystallinity gradients.

Owing to changes in stiffness and thermal expansion between plies, internal stresses may also be generated as a result of the chosen stacking sequence[23]. On a different scale, internal stresses can result from the mismatch of thermal expansion between matrix and fibre. Process-induced internal stresses may have undesirable effects on the performance of the end-product, reducing strength and initiating cracks or delamination, thereby leading potentially to premature failure of the composite. Internal stresses may also affect the dimensional stability of the part by generating post-processing distortions. Post-treatments such as annealing can be a solution to the problem of relaxing internal stresses in composite components, but the time and costs involved can be prohibitive.

6.3   Textile thermoplastic composite material forms

6.3.1   Thermoplastic matrix materials

Thermoplastic polymers differ from thermoset polymers in the way that the organic molecular units are bonded together. Thermoplastic monomer chains are held together by secondary intermolecular forces (e.g. van der Waals or hydrogen bonds) and mutual entanglements, rather than the covalent bonds (crosslinks) of thermoset polymers. Hence the intermolecular forces in thermoplastic polymers can be overcome by thermal energy and the polymer will reach a softened or molten state. Depending on the molecular structure, thermoplastic polymers are divided into the two groups of amorphous and semi-crystalline polymers.

Amorphous polymers have randomly oriented molecular chains, even in the solid state. The asymmetric chains (e.g. those with bulky side groups) inhibit arrangement of the chains into ordered crystalline structures. In semi-crystalline polymers, the chains are symmetric enough to be fitted into an ordered crystalline state. Due to the large molecular chain length, complete crystallinity (as in a low molecular weight substance) is not achieved and both crystalline and amorphous regions coexist in the solid. When a polymer is cooled from above the melt temperature (T_m), the chains lose mobility and begin to interact segment by segment until glass transition (T_g) or crystallisation takes place. At T_g the molecular chains are locked in place and the polymer assumes a 'glassy' state. The molecular orientation of amorphous polymers in the glassy state is random due to the irregular chain structure. Semi-crystalline polymers crystallise at temperatures well above the T_g and hence regions of amorphous and crystalline phase coexist in the temperature range between crystallisation and T_g. Crystallisation does not occur at a clearly defined temperature, but over a range, for example being shifted to a higher temperature with increased pressure, or reduced with an increased cooling rate.

As a thermoplastic polymer is heated above the T_m, a reduction in melt viscosity occurs. Given suitable impregnation techniques, this enables a variety of thermoplastic polymers to be used as matrix materials in textile composites, giving a wide range of thermal and mechanical properties (Table 6.1). Generally, the lower melting temperature polymers simplify processing, due to handling and drying issues. The following section describes a variety of approaches, shown schematically in Fig. 6.5, that are used to impregnate the reinforcing textile structure with the matrix polymer. Table 6.2 compares the mechanical properties of textile thermoplastic composites from a variety of sources, which should be taken as representative values rather than design data.

6.3.2   Bulk melt impregnation

The polymer is introduced (usually as a film or extruded) between layers of reinforcement[26]. The assembly is then pulled through a series of heated rollers that melt the matrix and reduce its viscosity sufficiently to enable impregnation. After a chilling roll, depending upon the thickness of the resulting laminate, the material is either collected on a rotating drum or sawn to a predetermined length. Double belt presses are often used (2–4 bar pressure) to achieve high-speed impregnation (5–8 m/min) and consolidation of semi-finished sheets[27], as is used commercially for GMTs[28] and textile thermoplastic composites[29]. Impregnation quality (void content) is dominated by the pressure–temperature history of the polymer and the speed of the lamination machine[30]. The significance of this depends upon the subsequent fabrication process (usually involving further void reduction) and the intended application. Melt impregnation is difficult or impractical for certain polymers due to their limited tolerance to the temperatures required for viscosity reduction. Thermal degradation, characterised by a reduction of molecular weight (where crosslinking does not occur), may initiate within a few degrees of the melt temperature, making this method of viscosity reduction impractical. While most systems[29] manufacture sheet product continuously, film stacking uses similar principles where fabrics are interleaved with polymer film and consolidated between heated dies or in a diaphragm-forming arrangement.

Table 6.1 Typical properties of unreinforced thermoplastic polymers used as matrix materials in textile composites (from MatWeb.com and adapted from ref. 24)

Unreinforced matrix	E_{ten} (GPa)	σ_{tens} (MPa)	$\varepsilon_{failure}$ (%)	ρ (g/m^3)	Izod impact (J/cm)	T_g (°C)	T_m (°C)	$T_{process}$ (°C)	€/kg
Polypropylene (PP)	1.4	31–42	100–600	0.91	31–42	–20	165	190–230	0.7
Polyethylene terephthalate (PET)	3.5	48–73	30–300	1.3	48–73	80	250	260–350	3.5
Polybutylene terephthalate (PBT)	2.5	56	50–300	1.3	56		235	240–275	3.2
Polyamide 6* (PA6)	3.0	35	min 50%	1.1	41–167	60	220	230–285	2.9
Polyamide 66* (PA66)	1.3	55	min 50%	1.1	51–96	70	260	270–325	3.1
Polyamide 12* (PA12)	1.1	40	min 50%	1.01	35–70	40	175	180–275	8.4
Polyphenylene sulfide (PPS)	3.4	48–87	1–4	1.35	48–87	90	285	310–335	8.3
Polyetherimide (PEI)	3.4	62–150	5–90	1.26	62–150	215	am	340–370	18–22
Polyvinylidene fluoride (PVDF)	1.0	24–51	12–600	1.76	43–80	–40	171	185–290	15
Polyetheretherketone (PEEK)	3.6	70–105	15–30	1.3	70–105	145	335	340–400	30–77

am = amorphous.
* As conditioned form

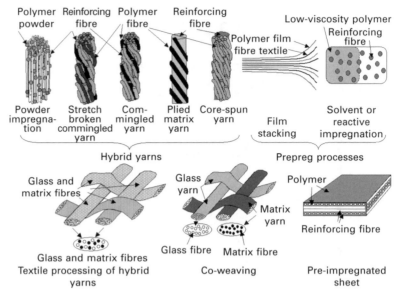

6.5 Techniques of combining thermoplastic polymers with reinforcing fibres for textile composites (adapted from Verpoest[25] and Wakeman et al.[8]).

6.3.3   Solvent impregnation

Solvent impregnation is an alternative to melt impregnation for generally amorphous thermoplastics, such as those used by Ten Cate[31]. Polymers suited to solution techniques include thermoplastic polyetherimide (PEI), polysulfone(PSU), polyethersulfone (PES) and polyphenyl sulfone (PPS) (http://www.matweb.com/reference/abbreviations.asp). The processing of a low-viscosity solution (*c.* 10wt% of polymer) facilitates impregnation and wetting of fibre tows or woven fabrics. However, handling problems occur including the containment of volatile organic compounds (VOCs) and the removal of residual volatiles (the final 5%) prior to fabrication[32]. Resistance to solvents is often poor and a sensitivity to stress cracking can result. When choosing composites based on amorphous resins, susceptibility to specific solvents needs to be assessed. Nevertheless, such materials are used in a range of engineering applications[31]. The most successful are those based on PEI, which has excellent fire, smoke and toxicity characteristics. The Ten Cate[31] CETEX® range of PEI-based composites has been used in a variety of aircraft applications with Airbus and Fokker.

6.3.4   Commingling

In addition to reducing the viscosity term by heating, minimisation of mass transfer distances also aids impregnation. Commingling is an established

Table 6.2 Suppliers and properties of textile thermoplastic composite materials

Material form		Reinforcement	Matrix	ρ (g/m^3)	σ_{tens} (MPa)	E_{ten} (GPa)	$\varepsilon_{failure}$ %	σ_{flex} (MPa)	$E_{flexural}$ (GPa)
Sandwich material	M_f%								
Quadrant	40	Glass	PP	1220	248	10.5	1.8	280	14.5
GMTex 4/1 (commingled fabric skins, GMT core)	40	0–90° glass	PP	1200	120	4.9	1.9	160	7.2
GMTex 1/1									
Experimental GF weave 4/1 + random glass core	30	Random core, fabric skins	PET	1620	184/82	12/6.9	1.8/1.5	–	–
Experimental 75% wt 4/1 commingled glass/PP + 500:m PP core	<75	PP film + fabric skins	PP	–	377	25	–	410	21
Aligned fibre materials	V_f%								
Plytron GN 638T (melt impregnated, surface = 0° direction)	35	[(0,90)$_2$]s	PP	1480	360	16	2.5	300	16.8
GN 638T (melt impregnated) tape	35	UD glass	PP	1480	720/11	28	1.9	436	21
Tech textiles Thermopreg non-crimp interdispered fibres, GF	35	Stitched 0/90 fabric	PP	1700	256	13	–	170	–
Vetrotex Twintex, commingled GF/PP	35	bal. 2 × 2 twill weave	PP	1490	300	13	–	275	12

Table 6.2 (Continued)

Material form		Reinforcement	Matrix	ρ (g/m³)	σ_{tens} (MPa)	E_{ten} (GPa)	$\varepsilon_{failure}$ %	σ_{flex} (MPa)	$E_{flexural}$ (GPa)
Vetrotex — Twintex, commingled GF/PP	35	4/1 weave	PP	1500	400	24/8	–	380/160	18/6
Twintex, commingled GF/PP	50	UD	PP	1750	700	38	–	400	32
Twintex, commingled GF/PET	50	1/1 weave	PET	1950	440	25	–	600	23
Twintex, commingled GF/PET	50	UD	PET	1900	870	37.6	–	1025	37.7
Experimental — GF/PET	48	Warp knit 0°	PET	–	87/6.6	8.2/3.5	–	747/25	35/4.6
GF/PBT	56		PBT	–	–	–	–	810	40.5
Porcher Ind. — Power impregnation at fabric level	52	2/2 twill carbon fabric	PA 12	–	–	–	–	603	44
Vestopreg	52	Bal. weave	PA 12	1850	350/350	26/26	1.6/1.6	500	22
Comfil — Commingled GF yarn	44	Twill 3/3	PET	–	260	22	–	–	–
Commingled GF yarn, modified PET with lower T_m	41	Twill 3/3	PET	–	280	20	–	–	–
Schappe Techniques — Stretch broken commingled fabric at 20 °C	56	5HS carbon fabric	PA 12	1440	801	61	1.3	606	52

Table 6.2 (Continued)

Material form	Reinforcement		Matrix	ρ	σ_{tens}	E_{ten}	$\varepsilon_{failure}$	σ_{flex}	$E_{flexural}$
---	---	---	(g/m³)	(MPa)	(GPa)	%	(MPa)	(GPa)	
EMS Grivory	Stretch broken commingled fabric at 80 °C	5HS carbon fabric	56	PA 12	1440	631	58	1.2	332	42
	PA12/carbon (anionically polymerised)	2/2 twill carbon fabric	54	PA 12	1430	790	62.6	1.3	–	–
Cyclics	Cyclic CBT	UD carbon	–	PBT	1600	1565	115	–	1310	–
Dow	Thermoplastic polyurethane	UD glass (76% M_f)	–	ETPU	1910	1000	45	2.2	1150	45
Bond Laminates	TEPEX 101-glass FG290 (melt impregnated)	Glass twill 2/2	47	PA 66	1800	380	21	2.0	570	22
	TEPEX 201-carbon C200 (melt impregnated)	Carbon twill 2/2	48	PA 66	1500	755	52	2.1	745	44
BP	Curv (PP/PP composite sheet)	Polypropylene	–	PP	900	180	5	–	–	–
Gurit	SUPreM (powder impregnated UD 1 × 13 mm tape)	Carbon, UD	60	PEEK	–	2800	142	1.9	2000	120
TenCate	CETEX PEI-glass	7781 glass fabric	50	PEI	1910	465	30	–	625	26
	CETEX PEI-carbon	5H satin carbon fabric	50	PEI	1510	665	56	–	830	47

Table 6.2 (Continued)

Material form		Reinforcement	Matrix	ρ (g/m³)	σtens (MPa)	Eten %	εfailure (MPa)	σflex (GPa)	Eflexural
TenCate	CETEX PPS–carbon	50	5H satin carbon fabric	PPS	1550	660	54	–	–	–
Cytec	APC-2 (melt impregnated) AS4 carbon	63	UD carbon	PEEK	–	2400/82	151/9.9	–	2280	138

UD, unidirectional; GMTex, textile reinforced glass mat thermoplastic; 500:m PP core, core of polypropylene 500 micrometers thick textile process for intimately blending reinforcing fibres and matrix in a fibrous form. With the short flow lengths given by the hybrid yarn structure, the secondary consolidation and impregnation phases may occur at substantially lower pressure than those required for conventional melt impregnation. Since the hybrid yarns are both flexible and relatively stable, they are suitable for further textile processing (e.g. weaving and braiding) and the arising fabrics can be formed conveniently at room temperature. Also, given that the yarns are (generally) continuous, the reinforcement to matrix ratio can be controlled closely via the filament size and count of the two yarns. In general, the quality of the laminate is controlled by the intimacy of the blend and the uniformity of the hybrid yarn.

A patented commingling process[33, 34] has been developed by Saint-Gobain Vetrotex for glass and polypropylene yarn (Twintex®). The manufacturing route consists of drawing and sizing glass fibres (as normal), which are then passed to an annular drawing head that is supplied with molten polypropylene from an extruder. Under pressure, the molten matrix passes through orifices that are distributed over an annular die plate and then drawn into multiple continuous filaments that converge with the glass fibres to produce commingled yarn at rates of 10–30 m/s. Blowers control air turbulence. Several strategies can be adopted to maintain yarn integrity including the use of film formers or encapsulants. Where photoactive size is used, the application of ultraviolet radiation prior to the winding phase triggers a chemical transformation of the size to improve the bonding of the composite yarn. This is especially important when downstream processes such as braiding are planned, since any rewinding operation tends to separate the constituent fibres depending on their different moduli.

An alternative approach patented by Schappe blends stretch-broken carbon fibres with polyamide 12 (PA12) staple fibres (EMS-GRIVORY, a business unit of EMS-CHEMIE AG) with the addition of a wrapping filament[35], which is available commercially. Multifilaments of reinforcement fibre (typically carbon) are stretch broken by passing yarns through rollers, where the roller speed varies between zones, producing a strip of discontinuous fibres of a controllable length (70 mm average). These fibres are then transformed on standard spinning equipment into the stretch-broken yarn[36]. The staple matrix fibres and the stretch-broken reinforcement are then brought together via a machine having an additional stretching zone that produces an inhomogeneous strip of reinforcing and matrix fibres. Each strip is then combined with ten other strips on a similar machine and repeated a nominal four times. The strip obtained is then wrap-spun on a long fibre laminating system before final wrapping of the parallel fibres by a continuous multifilament of the same matrix as the commingled yarn. This system of yarn production eliminates the formation of fibres into helices that occurs during conventional twisted yarn spinning techniques. A higher degree of mingling is claimed for this process than for competing methods. Additionally, the discontinuous fibres provide additional compliance during forming.

A final approach to commingling[37], developed for carbon and polyetheretherketone (PEEK) systems, involves separating the two yarns into open ribbons. The PEEK yarn cannot be opened via an air curtain and is instead passed around polytetrafluoroethene (PTFE) rollers to induce a charge of 6 kV, causing the carbon filaments to repel. The charged bundle is then drawn into a flat ribbon. The carbon yarn is opened with an air curtain into a ribbon that is then combined with the PEEK ribbon at a commingling bar at rates of approximately 0.5 m/s. This, it is suggested, gives a homogeneous distribution. From the commingling bar, the hybridised fibres are coated with size and wrapped on a take-up roll to form the final product. Mechanical separating elements that would limit commingling speed are eliminated.

6.3.5   Powder impregnation

The powder-coating approach blends fibres and matrix on an intra-bundle basis[38–40] and is arguably more versatile than commingling in that powders may be more readily available than thermoplastic fibres. The process deposits and distributes the powder, most variants of which involve a suspension of particles in a fluid stream. The solid particles adhere to the fibres mainly by electrostatic forces. Success has been reported for fluidised bed processing, aqueous slurries and electrostatic methods of coating. In each case, the yarns are spread prior to coating to reduce the mass transfer distances. Individual yarns are then passed through an oven to sinter the powder on the fibre surface, creating resin bridges and a semi-impregnated tow that is then fully impregnated during the final consolidation stage. An advantage cited for powder impregnation is the dominant flow direction along the fibre axis that is in the direction of higher permeability (opposed to flow normal to the fibre with commingled yarn). Ideally, particles should be at a minimum the same diameter as the fibres, with the consolidation time proportional to the sixth power of the particle radius. The FIT (fibre-impregnated thermoplastic) process, initially developed by Atochem, impregnated fibre systems with polymer powder before extruding the same polymer as a sheath around the bundle. The yarn remained sufficiently flexible for textile processing. Commercial products[40, 41] include unidirectional and cross-ply tapes and broad goods. Fibre contents range typically from 30% to 65%.

An alternative commercial approach involves powder coating unidirectional (UD) tows directly or fabrics after weaving[42]. In the case of fabrics, they are passed from rollers in a continuous (e.g. 10 m/min) process under a first powder dispenser followed by infrared heaters before a second powder dispenser and a final set of infrared heaters. The material is removed on a take-up roll and fabrics of up to 1000 g/m^2 (glass) and 600 g/m^2 (carbon) have been processed successfully at V_f of up to 52% with polymers including PP, PA6, PA12, PPS, PEI and PEEK. Typical particle sizes are 25–100 μm. As with competing methods, the coated fabrics may then be consolidated further, laminated alternately, or the initially set product heated in a frame and stamped directly.

6.3.6   Co-weaving and warp-knitting techniques

Other techniques, which may be economic in specific circumstances, adapt existing textile technology and include co-weaving, inter-dispersed fabrics, split-film knits and plied matrix systems[25]. Co-weaving combines yarns of the two fibres into a form possessing good drape characteristics, although an uneven reinforcement to matrix distribution may occur with generally longer impregnation times compared with a commingled system for a given set of processing parameters. The plied matrix technique[43] uses a polymer fibre woven around a bundle of reinforcement fibres offering improved distribution over the co-weaving technique. Glass and PP yarns have been inter-dispersed on line during fabric formation to produce drapable stitch bonded fabrics[44].

Another example is the warp knitting of polymer films with weft insertion of reinforcement fibres[45, 46]. Split films are cheaper than conventional (spun) polymer fibres and are cut lengthwise into tapes that can be processed in a similar way to yarns. Weft-insertion warp knitting has been used in this way to produce multi-axial, non-crimp glass–PET (polyethylene terephthalate) fabrics. Typically, the split films are inserted in the weft, although there is no particular reason why they could not be warped also. A light stitching yarn (typically PET) provides fabric stability in the usual way. During processing, the textile is restrained during the preheat phase to resist shrinkage of the oriented polymer chains in the stretched ribbon as a return to the random coil orientation occurs.

6.3.7   *In-situ* polymerisation

An alternative solution is to impregnate the fibre bed using low-viscosity monomers (viscosities in the mPas range) with subsequent *in-situ* polymerisation that commences with activation of the monomer. Hence impregnation must occur before the viscosity reaches a limiting level. When polymerisation is complete, depending on the material and time–temperature transformation relationship, the part is cooled and the component released. Based on these principles, three main systems based around reaction injection moulding (RIM) technology have been adapted to impregnate textile composite structures via resin transfer moulding (RTM) or pultrusion processes, where monomer, activator and initiator systems are injected into a mould where polymerisation is completed in a few seconds or minutes, with the polymerisation times generally shorter than for thermoset resin cure cycles[47].

Polyurethanes

When mixed, isocyanates and alcohols react to form polyurethanes, which are found as cross-linked thermoset or linear thermoplastic resins, depending on the selected molecular chemistry and the polymerisation process[48]. Dow has developed thermoplastic polyurethanes for a reactive pultrusion process, known commercially as 'Fulcrum®'[49]. Here the thermoplastic polyurethane resin, which can be repeatedly processed, shows a rapid molecular weight reduction when heated into the processing range. The lower molecular weight melt also has a low viscosity, enabling impregnation of glass fibres in the pultrusion process. After the impregnation stage, cooling rebuilds the molecular weight to a similar high level as the starting point. Reactive pultrusion line speeds of 10 m/min have been achieved. Additionally, the urethane chemistry is reported to give good fibre–matrix adhesion. Materials can be recycled via injection moulding.

Anionic polymerisation of polyamides

Polyamides, denoted by the number of carbon atoms in the monomer unit, can be obtained by the ring-opening polymerisation of lactams[50–52], representing more than 50% of polyamide production. Anionic mechanisms, enabling fast polymerisation, have been applied commercially (NyRIM) by DSM to polyamide via the RIM technique. This has been used for manufacturing articles via gravity casting, rotational moulding and RIM, with different initiators and activators. This material system has also been used to impregnate glass fibre textile structures in a RTM process, giving a glass fibre PA6 composite.

Traditional anionic polymerisation required the combination of two volumes of lactam preblended with catalyst and activator, resulting in a short pot-life where the materials slowly polymerised in the tanks. This was overcome through the development of a liquid activator system by EMS Chemie[50] that contains both activator and catalyst in solution. This has enabled the processing of laurolactam by an anionic ring-opening polymerisation reaction to produce either a fibre reinforced PA12 part directly[53] or a thermoformable pre-preg[54, 55]. Polymerisation can be completed within minutes, with the type and concentration of initiator and activator together with the temperature influencing polymerisation kinetics and the final molecular weight. At 200 °C, the viscosity of this material prior to polymerisation is 1 mPa s whereas the molten PA12 has a viscosity at this temperature of 50 000 Pa s.

Cyclics technology for PBT, PET and PC

An alternative reactive thermoplastic system is produced by 'cyclics' technology[56]. Reinforcement is impregnated with a low-viscosity thermoplastic resin (prepolymer) that subsequently reacts in the presence of heat and a catalyst to increase its molecular weight via conversion of its short molecular chains to a linear structure in a ring-opening polymerisation reaction. This technology has been demonstrated for polybutylene terepthalate (PBT) and polycarbonate (PC). Two different product forms are available: one-part systems that have the resin and catalyst premixed, and two-part systems where the resin and catalyst are supplied separately and mixed during processing. The prepolymer (CBT^{TM} Resins) starts as a solid at room temperature, melts at 150 °C to a low viscosity of 150 mPa s, reducing further to below 20 mPa s at 180 °C. When mixed with specific tin or titanium catalysts, the PBT rings in cyclical form open and connect (i.e. polymerise) to form high molecular weight PBT thermoplastic. Full polymerisation is reported to occur from tens of seconds to many minutes, depending on the temperature and type of catalyst used.

6.4   Processing routes

6.4.1   Process windows for textile-based thermoplastic composites

A typical processing cycle for a composite with a thermoplastic matrix consists of heating the constituents above the matrix T_m, applying pressure for a time period to ensure appropriate consolidation, and finally cooling the material while maintaining the pressure. From a manufacturing perspective, considering equipment investment cost and production rate, it is desirable to process the part at the lowest pressures and with the shortest cycle times. Reducing cycle times requires exposing the composite to high heating rates prior to final shaping and high cooling rates during the solidification stage of the processing cycle.

Optimal processing conditions, in terms of pressure and cooling rate, may be defined by a processing window (Fig. 6.6)[1]. The lower cooling rate limit is fixed by economic constraints, since long cycle times are costly. At high cooling rates, a considerable thermal gradient is imposed over the thickness of the composite part, which leads to both morphological skin–core effects and thermal skin–core stresses. If the polymer is exposed to excessively high temperatures during an extended time, thermal degradation mechanisms become increasingly important (and undesirable). Relations between different defect mechanisms, often governed by the internal stress build-up (voiding, microcracking) and cooling rates have been observed. An increased forming pressure, corresponding to the upper right boundary of the diagram, may

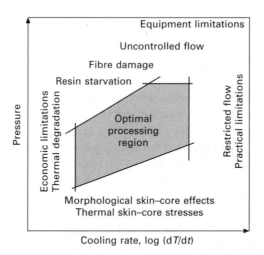

6.6 Generic process window for thermoplastic textile composites (adapted from Månson *et al.*[1]). T = temperature, t = time.

suppress this defect initiation. The forming pressure upper boundary will be governed by the practical limit of forces applied by the forming equipment, as well as by the increased tooling cost for high-pressure processes. High forming pressures may also cause damage to the fibre bed (and changing permeability) as well as leading to local resin fraction reductions due to a high degree of in-plane macro-scale resin flow from the mechanically locked fibre bed.

Based around these boundary conditions, a wide variety of processing routes have been developed to process thermoplastic composites. These offer different combinations of cycle time, material suitability (semi-finished products such as partially impregnated sheets or as textile structures composed of, for example, commingled yarns) and the investment in capital equipment needed.

6.4.2   Low-pressure processing techniques

A range of low-pressure processing techniques have been developed that rely on the intimate fibre and matrix distribution before processing, either in a 'dry' commingled yarn material, or a preconsolidated laminate. The lower pressures reduce the investment in tooling and forming equipment, which is an important consideration for lower manufacturing volumes.

Vacuum forming of thermoplastic composites

Thermoplastic composites can be processed by vacuum forming in a similar manner to thermoset composites. Vacuum forming also enables thermoplastic composites to be used in a hand lay-up type process. The produced part has only one surface with a finish that reflects the surface of the mould. Cycle times are a function of the heating route and sample size, ranging from 60 s[57] for carbon fibre (CF) reinforced PEEK, to 15 mins for glass fibre (GF)/PP plates[58] and to 24 h for a 7 m rigid inflatable boat[59]. With a two-shift pattern, a typical production volume is 30 parts per day[60].

Taking the example of commingled yarn-based textiles, the first process step is to position the material into a mould. Techniques employed to hold 'dry' commingled fabrics to vertical mould walls and to subsequent layers include the use of hot air guns to 'tack' layers together and, for polypropylene-based materials, a pressure-sensitive polypropylene-based adhesive that is fabric coated onto the commingled textile, together with low molecular weight polyolefin spray contact adhesives (e.g. 3M-77). Core materials, which resist collapse at the composite process temperature and pressure, can also be incorporated, with typical materials including foams (e.g. Airex KapexC51), aluminium honeycomb and balsa wood. Release agents are often used to facilitate part release.

For well-consolidated parts (low porosity), a perforated peel ply (100 g/m^2) and breather felt (non-woven nylon or nylon-PET of 100–300 g/m^2) are used beneath the vacuum bag, which enables the air to be fully evacuated. The peel ply is in direct contact with the thermoplastic composite to prevent adhesion between the breather felt and the moulded material. The vacuum bag can consist of several materials, which for the case of polypropylene-based composites must have a melting temperature above 200 °C. Depending upon production volumes, a flexible nylon vacuum bag (50–75 µm) or a pre-shaped reusable silicone rubber bag is sealed against the surface of the tool. Reusable silicon membranes can be used to replace films, reducing consumable costs and the time needed to position the vacuum system. Typically only 100–150 parts total production volume are needed to offset the higher initial costs. Preformed silicon rubber forms enable a male tool to be used.

With a vacuum approaching 1 bar being applied for the whole processing cycle, the tool and material are heated under vacuum to the polymer's processing temperature (e.g. 185–190° C for PP), which is maintained for typically 10–20 min (dependent on the textile fibre and matrix distribution) for full impregnation. The tool is then cooled to below 70 °C (to reduce distortion and ease handling) with the vacuum pressure still applied, after which the vacuum bag is removed and the part released.

As this is an isothermal process where the mould temperature is cycled above and below the composite matrix T_m, the overall mould mass that is heated and cooled needs to be minimised. The mould material chosen depends on the number of parts to be produced, the cycle time desired, and the surface finish needed. Four general types of mould are used: high-temperature epoxy composite (3000–6000 parts/mould), welded metal sheet moulds (5000–

10 000 parts/mould), cast or machined aluminium (10 000–30 000 parts/ mould) and electroformed nickel shell (50 000–100 000 parts/mould)[60].

Several methods are available to heat the evacuated tool depending on the size of the part, with the important requirement of a homogeneous temperature distribution. Heated tools avoid the need for a heated enclosure but are part-specific and generally used for part areas of 1–3 m² (Fig. 6.7a). However, tool durability is reduced due to differential expansion between the different mould materials and heating systems. Ovens offer a clean heating route, enabling a variety of components, including large structures, to be made from one heating system (Fig. 6.7b).

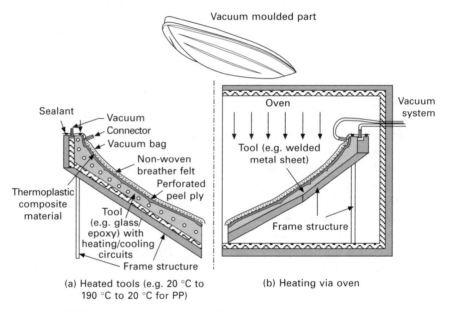

(a) Heated tools (e.g. 20 °C to 190 °C to 20 °C for PP)

(b) Heating via oven

6.7 Vacuum bag processing of thermoplastic composites.

As an example of the vacuum moulding process, a rigid inflatable boat hull[59] has been produced from commingled glass and PP textiles. A mould material was needed that would withstand 500 production cycles. The contoured base is a cast mould equipped with heating and cooling systems. The texture and colour of the component can be changed using thermoplastic films that are placed on the mould surface. For such a production volume, neither metal nor nickel-shell tooling was cost effective and hence composite tooling was developed using a high-temperature divinylmethane diisocyanate tooling resin. This produces a high-quality gel coat surface and can be highly filled with metallic powder to improve the thermal conductivity. Where appropriate, a heating system can be integrated into the mould by either laminating in PTFE-coated resistance wires (e.g. 2500 W/m²) or copper tubes for hot oil.

In the case of the electrically heated mould, the resistance wires are fabricated into a grid that is draped just beneath the mould surface. Using an oil heating system, 40 min were required to heat commingled fabric across the boat structure to melt temperature. Pressure was applied for a further 40 min before the cooling cycle of 20 min where cold water was passed through the integrated copper tubes, giving a total process cycle time of approximately 100 min. This was reduced to 45 min for the electrically heated mould.

Autoclave forming of thermoplastic composites

In an extension of the vacuum moulding process, autoclaves offer similar advantages to ovens during the heating stage (heating of the assembled mould, textile composite and vacuum bag) with the ability to apply additional external pressure from 3.5 to 7 bar. This reduces void content and offers closer thickness tolerances, but reduces the number of parts made per day (Fig. 6.8)[61]. Forming is performed by raising the autoclave temperature to above the matrix T_m followed by an increase of the autoclave pressure to effect forming and finally cooling under consolidation pressure. During forming, the mould is subjected to hydrostatic pressure and the only force tending to distort the mould is the reaction to the diaphragm stretching, which in the case of polymeric diaphragms is very low. Thin section moulds with low thermal mass can be used.

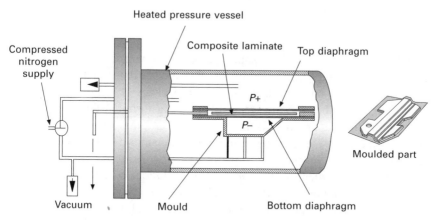

6.8 Autoclave processing (double diaphragm forming) of thermoplastic composites (adapted from Mallon *et al.*[61]).

Diaphragm forming

Diaphragm forming is a technique that consists of placing the laminate between two thin plastically deformable sheets, known as diaphragms. The diaphragms, when clamped around the mould edge, maintain biaxial tension on the laminate during deformation, consequently restricting laminate wrinkling, splitting and thin spots. The normal procedure consists of heating the whole mould, laminate and diaphragms to above the laminate polymer T_m, generally limiting process cycle times to 30 min or more. The combination of air pressure above the diaphragm, and a vacuum drawn from beneath the diaphragm forces the laminate to the mould geometry enabling the production of complex parts. Diaphragm forming can be performed either using compressed air or an autoclave. As an alternative to using two diaphragms, a single diaphragm placed over the preform can effect forming in a similar manner to double diaphragm forming (Fig. 6.9).

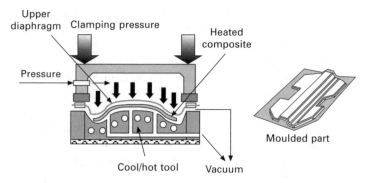

6.9 Isothermal single diaphragm forming of thermoplastic composites.

The diaphragms most commonly used are either polymeric or metallic[62]. Metallic diaphragms, such as superplastic aluminium, have been used for forming APC-2 (CF/PEEK) laminates. Polyimide films, such as Upilex-R (0.13 mm thick), have good elongation properties at temperatures up to 400 °C and have been used extensively. During forming, the laminate can slide within the diaphragms. As the diaphragms are stretched, surface friction between the laminate and the diaphragms transmits tensile forces to the laminate, creating out-of-plane support and suppressing wrinkling.

Recent developments have applied this technique to commingled glass and polypropylene textiles where panels have been produced using 0.5–1.0 mm thick silicone sheets. The composite fabrics and two silicon sheets are preheated in an initial stage between hot platens (210 °C) before rapid transfer to the forming unit and the application of positive pressure (7 bar) to form the material into a 'female' tool (Fig. 6.10). The silicon membranes are reusable, for between 500 and 1000 parts in polypropylene. This reduces cycle times to typically 2 min, as the temperature of the one-sided mould does not need to be thermally cycled.

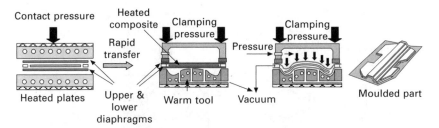

Contact pressure | Heated composite | Clamping pressure | Clamping pressure

Rapid transfer

Pressure

Heated plates | Upper & lower diaphragms | Warm tool | Vacuum | Moulded part

6.10 Non-isothermal double diaphragm forming of thermoplastic composites – external heating in hot plates and rapid transfer.

Bladder inflation moulding

The bladder inflation moulding (BIM) technique enables production of composite components with hollow geometries that have both high geometrical complexity and high intrinsic stiffness[63–66]. BIM uses an outer mould and inner bladder to define the shape of the part. The low process pressures (~10 bar) permit the use of simple tooling, giving a process for producing thin-walled hollow composite parts with evolving cross-sections. Previous work has demonstrated bladder inflation moulding for small generic components, bicycle frames and handlebars, tennis rackets, suspension arms and pressure vessels[65]. Two process variants of the BIM technique have been developed: (i) an isothermal process where the mould temperature and thereby the composite temperature is cycled above and below the composite matrix T_m (Fig. 6.11) and (ii) a non-isothermal process where the thermoplastic composite textile is preheated above the matrix T_m outside of the mould and then transferred rapidly to the cool mould for bladder moulding (Fig. 6.12).

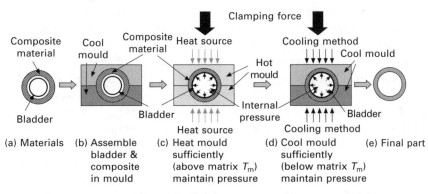

Clamping force

Composite material | Cool mould | Composite material | Heat source | Cooling method | Cool mould

Hot mould

Internal pressure

Bladder | Bladder | Heat source | Cooling method | Bladder

(a) Materials | (b) Assemble bladder & composite in mould | (c) Heat mould sufficiently (above matrix T_m) maintain pressure | (d) Cool mould sufficiently (below matrix T_m) maintain pressure | (e) Final part

6.11 A cross-sectional sequential view of the isothermal bladder inflation moulding process.

The isothermal BIM (isoBIM) process (Fig. 6.11) consists of assembling a thermoplastic composite braid around a bladder, which is pressurised

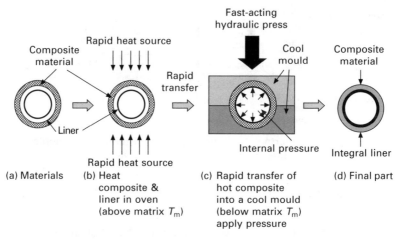

6.12 A cross-sectional sequential view of the non-isothermal bladder inflation moulding process.

internally via compressed gas. The bladder elongates elastically (silicon rubber) or plastically (thermoplastic extrusion blow mouldings) under the influence of the applied internal pressure and the braid is compressed between the bladder and a female mould. By heating the mould, the braid temperature is increased sufficiently above the matrix T_m in order to lower the matrix viscosity while maintaining an internal pressure inside the bladder to drive impregnation and consolidation. At process temperature, the pressure is maintained for the required period, after which the mould is cooled below the matrix T_m for solidification and crystallisation to occur. The internal pressure is then released and the part removed. In one example[65], eight-litre vessels (160 mm diameter) were produced from carbon fibre and PA12 commingled yarns in 12–15 min. The PA12 bladder, which formed an integral liner, was produced via extrusion blow-moulding.

To further reduce cycle times, suiting higher manufacturing volumes, a non-isothermal BIM (non-isoBIM) process has been developed[65, 66] that eliminates the tool thermal cycling needed for the isoBIM process (e.g. 20 °C to 260 °C to 80 °C) (Fig. 6.12). The assembled PA12/CF commingled yarn and PA12 bladder were heated directly in a combined forced hot air convection (heating the bladder) and IR oven (heating the external braid to higher temperatures) and then transferred rapidly using a sled and rail system into a cool mould, maintained below the polymer melt temperature. A hydraulic press was used to close the tool and internal pressure was applied inside the bladder via a rapid pneumatic connection system to produce the part. This reduced in-mould cycle times to 140 s for a generic pressure vessel component of 85 mm diameter and 2 mm wall thickness.

Reactive thermoplastic RTM

RTM techniques that are commonly used for thermosetting resins are also applicable to reactive thermoplastic resins. Thermoplastic RTM (TP-RTM) has been used to produce both generic plates[54, 55] and sections of a car floor pan[53]. As an example, TP-RTM of a PA12 material consisted of the following steps, which are shown schematically in Fig. 6.13. The injection unit consists of two tanks, two gear pumps and two pipes conveying respectively the activator and the monomer from the tanks to the mixing head. The monomer is in the molten state; therefore tank, pump and pipes are maintained at 180 °C during processing. The liquid system remains at room temperature to avoid degradation. Both materials need to be held under nitrogen. Immediately prior to injection, mixing of the molten monomer and the liquid activator system (2.5%) occurs in a static mixing head.

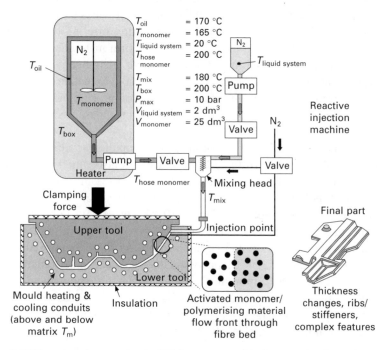

6.13 Reactive thermoplastic RTM process, with the reactive injection machine illustrated for APLC12 materials (PA12 final polymer).

This particular monomer and activator system requires isothermal processing for economic processing. The mould temperature needs to be cycled from above the monomer T_m of 150 °C to a peak ranging from 200 °C to 250 °C (with higher temperatures increasing the polymerisation rate), followed by cooling sufficiently below the PA12 polymer T_m of 175 °C for demoulding.

An aluminium mould was used that enabled the production and study of 450 mm × 550 mm plates of 1–3 mm thickness. Using oil heating, less than 30 min was required to reach a mould temperature of 200 °C, and with water cooling, temperatures were reduced from 200 °C to 50 °C in 5 min.

The TP–RTM cycle consisted of the following steps. First, a release agent compatible with the anionic polymerisation was applied to the mould and two layers of a dried 5-harness satin carbon weave (440 g/m^2) laid in the mould cavity [0/90]$_s$. Nitrogen was then flushed through the closed mould to remove oxygen and residual moisture. As the mould temperature exceeded 180 °C, activated monomer was injected into the mould. When the mould cavity was full with polymerising material, inlets and outlets were closed and heating maintained to enable complete polymerisation. The polymerised material then solidified during cooling, before final demoulding. The whole processing cycle required less than 60 min, with this being reduced to 15 min for a tool with faster heating circuits. This process has been used to produce a section of an automotive floor pan structure, offering the advantages of increased toughness and the ability to post form materials[53].

Injection-compression techniques have also been used to control crystallisation-based shrinkage and associated void formation to reduce porosity levels from 10% to 1%. Injection-compression processing applies a typical surface pressure of 55 bar during the cooling phase, using a shear edge tool.

The polymerisation of cyclic butylene terephthalate oligomers to form a PBT material system (offered by Cyclics[56]) is also suited to TP-RTM. The prepolymers can be processed without cycling the mould temperature and will melt, polymerise and solidify when processed between 180 and 200 °C. With a prepolymer melting point of 150 °C, a processing temperature (mould temperature) of 180–190 °C is reported to give a compromise between higher rates of polymerisation and crystallisation. As the polymerised PBT has a melt temperature of 220 °C, the mould does not need to be cooled before part release. Cyclics® Corporation[56] claims to have demonstrated the use of CBT^{TM} resins in the following textile composite processes: compression moulding, vacuum bagging, resin film infusion (RFI), RIM structural reaction injection moulding (SRIM), RTM and pultrusion.

6.4.3   High-pressure processing

Press-based processes, often involving separate preheating of the composite and then moulding in a cool tool, require higher forming pressures due to the short residence time in the tool under pressure during which final impregnation and consolidation need to occur. These processes have evolved to reduce costs and decrease cycle times for high-volume manufacturing. The alternative (isothermal) technique of cycling the tool temperature is less suited to high production volumes, requiring large transfers of energy and resulting in generally longer cycle times and higher operating costs.

During non-isothermal processing, drapable textiles are still essential for maximum formability, but there is no longer a need for the composite product to have tack or drape at room temperature (forming occurs above matrix T_m), and a degree of mechanical handling integrity is desirable. The choice of fully impregnated, partially impregnated or unconsolidated material forms is an interaction of economics (eliminating pre-consolidation costs *vs.* potentially lower stamping times) and mechanical properties (cycle time *vs.* void content) and of practical handling considerations (degree of automation). Pre-consolidation eliminates the need to heat multiple layers of fabric, which must be assembled before stamping. Where fully pre-consolidated materials are used, the final stamping stage forces the material to conform to the mould geometry and (ideally) to eliminate any deconsolidation introduced during preheating. A comparison of stamp-forming with unconsolidated (dry commingled textile), partially pre-consolidated (40% porosity) and fully pre-consolidated CF/PA12 commingled yarn material showed increased laminate properties and reduced porosities for increasing levels of pre-consolidation, with the effect strongest at lower forming temperatures[54].

The following processing techniques, described after a brief examination of common preheating and blank holder technologies, are essentially all stamp-forming variants. After heating of the material to processing temperature, the primary forming mechanisms are the simultaneous phases of[9] intra-ply shearing to facilitate fabric deformation, inter-ply shearing for shape accommodation (for two or more plies), viscous-friction between the plies and tool surfaces, pressure application for consolidation and removal of heat to permit solidification of the matrix.

Preheating technology for stamp-forming processes

Established industrial preheating techniques include infrared, hot air and air impingement heating, with contact heating, microwave, radio frequency and resistance heating (for carbon materials) under development[67–69]. Infrared heating[70] offers several advantages, including fast absorber heat-up times, programmable heating, controllability and higher energy efficiency. Problems occur with surface and through-thickness temperature inhomogeneities, which are oven design issues[71]. Circulated hot air ovens[72] are best used for porous materials or pre-consolidated sheets that strongly deconsolidate. They have longer response times compared with infrared, but limit degradation from high flux levels at the composite surface and generally give a more even surface temperature distribution. Additionally, inert gases can be used to further reduce degradation. Air impingement ovens direct hot air at high velocity towards the composite blank and limit the upper temperature.

Infrared oven emitters radiate electromagnetic energy in the infrared region, with the laminate reflecting, absorbing or transmitting this energy. Heat transfer occurs when radiation is absorbed, increasing the random molecular energy. For example, carbon black pigmented PP shows minimal transmission, with 3% reflection and 97% absorption, irrespective of the wavelength, over the range of 0.8–3 μm[73]. This analysis needs to be performed for the specific polymer in question. Therefore, only as much energy can be delivered to the surface as can be transferred into the blank, without the surface overheating. The advantages of short wave emitters for PP are therefore limited to reducing response times and hence increasing the control potential.

A preheat processing window (Fig. 6.14) can be established as a period where the mid-plane of the laminate has reached the desired processing temperature for impregnation but without degradation at the outer surfaces. Different textile composite forms exhibit different heat transfer characteristics, for example, the range of pre-consolidation levels between a 'dry' commingled textile and a fully pre-impregnated laminate. Additionally, fully impregnated materials (fibre architecture dependent) will deconsolidate during heating[74] due to release of strain energy in the reinforcing fibres, the condensation of dissolved gas into voids of finite size[17] and fibre to matrix adhesion effects. Hence during heating, the centre of the composite can become insulated because of its lofted structure. High matrix viscosities require heating to the maximum temperature possible to facilitate impregnation, while avoiding oxidative degradation, which results in chain scission of the macromolecules, and hence a reduction in the molecular weight. Matrix degradation occurs as a function of both temperature and time at temperature and frequently phosphite

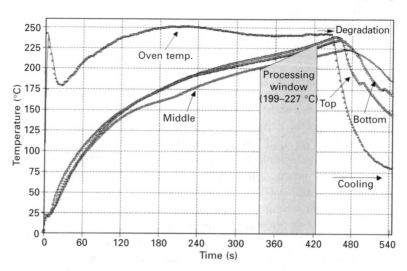

6.14 Example of a preheat window for PP-based thermoplastic composites.

antioxidants are used[75] to minimise these effects during both preheating and the service life of subsequent products. As an example, gel permeation chromatography (GPC) has been used to define the upper processing window limit (with a given stabiliser package) for glass and PP laminates[73]. Higher laminate temperatures reduced the time to the onset of degradation. For a standard infrared oven and for a laminate temperature of 220 °C, the preheat time adopted for the fabric blank of 145 s was within the onset of degradation at 200 s. At laminate temperatures of 180 °C, preheat times of greater than 10 min were possible.

Engineering polymers, such as PEI, require pre-drying before the main heating cycle, for example 4 h at 150 °C, to remove residual moisture. This is not necessary for commodity polymers such as PP.

Blank-holders and membrane forces

Blank-holders are a common feature of the stamp-forming processes discussed below, performing three principal functions. Due to thermoplastic memory effects, the matrix phase of certain composite textiles shrinks during preheating, requiring a blank-holder to hold the material. Additionally, to maintain fibre orientations correctly in the final part, an automated route is needed to transfer the hot and hence conformable material from the oven to the tool. Manual transfer is unsuitable because of the rapid convective heat loss and also localised conductive heat loss where the operator grips the laminate. Blank-holders thus restrain the material during preheating and also enable rapid (<10 s) transfer to the cool mould, which has been minimised in practice to 2–3 s[10, 29].

To maximise fabric drapability and reduce forming faults during moulding, a membrane stress should be applied to the fabric by a blank-holder, permitting higher shearing angles without visible wrinkling. For example, a 20 MPa membrane stress increased fabric shear angles from 35° to 60° without visible wrinkling[10]. As the fabric conforms to the tool, it is pulled through the blank-holder when the force needed to overcome the viscous friction at the fabric/blank-holder contact interface is less than the force needed to continue intra-ply shearing. The clamping load is transmitted to the fibre through the shear action of the resin, which will act as a lubricant depending on matrix temperature. Resistance at the tool/ply and ply/blank-holder interfaces is a function of the matrix film viscosity and Coulomb friction from the reinforcing fibre and tool contact. Increased pullout velocities and normal pressures increase the shear stress. Blank-holder temperatures should be lower than the material bulk[76]; for PP-based composites, a blank-holder temperature of 125 °C improved quality versus 50 °C. An alternative to planar blank-holders is to clamp the fabric between rollers at selected zones, with the knurled roller surface fixing but not damaging the sheets[10]. The torque resistance of the rollers inhibits draw-in of the material into the mould, with the normal pressure on the rollers controlled via pneumatic cylinders to vary retaining forces as a function of draw depth. This reduces heat transfer from the hot sheet to the blank-holder by the small contact surface of the rollers, reducing residual stresses where high membrane tensions are used.

Elastomeric tool forming

The upper tool used in this process consists of an elastomeric pad that can be contoured to match the shape of the lower mould for forming complex shaped components (Fig. 6.15). Elastomeric moulds deform elastically to a rigid lower mould under relatively low loading, but with low compressibility (constant volume). In the locked situation, elastomeric tools behave near hydrostatically, where an increase in the applied force on the mould transmits to a general increase in surface pressure on the whole laminate. Consolidation quality is influenced less by the part geometry than where two rigid moulds are used. Rubber moulds can be created from polyurethane or cast silicon rubber types (e.g. GE RTV630, offering high strength and tear resistance with inherent release capabilities). The laminate is preheated to above the matrix T_m and transferred rapidly into the mould. Blank-holders are used to control wrinkling. Closing of the mould deforms the laminate into the required shape and subsequently the component cools and solidifies within the mould. Marginally longer cycle times are needed than for matched metal moulds because of the lower thermal conductivity of the rubber mould. By changing the rubber tool geometry, the pressure distribution around the mould can be varied. The flexible nature of the mould accommodates composite material thickness variations caused by deformation of the laminate during forming.

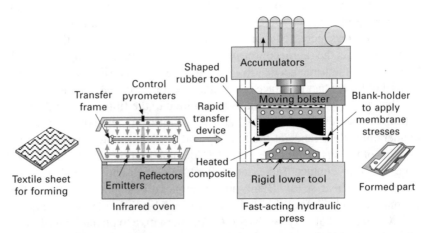

6.15 Use of shaped elastomeric tooling for the stamping–forming of textile composites.

Elastomeric moulds are generally less suited to very high production volumes due to the reduced tool life compared with metallic moulds (contact with hot laminates and a low rubber thermal conductivity)[12].

The rubber pad may also be in the form of a rubber block (Fig. 6.16). This is generally suited to forming only simple shaped components, due to the limited local deformation of the rubber pad. Rubber block pressing provides high normal forces and surface tension but these are not uniformly distributed over the laminate. The local forming forces are determined by the shape of the die only, which counteracts closure of the press. Such processes have been widely used in production for Airbus and Fokker aerospace applications, in PEI and PPS-based textile composites (CETEX®).

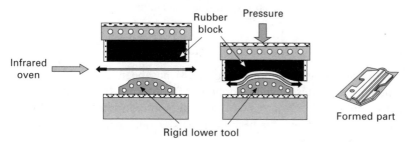

6.16 Deformable rubber block tooling for the stamping–forming of textile composites.

Hydroforming

The hydroforming process (Fig. 6.17) utilises fluid pressure (hydraulic oil) behind a rubber diaphragm to force a sheet of material into the shape of an adjoining mould. The cut textile preform is first located on the diaphragm that is fixed to the fluid filled chamber. The draw-cap is fixed in position and the punch lowered to form the part against the resistance of the fluid pressure. Both the punch and the fluid are held at the appropriate temperature and, after the required hold time, the pressure is released, the punch withdrawn and the part removed. The main advantage of hydroforming is the application of hydrostatic pressure and a uniform normal pressure over the laminate surface, during both forming and consolidation phases. This facilitates the deformation mechanisms of inter-ply and intra-ply shear while at the same time squeeze flow effects are not restricted. As a result, the formed part will not suffer from damage in high spots and sharp corners in the same way as can occur in matched die forming.

Hydroforming, like diaphragm forming, eliminates the need for mutually conforming dies and consequently the tooling costs are relatively low. Elastomers used in hydroforming are generally suited only to lower temperatures and hence, for higher temperatures, additional rubber sheets

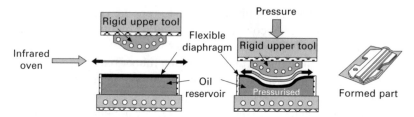

6.17 Hydroforming of textile composites.

have been placed over the preform before forming to prevent diaphragm rupture. Compared with diaphragm forming, hydroforming offers fewer possibilities to control forming forces on the laminate, but higher hydrostatic pressures can be imposed on the laminate surface.

Rapid stamp-forming with matched rigid moulds

The rapid stamping technique (Fig. 6.18) has been extended from more traditional techniques such as diaphragm forming for high-volume or cost-sensitive industrial sectors, notably the automotive industry. Metallic tools are necessary to achieve the high heat transfer rates needed to reduce cycle times. The process consists of preheating the textile in an oven to above the matrix T_m before rapid transfer to a cool tool where forming and consolidation take place. Blank-holders are used to control wrinkling. With consideration of the following issues, textile composites can be processed with matched metal tools using the techniques commonly used for GMTs:

- avoid obstructing deformation of the laminate when local pressure is applied too early;
- thickness differences due to intraply shearing must be predicted precisely (dependent on the fabric type and the initial orientation in the tool);
- the final consolidation will be different over the product due to different surface pressures during the consolidation phase (e.g. wall sections with lower normal pressures);
- fabric waste management and recycling.

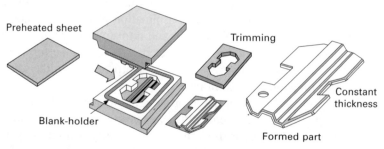

6.18 Non-isothermal stamping of thermoplastic textile composites.

The required consolidation pressure depends on the tolerated porosity level, but 5–15 bar is typical[3, 29], and pressures above 40 bar (e.g. for GF/PP and CF/PEI) may promote resin percolation to the outer plies[9, 77]. The forming rate has important effects during processing. Reported stamping rates include 25 and 210 mm/s for PP and PPS systems[78], and 12 to 166 mm/s for glass or carbon and PA12[10, 54]. Higher stamping rates reduce material cooling and were found to increase hemisphere quality for CF/PA12 materials. Examination of four materials at rates of 10–100 mm/s, with right angle and cylindrical dish tools, showed that higher forming rates limited inter-ply slip, resulting in increased surface fibre buckling[79]. The use of a rapid approach, followed by a relatively slow forming and consolidation period, has also been suggested. Mould temperatures are material (matrix) dependent and are recommended as: 80 °C (PP), 130–160 °C (PA), >160 °C (PPS), 230–280 °C (PEI) and 230–300 °C (PEEK). Cool moulds cause matrix solidification prior to complete part drape (limiting inter-ply slip) whereas materials conform fully at higher forming rates with warmer tools.

Continuous compression moulding

This is a derivative of the stamp-forming process where materials are fed into a preheat oven and then into an intermittent press with a heating and cooling zone[80]. The process is normally automated. The material is pulled through the tool when the press is open and feed stops as the section of material in the tool is compressed. The tool then opens again and the material is advanced, by a step size smaller than the part size. Shaped flat sections, 'U' sections, 'I' sections and 'II' sections can be produced together with closed hollow sections. Shapes with a constant radius (in one degree of freedom only) can also be formed. The lay-up of the composite sheet used to feed the process can be varied, such that UD fibres can be located on the flange area, and ±45° fibres on the wall areas of an 'I' beam. Pressures of 25 bar are typically applied, and feed rates of 30–80 m/h are common. Part thickness of 5 mm is possible and 30 mm thick sandwich structures can be made. The process is shown schematically in Fig. 6.19.

6.5   Novel thermoplastic composite manufacturing routes

6.5.1   Co-compression moulding of textile preforms with a flowable core

In addition to adding thickness changes and functional parts (ribs and bosses) to a structural item, the compression moulding of fabrics is often made easier by incorporating compliant polymer film or GMT interleaves between

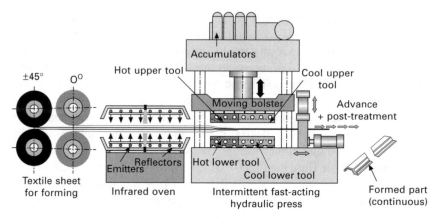

6.19 Continuous compression moulding for the stamping–forming of textile composites.

layers of fabric to overcome problems of thickness variations and areas of poor consolidation (Fig. 6.20)[13,14]. While the flow-moulding core can reduce the overall V_f, it has a minimal effect on flexural stiffness. The polymer or fibre suspension flows in an open channel at the inter-ply region (insulated from the relatively cold tool by the outer plies), yielding pseudo-hydrostatic consolidation. This has been demonstrated[14] for an automotive door module, producing a fourfold increase in modulus over monolithic GMT, with no cycle time penalty. An example of a material commercially available from Bond Laminates[27, 28] is 'flow-core' and Quadrant GMTex® where commingled fabric skins have been combined in a double belt press with a GMT core, which are now in series production for automotive applications. Another

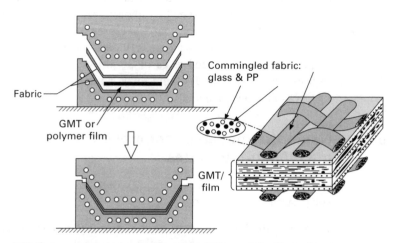

6.20 Co-compression moulding of textile composites with a compliant core.

solution is the extrusion of long fibre thermoplastics (LFTs) between commingled fabric skins prior to compression moulding[81]. This co-compression moulding technique gives a solution for applications that do not need textile reinforcement to the edge of the component, where the area of aligned fibre materials is smaller than the tool, eliminating waste and the need for blankholders during preheat and transfer stages. This has also been applied to more conventional sandwich structures using either polymeric foam or honeycomb cores[81].

6.5.2   Over-injection moulding of stamped preforms

A hybrid moulding process, developed by EPFL-LTC[83–85], that enables composite stampings to be over-injection moulded (by direct recycling of waste prepreg material) is shown in Fig. 6.21. The combination of sheet stamping with over-injection moulding facilitates the combination of the intrinsic stiffness of sheet forming with the ability to include a high degree of integrated functionality via the over-injected polymer. The key issue behind the process is to achieve interfacial healing between the hot, over-moulded polymer and the warm textile composite stamping, with the requirement that the average temperature of the two is above the polymer T_m[86–88]. In practice, a standard injection moulding tool and melt temperatures (e.g. 50 °C and 240

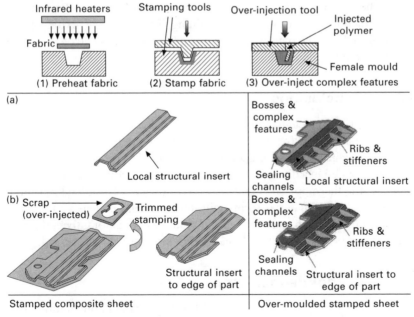

6.21 Over-injection moulding of stamped textile composites:
(a) rectilinear stamped sheet and (b) net-shape stamped sheet.

°C respectively for PP) can be used where the stamped insert is preheated off-line to a temperature below the polymer T_m (e.g. 150 °C for PP). This maintains the rigidity of the insert during transfer into the over-moulding tool and also gives the thermal requirements needed for *in-situ* fusion bonding. Ejector pins can be used to hold the sheet in position while the injection moulding tool fills. The pins are removed when the tool is full but before packing pressure is applied.

Figure 6.21(a) illustrates a route where the area of stamped sheet would be inside the edges of the part. Preconsolidated sheet would be heated and then hot draped to the tool geometry using a low-cost shaping tool. The preform would be reheated and then placed locally in the mould where it is needed. Hence a separate stamp-forming stage would not be used, reducing tool and press costs. The tool is then closed ready for the injection moulding cycle. Final consolidation of the sheet occurs from the over-moulding process. Thus the need for an external blank-holder is eliminated and rectilinear fabric sections can be used in the part. A limitation here is that sheet would not span the whole component, notably at non-linear edge regions. This approach has been used commercially for GF/PP material by PlasticOmnium to produce a rear bumper for General Motors.

Figure 6.21(b) shows a generic component where the stamped part covers the entire hybrid component surface. Here the textile composite is stamp formed in the standard manner. The stamping scrap, typically 30% of the part, is then ground and compounded with virgin polymer to adjust the fibre volume fraction for over-injection moulding. The net-shape textile structure is then placed into the over-moulding tool and ribs, bosses and other complex features are over-moulded.

6.6   Conclusions

This chapter has presented an overview of the interacting material/process/ property relationship for thermoplastic textile composites for a variety of conversion techniques. A wide choice of raw material forms and final processing techniques suited to a diverse range of applications is commercially available. Hence textile thermoplastic composites have evolved considerably from the first generation of material forms with the slow and expensive techniques developed for the aerospace industry to a state where automotive production volumes can be considered or large marine structures be processed.

6.7   Acknowledgements

The authors wish to acknowledge the following for their collaboration in the surrounding research and industrial implementation: Dr P-E. Bourban, P. Vuilliomenet, F. Bonjour, N. Bernet, A. Luisier, O. Zinger, N. Weibel and Dr

V. Michaud. The author also wishes to thank the following companies for information: Bond Laminates, EMS-GRIVORY, Gurit, Plytron, Flex Composites, Porcher Industries, Saint-Gobain Vetrotex, Schappe Techniques and TenCate.

6.8   References

1. Månson J.-A.E., Wakeman M.D. and Bernet N., 'Composite processing and manufacturing – an overview', in *Comprehensive Composite Materials*, Ed. A. Kelly and C. Zweben, vol. 2, 577–607, Elsevier Science, Oxford, 2000.
2. Klinkmuller V., Um M.K., Friedrich K. and Kim B.S., 'Impregnation and consolidation of different GF/PP co-mingled yarn', *Proceedings ICCM-10*, Whistler, B.C., Canada, vol. 3, 1995.
3. Bernet B., Michaud V., Bourban P.E. and Månson J.-A.E., 'An impregnation model for the consolidation of thermoplastic composites made from commingled yarns', *J. Compos. Mater.*, 1999, 33(8), 751–772.
4. Wakeman M.D. and Rudd C.D., 'Compression moulding of thermoplastic composites', in *Comprehensive Composite Materials*, Ed. A. Kelly and C. Zweben, vol. 2, 915–963, Elsevier Science, Oxford, 2000.
5. Van West B.P., Pipes R.B. and Advani S.G., 'The draping and consolidation of co-mingled fabrics', *Composites Manufacturing*, 1991, 2(1), 10–22.
6. Cain T.A., Wakeman M.D., Brooks R., Long A.C. and Rudd C.D., 'Towards an integrated processing model for a commingled thermoplastic composite', *11th International Conference on Composite Materials*, Fisherman's Bend, Gold Coast, Australia, 1997.
7. Svensson N. and Shishoo R., 'Manufacturing of thermoplastic composites from commingled yarns – a review', *J. Thermoplastic Composite Materials*, 1998, 11(Jan), 22–56.
8. Wakeman M.D., Cain T.A., Rudd C.D., Brooks R. and Long A.C., 'Compression moulding of glass and polypropylene composites for optimised macro- and micro-mechanical properties – 1. commingled glass and polypropylene', *Composites Sci. Technol.*, 1998, 58, 1879–1898.
9. Wilks C.E., PhD Thesis, Department of Mechanical Engineering, University of Nottingham, 1999.
10. Breuer U. and Neitzel M., 'The challenge of stamp forming high-strength thermoplastic composites for transportation', *42nd International SAMPE Symposium*, Anaheim, CA, 1997.
11. Long A.C. and Rudd C.D., 'A simulation of reinforcement deformation during the production of preforms for liquid moulding processes', *Proc. Inst. Mech. Eng. J. Eng. Manufacture*, 1994, 208, 269–278.
12. Robroek L., 'The development of rubber forming as a rapid thermoforming technique for continuous fibre reinforced thermoplastic composites', PhD Thesis, Structures and Materials Laboratory, Delft University of Technology, 1994.
13. Wakeman M.D., Cain T.A., Rudd C.D., Brooks R. and Long A.C., 'Compression moulding of glass and polypropylene composites for optimised macro- and micro-mechanical properties. Part 3: Sandwich structures of GMTS and commingled fabrics', *Composites Sci. Technol.*, 1999, 59, 1153–1167.
14. Wakeman M.D., Cain T.A., Rudd C.D., Brooks R. and Long A.C., 'Compression

moulding of glass and polypropylene composites for optimised macro- and micro-mechanical properties. Part 4: Technology demonstrator – a door cassette structure', *Composites Sci. Technol.*, 2000, 60, 1901–1918.

15. Janeschitz-Kriegl H., Fleischmann E. and Geymayer W., 'Process-induced structure formation', *Polypropylene Structure, Blends and Composites, Vol. 1, Structure and Morphology*, J. Karger-Kocsis, Chapman & Hall, London, 1995.

16. Tornqvist R., Sunderland P. and Månson J-A.E., 'Non-isothermal process rheology of thermoplastic composites for compression flow moulding', *Composites Part A: Appl. Sci. Manufact.*, 2000, 31, 17–927.

17. Leterrier Y. and G'Sell C., 'Multilayer plug flow modeling of the fast stamping process for a polypropylene/glass fiber composite', *Polymer Composites*, 1996, 17, 2, 231–240.

18. Chapman T.J., Gillespie J.W., Pipes R.B., Månson J-A.E. and Seferis J.C., 'Prediction of process-induced residual stresses in thermoplastic composites', *J. Composite Materials*, 1990, 24, 616–642.

19. Wijskamp S., Lamers E., Akkerman R. and Van de Ven E., 'Warpage of rubber pressed composites', *ECCM-10*, 3–7 June 2002, Brugge, Belgium.

20. Lawrence W.E., Månson J.-A.E. and Seferis J.C., 'Thermal and morphological skin-core effects in processing of thermoplastic composites', *Composites*, 1992, 21(6), 475–480.

21. Ye L. and Friedrich K., 'Interlaminar fracture of commingled fabric based GF/PET composites', *Composites*, 1993, 24(7), 557–564.

22. Curtis P.T., Davies P., Partridge I.K. and Sainty J.P., 'Cooling rate effects in PEEK and carbon-fibre PEEK composites', *Proc. 6th International Conference on Composite Materials*, London, England, 1987, vol. IV, pp. 401–412.

23. Sunderland P., Yu W. and Månson J.-A.E., 'A thermoviscoelastic analysis of process-induced internal stresses in thermoplastic matrix composites', *Polymer Composites*, 2001, 22(5), 579–592.

24. Muzzy J.D., 'Thermoplastics – properties', in *Comprehensive Composite Materials*, Ed. A., Kelly and C., Zweben, vol. 2, 57–76, Elsevier Science, Oxford, 2000.

25. Verpoest I., 'Composite preforming techniques', in *Comprehensive Composite Materials*, Ed. A. Kelly and C. Zweben, vol. 2, 623–669, Elsevier Science, Oxford, 2000.

26. Peltonen P., Lahteenkorva K., Paakkonen E., Jarvela P. and Tormala P., 'The influence of melt impregnation parameters on the degree of impregnation of a polypropylene/glass fibre prepreg', *J. Thermoplastic Composite Materials*, 1992, 5(Oct.), 318–343.

27. Wang X., Mayer C. and Neitzel M., 'Impregnation of thermoplastic composites manufactured bu double belt press technique', *Proc. Eleventh International Conference on Composite Materials*, Gold Coast, Australia, Woodhead Publishing, Cambridge, 1997.

28. Quadrant Plastic Composites, http://www.quadrant.ch/, 11 April 2003.

29. Bond Laminates, http://www.bond-laminates.de/, 11 April 2003.

30. Mayer C., Wang X. and Neitzel M., Macro- and 'microimpregnation phenomena in continuous manufacturing of fabric reinforced thermoplastic composites', *Composites Part A*, 1998, 29A, 783–793.

31. TenCate CETEX, http://www.tencate-ac.com, 11 April 2003.

32. Connor M., Toll S., Månson J.-A.E. and Gibson A.G., 'A model for the consolidation of aligned thermoplastic powder impregnated composites', *J. Thermoplastic Composite Materials*, 1995, 8(2), 138–162.

33. Roncato G. and Federowsky R., 'Process and device for producing a yarn or ribbon formed from reinforcement fibers and a thermoplastic organic material', US Patent Office, Pat. No. 5011523, 1991.
34. Roncato G., Federowsky R., Boissonnat P. and Loubinoux D., 'Apparatus for manufacturing a composite strand formed of reinforcing fibers and of organic thermoplastic material', US Patent Office, Pat. No. 5316561, 1994.
35. Guevel J., Francois M. and Bontemps G., 'Hybrid yarn for composite materials with thermoplastic matrix and method for obtaining the same', US Patent Office, Pat. No. 5910361, 1999.
36. Schappe Techniques, http://perso.wanadoo.fr/schappe/tpfl.html, 10 April 2003.
37. Stuart L.M., 'Apparatus for commingling continuous multifilament yarns', US Patent Office, Pat. No. 5241731, 1993.
38. Miller A., Wei C. and Gibson A.G., 'Manufacture of polyphenylene sulphide matrix composites via the powder impregnation route', *4th International Conference on Automated Composites*, Nottingham, The Institute of Materials, 1995.
39. Ramani K., Hoyle C. and Parasnis N.C., The American Society of Mechanical Engineers – Use of Plastics and Plastic Composites, *Materials and Mechanics Issues*, 1993, (46), 633–657.
40. Gurit Suprem, http://www.guritsuprem.ch/, 10 March 2003.
41. Flex composites, Delfweg 50, 2211 VN Noordwijkerhout, The Netherlands.
42. Porcher Industries, Division Verre France, Badinieres BP 501, 38317 Bourgoin-Jallieu, Cedex, France, 2000.
43. Lynch T., 'Thermoplastic/graphite fibre hybrid fabrics', *SAMPE J.*, 1989, 25(1), 17–22.
44. Hogg P. and Woolstencroft D., 'Non-crimp thermoplastic fabrics – aerospace solutions to automotive problems', *Advanced Composite Materials: New Developments and Applications Conference Proceedings*, Detroit, MI, ASM International, 339–349, 1991.
45. Kothe J., Scherieble G. and Ehrler P., 'Thermoplastic composites made of textile prepregs', *J. Thermoplastic Composite Materials*, 1997, 10(July), 316–333.
46. Baeten S. and Verpoest I., 'Optimisation of a GMT-based cold pressing technique for low cost textile reinforced thermoplastic composites', *Composites Part A: Appl. Sci. Manufacturing*, 1999, 30, 667–682.
47. Bourban P.E., 'Liquid moulding of thermoplastic composites', in *Comprehensive Composite Materials*, Ed. A., Kelly and C., Zweben, vol. 2, 2.28, 965–977, Elsevier Science, Oxford, 2000.
48. Schmelzer H.G., Kumpf R.J., in *Journal of Macro-molecular science, Part A: Pure Appl. Chem.*, 1997, A34 (10), 2085–2101.
49. 'Fulcrum®', http://www.dow.com/fulcrum/, 26 April 2003.
50. Leimbacher R. and Schmid E., US Patent Office, Pat. No. 5837181, 1998.
51. Luisier A., Bourban P.E. and Månson J.-A.E., 'Time–temperature-transformation diagram for reactive processing of polyamide 12', *J. Appl. Polymer Sci.*, 2001, 81, 963–972.
52. Luisier A., Bourban P.E. and Månson J.-A.E., 'Initiation mechanisms of an anionic ring-opening polymerization of Lactam-12', *J. Polymer Sci.*, 2002, 40, 3406–3415.
53. Verrey J., Michaud V. and Månson J.-A.E., 'Processing complex parts with thermoplastic RTM techniques', *24th International SAMPE Europe Conference*, 1–3 April 2003, Paris, France.
54. Wakeman M.D., Zingraff L., Kohler M., Bourban P.E. and Månson J.-A.E., 'Stamp-

forming of carbon fibre/PA12 composite preforms', *Proc. Tenth European Conference on Composite Materials*, 3–7 June 2002, Brugge, Belgium.

55. Zingraff L., Bourban P.E., Wakeman M.D., Kohler M. and Månson J.-A.E., 'Reactive processing and forming of polyamide 12 thermoplastic composites', *23rd International SAMPE Europe Conference*, 9–11 April 2002, Paris, France.

56. Rösch M. and Eder R.H.J., 'Processing cyclic thermoplastic polyester', *9th International Conference on Fibre Reinforced Composites*, University of Newcastle, UK, 26–28, March 54–62, 2002.

57. Olson S., 'Manufacturing with commingled yarns, fabrics, and powder prepreg thermoplastic composite materials', *SAMPE J*, 26(5), 31–36, 1990.

58. Gil R.G., Long A.C. and Clifford M.J., 'Vacuum consolidation of glass/PP commingled fabrics', *Proc. 4th Congreso Nacional de Materiales Compuestos (MATERIALES COMPUESTOS '01)*, Spanish Association of Composite Materials, Gijon, Spain, Nov. 2001, 773–780.

59. McMillan A., Boyce G., Pohl C. and Michaeli W., 'Processing of large surface area components from thermoplastic composites', http://www.twintex.com/docu/tw_processing.html, 10 April 2003.

60. http://www.twintex.com/fabrication_processes/tw_process.html, 'Twintex vacuum moulding manual', 10 April 2003.

61. Mallon P.J. and Ó Brádaigh C.M., 'Development of a pilot autoclave for polymeric diaphragm forming of continuous-fiber reinforced thermoplastics', *Composites*, 1988, 19(1), 37–47.

62. Monaghan M.R., Ó Brádaigh C.M., Mallon P.J. and Pipes R.B., 'The effect of diaphragm stiffness on the quality of diaphragm formed thermoplastic composite components', *J. Thermoplastic Composite Materials*, 1990, 3, 202–215.

63. Weibel N.D., Lutz C., Wakeman M.D. and Månson J.-A.E., 'High rate bladder moulding of thermoplastic composites', *Proc. 21st International SAMPE Europe Conference*, Paris, France, 18–20 April 2000, 317–327.

64. Bernet N., Michaud V., Bourban P.E. and Månson J.-A.E., 'Commingled yarn composites for rapid processing of complex shapes', *Composites Part A: Appl. Sci. Manufact.*, 2001, 32A (11), 1613–1626.

65. Wakeman M.D., Vuilliomenet P. and Månson J.-A.E., 'Isothermal and non-isothermal bladder inflation moulding of thermoplastic composite pressure vessels: a process/cost study', *24th International SAMPE Europe Conference*, 1–3 April 2003, Paris, France.

66. Patent W.O., 02/072337 A1 'Non-isothermal method for fabricating hollow composite parts', Ecole Polytechnique Fédérale de Lausanne 2001.

67. Anon, 'GMT front end for the VW Golf', *Reinforced Plastics*, Feb 1992, 43–54.

68. Mahlke M., Menges G. and Michaeli W., 'New strategies to preheat glass mat reinforced thermoplastics (GMT)', *ANTEC'89*, 1989, 873–875.

69. O'Brien K. and Kasturi S., 'A computational analysis of the heating of glass mat thermoplastic (GMT) sheets by dual beam microwave sources', *Polymer Composites*, 1994, 15 (June), 231–239.

70. Sweeny G.J., Monaghan P.F., Brogan M.T. and Cassidy S.F., 'Reduction of infra red heating cycle time in processing of thermoplastic composites using computer modelling', *Composites Manufacturing* 1995, 6(3–4), 255–262.

71. Brogan M.T. and Monaghan P.F., 'Thermal simulation of quartz tube infra red heaters used in the processing of thermoplastic composites', *Composites: Part A*, 1996, 27A, 301–306.

72. Chilva T.E., Oven for heating and stacking thermoplastic fiber reinforced resin sheets, US Patent Office, Pat. No. 4767321, 1988.
73. Wakeman M.D., 'Non-isothermal compression moulding of glass reinforced polypropylene composites', PhD Thesis, University of Nottingham, March 1997.
74. Davis S. and Mcalea K., 'Stamping rheology of glass mat reinforced thermoplastic composites', *Polymer Composites*, 1990, 11(6), 368–378.
75. Lemmen T. and Roscoe P., 'Relation of phosphate structure to polymer stability as determined by physical and analytical methods', *J. Elastomers Plastics*, 1992, 24(2), 115–128.
76. Okine R.D., Edison N.K. and Little O.C., 'Properties and formability of a novel advanced thermoplastic composite sheet product', *Proc. 32nd International SAMPE Symposium*, Anaheim, CA, April 1987, 1413–1425.
77. Hou M., PhD Thesis, University of Kaiserslautern, 1993.
78. Bigg D. and Preston J., 'Stamping of thermoplastic matrix composites', *Polymer Composites*, 1989, 10 (August), 261–268.
79. Williams R. and Mallon P.J., 'The optimisation of the non-isothermal press-forming process for fibre reinforced thermoplastics', *Proc. Fourth International Conference on Flow Processes in Composite Materials*, Aberystwyth, Wales, University of Wales, 1996.
80. Advanced Composites and Machines GmbH, Robert Bosch-Str.5, Markdorf, Germany.
81. Henning F., Tröster S., Eyerer P. and Kuch I., 'Load oriented one-step-TWINTEX®-sandwich-structure for large scale production of automotive semi-structural components', *59th ANTEC*, 6–10 May 2001, Dallas, Texas, USA.
82. Rozant O., Bourban P.E. and Månson J.-A.E., 'Manufacturing of three dimensional sandwich parts by direct thermoforming', *Composites Part A: Appl. Sci. Manufact.*, 2001, 32A(11), 1593–1601.
83. Patent E.P. 0 825 922 B1, 'Process for equipments for the manufacture of polymer and for composite products', Ecole Polytechnique Fédérale de Lausanne, 1995.
84. Wakeman M.D., Bonjour F., Bourban P.E., Hagstrand P.O. and Månson J-A.E., 'Cost modelling of a novel manufacturing cell for integrated composite processing', *23rd International SAMPE Europe Conference*, 9–11 April 2001, Paris, France.
85. Bourban P.E., Bogli A., Bonjour F. and Månson J.-A.E., 'Integrated processing of thermoplastic composites', *Composites Sci. Technol.*, 1998, 58, 633–637.
86. Bourban P.E., Bernet N., Zanetto J.E. and Månson J.-A.E., 'Material phenomena controlling rapid processing of thermoplastic composites', *Composites Part A: Appl. Sci. Manufact.*, 2001, 32, 1045–1057.
87. Zanetto J.E., Plummer C.J.G., Bourban P.E. and Månson J.-A.E., 'Fusion bonding of polyamide 12', *Polymer Eng. Sci.*, 2001, 41(5), 890–897.
88. Smith G.D., Plummer C.J.G., Bourban P.E., Månson J.-A.E., 'Non-isothermal fusion bonding of polypropylene', *Polymer*, 2001, 42(14), 6247–6257.

7

Modeling, optimization and control of resin flow during manufacturing of textile composites with liquid molding

A GOKCE and SG ADVANI,
University of Delaware, USA

7.1 Liquid composite molding processes

Liquid composite molding (LCM) processes are commonly used to manufacture polymer–textile composites because of their low equipment and tooling costs, low pressure requirements, short cycle times and ability to yield net-shape parts [1]. Some variations of LCM processes are resin transfer molding (RTM), vacuum-assisted resin transfer molding (VARTM), resin infusion process (RIP) and Seemann composites resin infusion molding process (SCRIMP).

LCM processes include three basic steps (Fig. 7.1) [2]:

1. The textiles, which constitute the reinforcing component of the composite part, are cut from large rolls to the dimensions of the composite part to be manufactured. The textile sheets are placed into the mold. The mold

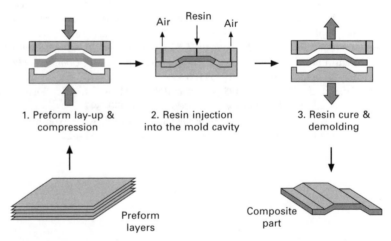

7.1 Three basic steps of resin transfer molding, a liquid composite molding process.

is closed and sealed. If the mold is single-sided as in SCRIMP and VARTM, it is covered with a plastic bag instead of a rigid tool. In RTM, which involves matching dies, the mold is closed and the mold cavity is sealed by fastening the mold with nuts and bolts, or by applying pressure with a press.

2.   The resin is injected into the mold cavity through the gates. In vacuum-assisted processes such as SCRIMP and VARTM, the air is removed from the mold cavity prior to the resin injection. Atmospheric pressure drives the resin through the preform and the maximum pressure at the injection gates is one atmosphere. In RTM, high pressures can be applied at the gates. Resin advances through the fibrous preform owing to the injection pressure, pushing air out of the mold through the vents.

3.   After the injection step is complete, the resin in the mold cavity cures through a series of chemical reactions to form a solid part.

The textile sheets that form the reinforcements placed in the mold before resin injection are made of fibers. The textile properties are determined by the fiber material and the textile architecture. In most of the textiles, glass fibers are used [3, 4]. Carbon and aramid (Kevlar) are among the other fiber types used. The fiberscan be either used in single strands as in *chopped* and *continuous strand random mat*, or can be used to form tows containing 200 to 4000 fibers, which are then used to make textile sheets (also known as preforms). Three types of textiles are illustrated in Fig. 7.2 a random mat, a preform that is woven using carbon fiber tows and a preform with loosely stitched glass fiber tows. Several types of preforms can be produced by stitching, weaving and braiding fiber tows.

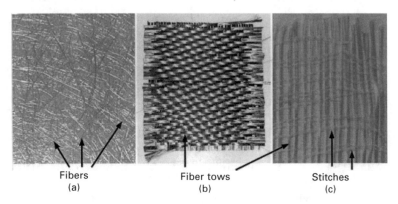

Fibers   Fiber tows   Stitches
(a)   (b)   (c)

7.2 (a) A random mat. (b) A preform that is woven using carbon fiber tows. (c) A preform that is manufactured by loosely stitching fiber tows.

Manufacturing an acceptable composite part depends on successful completion of the three steps shown in Fig. 7.1. The mold filling step is studied exclusively in this chapter [5–8]. Flow through porous media is modeled in the following section. In section 7.3, liquid injection molding simulation (LIMS), software that is used to simulate resin flow in the mold cavity, is introduced and its utilization in design, optimization and control studies is summarized. A method to optimize injection gate location is presented in section 7.4. Section 7.5 introduces the disturbances associated with the mold filling process and explains the major disturbance, racetracking, in detail. Section 7.6 explains the use of feedback control to offset the adverse effects of racetracking on the mold filling process. In Section 7.7, a passive control method to minimize the defect content in the composite parts is presented. The chapter is summarized and potential future research directions are discussed in section 7.8.

7.2   Flow through porous media

In LCM processes, the resin flows through a complex network of channels and paths that are formed by the architecture of the textile sheets in the mold cavity. The diameters of these channels are several orders of magnitude smaller than the gap-width of the mold. Thus, in addition to the stresses due to the presence of textile fibers, the capillary and surface tension effects may become significant. The resin flow through the fiber textiles can be characterized as a complex, dual-scale flow as illustrated in Fig. 7.3. The advance of resin through the air channels between tows is a macro-scale flow. The penetration of resin through the fibers in the tows is a micro-scale flow.

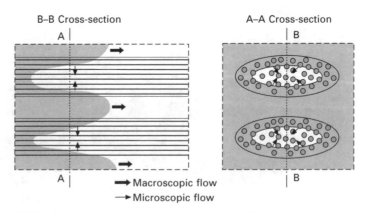

7.3 Dual-scale flow through the porous medium.

While it is theoretically possible to analyze this complex process in detail, the magnitude of immense computational power required to solve the problem forbids any attempt. Moreover, such a detailed study is not necessary as one is only interested in the macroscopic flow rate–pressure relationship during the impregnation of resin. Thus, one would have to average the microscopic stresses to find the macroscopic relationship. The textile preforms are made up of repeated patterns. Instead of analyzing the entire mold geometry, the repetitive pattern can be identified as a unit cell and the flow can be studied in detail. The results of this microscopic study can be used determine the resistance to flow through the unit cell in an average sense. Permeability of a textile preform is defined as the ease of resin flow through the preform. Note that the permeability is an inverse measure of the flow resistance. Extensive research has been conducted to predict and measure the permeability of textiles with various architecture. These studies can be grouped in three major categories as analytical [9–15], numerical [16–22] and experimental [22–28].

The permeability is used along with Darcy's law to model the complicated flow through the porous media in an average sense:

$$\langle u \rangle = -\frac{K}{\eta} \cdot \nabla \langle P \rangle_f \qquad\qquad 7.1$$

where $\langle u \rangle$ is the resin velocity vector averaged over the volume, K is the positive definite permeability tensor of the textile preform, η is the viscosity and $\nabla \langle P \rangle_f$ is the pressure gradient averaged over the fluid volume. In eqn 7.1, resin is modeled with η and the preform is modeled with K. Note that resin flow through the porous medium is assumed be an isothermal, Newtonian flow, thus resin viscosity is taken to be constant. The resin is assumed to be incompressible, hence the continuity equation can be written as

$$\nabla \cdot \langle u \rangle = 0 \qquad\qquad 7.2$$

Inserting eqn 7.1 into eqn 7.2, the mold-filling process can be written as an elliptic partial differential equation as:

$$\nabla \cdot (K \cdot \nabla \langle P \rangle_f) = \nabla K \cdot \nabla \langle P \rangle_f + K : \nabla (\nabla \langle P \rangle_f) = 0 \qquad\qquad 7.3$$

which governs the pressure field in the mold region that is covered with resin. Equation 7.3 is subject to the following boundary conditions (BC):

Neumann BC at the mold walls: $n \cdot (k \cdot \nabla_f) = 0$ \qquad 7.4

Dirichlet BC at the flow front: $P = 0$

and at the injection gates: $P = P_i$,

or: $Q = Q_i$,

where n is the normal to the mold wall and Q is the flow rate. The flow rate boundary condition at a gate can be converted to pressure boundary condition through Darcy's law. Note that the boundary condition at a gate is applied along the boundary of a hypothetical area around the gate in two-dimensional (2D) flow, and over the surface of a hypothetical volume around the gate in three-dimensional (3D) flow.

Figure 7.4 shows domains and boundaries of interest in the mold-filling process. The entire mold cavity forms the mold domain. The region filled with the resin is the resin domain and eqn 7.3 applies in this domain only. As the resin is injected into the mold domain from the gates, the resin domain grows. Since the resin domain changes with time, the pressure distribution in the domain changes as well. Also, varying the boundary conditions at the injection gates modifies the pressure distribution in the resin domain. Therefore, although not explicit in eqn 7.3, the pressure field in the resin domain is a function of time. The resin flow through the porous medium is low Reynolds number flow, which explains the non-existence of inertia terms in eqn 7.3. Moreover, the flow can be considered to be in quasi-steady state at any given time step. Dirichlet boundary conditions apply at the free (moving) section of the outer boundary of the resin domain, which is called the flow front. Neumann boundary conditions apply at the common boundary of the mold domain and the resin domain.

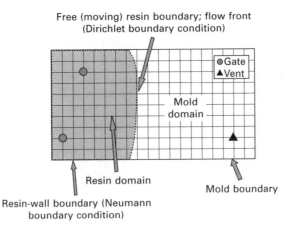

7.4 Resin and mold domains and boundaries.

The pressure gradient in eqn 7.1 is created by the pressure applied at the gates, P_i. The injection rate Q_i determines the wetted region expansion rate. The relationship between P_i and Q_i is a function of the permeability (or resistance) field in the wetted region. One either specifies the pressure P_i or the flow rate Q_i at a gate. As the problem is quasi-steady state, these values may change from one time step to the next time step.

The main function of a vent is to provide an escape passage to the air in the mold cavity. Mathematically, it enforces the Dirichlet boundary condition at the flow front. If a contiguous dry region in the mold cavity lacks a vent, the Dirichlet boundary condition in eqn 7.4 does not apply anymore along the boundary between the dry region and the resin-covered region. The pressure in the dry region increases owing to compression of air and prevents the resin from saturating the dry region. It is highly likely that air in such a region will not be able to escape the mold cavity and will cause a dry spot in the final composite product, which has to be rejected.

The process modeled by eqns 7.3 and 7.4 is a distributed parameter system, whose solution depends on the mold geometry, permeability field in the mold cavity and injection locations and conditions. Since analytical solution is difficult for 2D and 3D flows, researchers employed numerical methods to solve it. Various numerical methods have been employed to simulate the mold filling step [29–35]. LIMS [36], software developed at the University of Delaware to simulate resin impregnation through the porous media, is introduced in the following section. Note that LIMS can also simulate resin curing during the filling phase, which is not discussed here.

7.3   Liquid injection molding simulation

LIMS uses a finite element/control volume approach to simulate the resin flow in the mold cavity. The finite element mesh of the mold geometry, the resin and textile preform properties (η and K values), gate locations and injection conditions are input to LIMS as shown in Fig. 7.5. The area around a node is assigned to the node as its control volume. In Fig. 7.6(a), the nodes at the flow front are marked with solid circles and the control volumes

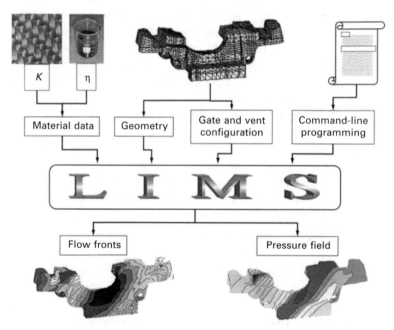

7.5 Inputs and outputs of LIMS.

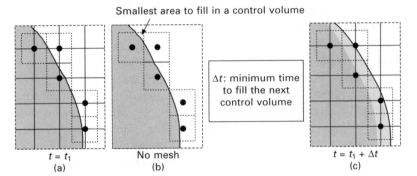

7.6 (a) Flow front location at $t = t_1$. Flow front nodes are marked with circles and the associated control volumes are circumscribed with dashed squares. (b) Mesh is removed to facilitate visualization. (c) Flow front location at $t = t_1 + \Delta t$. Note that the top left node became part of the resin domain.

around the nodes are circumscribed with dashed squares. The mold filling is simulated recursively as follows:

- *Step I.* $t = t_{new}$. LIMS solves eqns 7.3 and 7.4 using finite element methods to obtain pressure distribution $P(x)$ in the mold at $t = t_{new}$.
- *Step II.* The velocity field at the flow front u is determined using eqn 7.1.
- *Step III.* Under the current velocity field, the shortest time to fill a new control volume, Δt_{min} is found. In Fig. 7.6(b), the mesh is removed to facilitate visualization and clearly show the smallest area that needs to be filled to saturate the next control volume. Assuming a homogeneous permeability field, it is clear that the control volume on the top-left will take shortest time to fill. Thus, Δt is taken such that this control volume is filled at the next time step.
- *Step IV.* The current velocity field u is applied on the resin flow for Δt_{min}, the new fill state is estimated and the time is adjusted $t_{new} = t_{new} + \Delta t_{min}$. Figure 7.6(c) shows the flow front at the next time step. The discrepancy between the previous and current flow front locations is highlighted with a different shade in the figure. Note that the top-left control volume is completely filled. Also, three new nodes become part of the resin domain at this time step.
- *Step V.* Stop if all control volumes are saturated, otherwise go to step I.

7.3.1   Features of LIMS

Interaction with LIMS

The numerical solver in LIMS is coupled with a scripting language LBASIC. The communication between the user and the solver is provided via a console using LBASIC. The process parameters and commands to the numerical solver are entered from the console. Selected simulation output can be displayed on the console if desired. Also a graphical user interface (GUI) module is available to interact with LIMS, which provides *point-and-click* capability for most common steps. Using LIMS through the GUI module is easier, while the console and LBASIC provide the widest capabilities.

Interrogation of simulation at each time step

In LIMS, each Solve command executes a single loop of the five steps presented above. Therefore, one can intervene in the simulation at any time step and modify process and material parameters such as the preform properties, injection location and conditions, or vent locations. Moreover, one can retrieve data from the simulation at any time step. During the actual operation, the operator can intervene in the process and obtain necessary data.

Procedures

Procedures – batches of LBASIC commands – further facilitate the use of LIMS. One can create complex algorithms in LBASIC to try various filling scenarios, retrieve specific data and undertake control actions during mold filling, as illustrated below.

In Fig. 7.7, a step mold is partially filled and the injection has to be switched to a secondary gate. Three nodes, Gate A, Gate B and Gate C in the

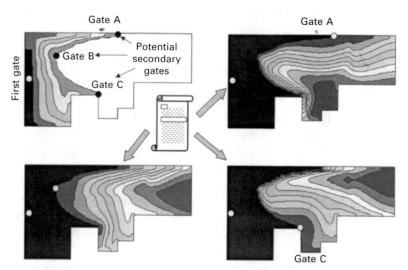

7.7 Filling contours at progressive times to demonstrate changes in filling patterns when injection gate is switched to secondary Gates A, B or C using the scripting procedures listed in Fig. 7.8.

figure, are chosen as the possible locations for the second injection location. Two procedures, *TestGates* and *SwitchGate*, are written to try each of these locations automatically. Initially, the first procedure fills the mold until resin arrives at Gate A and saves the mold filling state in the file *State.dmp*. Note that the resin arrives at Gates A, B and C simultaneously. Next, procedure *TestGates* calls the second procedure three times to try each of Gates A, B and C. Procedure *SwitchGate* loads the mold filling state from *State.dmp*, sets the second injection location and completes the mold filling simulation. The procedures are listed in Fig. 7.8 and the flow patterns at various time steps are illustrated in Fig. 7.7.

Tracers

A tracer simulates a sensor in the mold cavity. It follows the state of a given node and records the information either to a window on the screen or to a tab-limited ASCII file where each line contains the time and traced value. The tracers are valuable tools when one wants to use sensors in resin flow feedback control studies.

Virtual experimentation environment

LIMS can be called from other programs such as LabView and MATLAB® through dynamic linking, which provides seamless integration between the two environments and computational efficiency. There are two modes of data exchange between the LIMS subroutine and the master program:

1. *The text (command) data exchange*: in this mode, the master program behaves like a user sitting at the console, sending lines of commands to the LIMS subroutine and receiving lines of LIMS output. Buffers are used to store information during data exchange between the master program and the LIMS subroutine. Figure 7.9 illustrates schematically an instance of message traffic between MATLAB and LIMS. In the figure, the last command MATLAB put into the buffer is *print socurrenttime( )*, which requests LIMS to print the current simulation time, and the last output LIMS put into the buffer in response to an earlier request is *342 345.231 10000*, which can be explained as *<node number>, <current simulation time, i.e. the duration before resin arrives at Node 342>, <pressure at Node 342>*. The computational overhead that comes with the message traffic limits the amount of data that can be exchanged between the LIMS subroutine and the master program.

2. *The binary data exchange*: in binary mode, shared memory blocks (memory-mapped file) are used to exchange data between the LIMS subroutine and the master program (Fig. 7.10). A shared memory block

```
REM Syntax: TestGates FirstGate, GateA, GateB, GateC, GateType, GateStrength
proc TestGates
  defint i, j, FirstGate, GateA, GateB, GateC, GateType, Flag1,
  defdbl Time, GateStrength
  let FirstGate=argument(1)
  let GateA=argument(2)
  let GateB=argument(3)
  let GateC=argument(4)
  let GateType=argument(5)
  let GateStrength=argument(6)
  let Flag1=1
  setintype "in"
  read "onrsmall.in"
  setgate FirstGate, GateType, GateStrength
REM Fill until the specified limit
  do while (Flag1)
    solve
    for i=1 to sonumberfilled()
      let j=sonextfilled()
        if (j=GateA) then
          let Flag1=0
          setgate FirstGate, 0, 0
          setouttype "dump"
          write "State"
          let Time=socurrenttime()
        endif
    next i
  loop
  REM Switch injection from the first gate to gates A, B and C
  SwitchGate GateA, GateType, GateStrength, Time, "State_GateA"
  SwitchGate GateB, GateType, GateStrength, Time, "State_GateB"
  SwitchGate GateC, GateType, GateStrength, Time, "State_GateC"
endproc
```

```
REM Syntax: SwitchGate SecondGate, GateType, GateStrength, Time, OutFile
proc SwitchGate
  defint SecondGate, GateType, GateStrength
  defdbl Time
  defstring OutFile
  let SecondGate=argument(1)
  let GateType=argument(2)
  let GateStrength=argument(3)
  let Time=argument(4)
  let OutFile=argument(5)
  setintype "dump"
  read "State.dmp"
  settime Time
  setgate SecondGate, GateType, GateStrength
  do while (sonumberempty()>0)
    solve
  loop
  setouttype "tplt"
  write OutFile
endproc
```

7.8 Two procedures created to try three different filling scenarios.

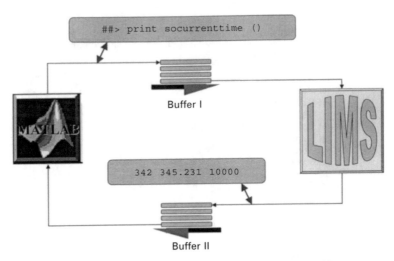

7.9 Text mode data exchange between MATLAB® and LIMS.

is essentially a data structure that contains all process information. Each shared block has a name and a certain organization, which allows the LIMS subroutine or the master program to access any specific data from the block: the fill factor of a certain node, for example. At any given time, either the master program or the LIMS subroutine can access the data block. Note that this exchange mode is binary and does not involve any actual data transfer between the master program and the LIMS

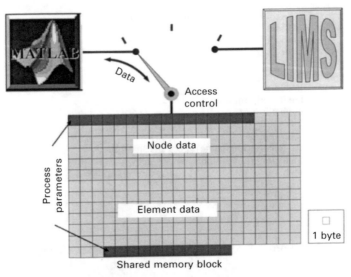

7.10 Binary mode data exchange between MATLAB ® and LIMS.

subroutine; in essence, two software programs share a certain portion of the RAM memory. Hence, it is very fast compared with text mode and is suitable for large volume data exchange such as providing data for a contour plot visualization.

The text mode is suitable for small amounts of data exchange and the binary mode is advantageous for large volumes of data exchange. If the same data are to be accessed repeatedly for a period of time, permanent memory can be used for data exchange. The majority of the study presented in this chapter was conducted in a virtual optimization and control environment created by the integration of LIMS and MATLAB® [37], a high-level programming language and an integrated technical computing environment that combines numerical computation, advanced graphics and visualization.

7.3.2   Process model in design, optimization and control

Several engineering practices, such as design, optimization and control, require a mathematical model of the physical system under study. A mathematical model imitates the process physics, and for a given set of process inputs, the model is expected to reproduce the same results as the actual system. For example, in the mold filling process, the inputs are the mold geometry, the gate locations, the injection conditions, the vent locations and the preform and resin properties, and the important outputs are the flow pattern, the fill time, dry spot formations and the pressure requirements.

In some cases an analytical mathematical model is available for the physical system under study:

$$F = f(x) \hspace{10em} 7.5$$

where x represents the process parameters and, f is an explicit function of x that maps process inputs x to the process output(s) F ($\in R$). Several design, optimization and control tools exist to address these types of situations. In other cases such as the mold filling process, an explicit function that relates the process inputs to the process outputs does not exist. Instead, LIMS is used as the mathematical model (also called the process model) of the mold filling process, which maps the process inputs to the process outputs. It provides a virtual testing environment, where one can try different designs, investigate different ideas and concepts, study input–output relations, etc. Common analytical tools in design, optimization and control fields do not apply in these cases, since an explicit form of f is not available. In the following two sections, the utilization of LIMS in the optimization and control studies is summarized.

Optimization studies

Several design choices exist for the mold in LCM processes. The optimization algorithms can help designers achieve the design that fits the performance criteria best. These algorithms require the mathematical model of the physical system to be optimized. When the mathematical model f is explicitly available, the optimal values of the process parameters x can be estimated using the standard procedures.

In the mold filling process, LIMS provides the input–output relationship. The cost functions are defined in terms of LIMS output. Several optimization algorithms that do not require an analytical process model are available for use with LIMS [38, 39]. These algorithms are usually iterative. At each iteration, the process is simulated with the design parameter values from the previous iteration. The cost function is evaluated using LIMS outputs. Next, the optimization algorithm calculates the new design parameter values, such as gate locations, for the next iteration. The purpose of all optimization algorithms is to converge towards the optimal solution by each iteration [40, 41].

Feedback control

During the manufacturing stage, feedback control may be necessary to offset the adverse effects of the disturbances and fill the mold successfully. It is straightforward to simulate the control hardware. Ability to interrogate the simulation at each time step allows one to simulate actuator actions such as opening/closing a gate and modifying the injection conditions. Sensors are simulated by using the LIMS output and the tracer data at each time step. The LBASIC procedures are extremely helpful for these purposes.

In the feedback control, once the sensor data have been retrieved from the LCM mold and processed, the optimal control action must be generated by the controller and sent to the actuators (the gates) to drive the resin flow in the desired direction. The controller usually will employ a process model to predict the responses to various actuator actions. For such methodology, LIMS can serve as the mold filling process model.

7.4   Gate location optimization

7.4.1   Problem statement

In this section, a gate location optimization problem is stated [39]. The cost function (also called the performance index) is expressed in terms of LIMS output. An assumption on the cost function f is verified using LIMS simulations.

The single gate optimization problem in LCM can be stated as follows:

Given a non-empty, closed set $D \subset R = \{$integer$\}$ and a continuous function $f : D \to R$, find at least one point $x^* \in D$ satisfying $f(x^*) \leq f(x)$ for all $x \in D$.

In LCM, x correspond to a node in the finite element mesh where a gate can be placed and set D consists of all possible x values. f is the cost function that needs to be minimized. Note that an analytical expression is not available for f. Instead LIMS output is used to estimate $f(x)$ value. Two assumptions are made about f.

Assumption I: $f(x)$, $x \in D$, is continuous.

Since f is not available, it is impossible to apply standard tests to prove this assumption. Instead, a case study is presented for verification using simulations.

The bottom picture in Fig. 7.11 is a simplified mesh of a bus chassis. There are two high fiber volume fraction regions in the geometry as illustrated in the figure. Flow resistance is much higher in regions with a high fiber volume fraction. It is desired to find the injection location for a single gate that will fill the mold in minimum time. The injection would be done under a constant pressure boundary condition. In order to achieve this physically, the resin reservoir is pressurized and held under constant pressure until the resin reaches the last vent. All nodes were tried as injection locations (exhaustive search) and corresponding fill times are shown with two surface plots in Fig. 7.11. In the plots, the z axis shows the fill time. The lower a node is on the z axis, the less time it takes to fill the mold. The middle picture in Fig. 7.11 shows the overall exhaustive search results. As expected, the nodes in the high-volume fraction regions are not suitable as the gate location: it takes much longer to fill the mold if the injection location is in these regions. Note that in the middle picture, the results of the high-volume fraction region dominate the plot; therefore the fill time variation in the remaining region cannot be further discriminated. In the top picture in Fig. 7.11, the results of the high-volume fraction region are removed from the plot, making the behavior of the remaining nodes visible. As expected, the injection locations in the middle of the geometry will fill the mold in shorter times. Note that plots in both middle and top pictures are continuous, hence they verify Assumption I.

Assumption II: df/dx is small and of the same order throughout the mesh surface, where dx represents spatial differential of node x in the mesh.

7.4.2   Branch and bound search

The branch and bound search (BBS) is used as the optimization algorithm [42, 43]. BBS converges on the optimal solution by dividing the solution set into smaller sets and eliminating the sets that are unlikely to include the

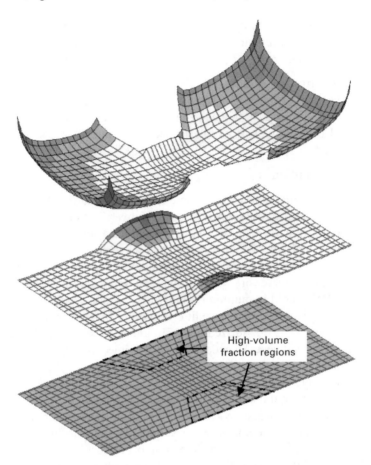

7.11 Bottom: simplified mesh of bus chassis with two high-volume fraction areas. *Middle*: Surface plot of exhaustive search results. The gate is desired at a location that will minimize fill time. High-volume fraction regions return high fill times and dominate the plot. *Top*: The results of high-volume fraction region are removed from the plot, which clearly shows that the gate should be placed at the center.

optimal solution. This process is repeated until it is no longer possible to partition a set or a stop criterion is met. In order to apply BBS to an optimization problem, it should satisfy two requirements.

Requirement I: Given a solution set, it should be possible to estimate a lower bound for the function f.

A lower bound for a solution set S_i, $i = 1, 2 \ldots n$, denoted as $\beta(S_i)$, satisfies the condition $f(x) \geq \beta(S_i)$, $x \in S_i$. The only way to find a lower bound for a set in the current case is to use LIMS with each node in the set, which amounts to impractical exhaustive search. Hence, in this study, the lower bound is replaced with a qualitative parameter as follows.

Owing to the physics of the problem, f can take in multiple values and return a single value: $f = f([x_1, x_2 \ldots x_n])$, $x \in D$, i.e. multiple gates can be used simultaneously to fill the mold, which will yield a single flow pattern that will return a single f value. Given a set S_i, the pseudo-average of f over the set is defined as $\bar{f}_{S_i} = f(S_i)$, i.e. all nodes in the solution set S_i are simultaneously designated as gates and the subsequent value of the cost function f is assigned to $\bar{f}_{S_i}$. A third assumption is derived from Assumption II as follows:

> **Assumption III**: For the sets at the same partitioning level, the differences between the pseudo-average $\bar{f}_{S_i}$ and the unknown lower bound $\beta(S_i)$ is of the same order: $\varepsilon_i \sim \text{avg}(\varepsilon_i)$, where $\varepsilon_i = |\bar{f}_{S_i} - \beta(S_i)|$.

According to Assumption III, $\bar{f}_{S_i} = \beta(S_i) \pm \varepsilon_i$. Since ε_i varies slightly for sets under evaluation, $\bar{f}_{S_i}$ can replace unknown $\beta(S_i)$ for the purpose of ranking the performances of various solution sets.

> **Requirement II**: There should be a way to partition a given solution set.

The solution set is partitioned according to geometrical guidelines (Fig. 7.12). Although there are standard partitioning methods in BBS, none has yet been proven to be applicable to the mold filling problem. A rule of thumb

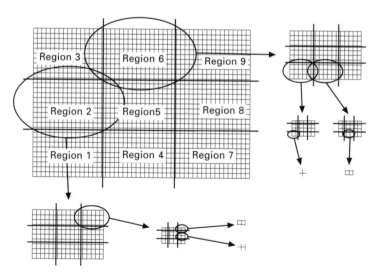

7.12 Recursive domain reduction in branch and bound search while converging on the optimal solutions.

is to partition the set in a way to validate Assumption III. A fine partitioning should be used for sets that include high gradients of f, whereas sets with low gradients may perform well with a coarse partitioning. A major cause of high gradients is significant variations to the permeability distribution within the area covered by the set. In this study, homogeneous, equal partitioning is used regardless of the permeability field in the mold.

Procedure

The algorithm of the BBS is explained step by step below. The nodes in the entire mesh are taken as the first solution set.

- *Step 1*: Partition the set(s) into subsets to form a new group of smaller sets for evaluation.
- *Step 2*: For each set in the group estimate $\bar{f}_{S_i}$.
- *Step 3*: Choose the set with $\bar{f}_{min} = min(\bar{f}_{S_i})$. Also choose the sets whose $\bar{f}$ values are close to $\bar{f}_{min}$ in order to compensate for the impreciseness in Assumption III.
- *Step 4*: Remove the remaining sets from the search space.
- *Step 5*: If the chosen sets include suffcient number of nodes for partitioning, go to Step 1. Otherwise, simulate mold filling for each node in the chosen set(s), take the best solution and Stop. In this study, partitioning is continued until there are four or fewer nodes in a chosen set.

A schematic explanation of the procedure is given in Fig. 7.12: A 3 × 3 partitioning scheme is used throughout the solution. In the first iteration, sets 2 and 6 give superior performance as compared with other sets and hence, are selected for the next level of subdivision. In order to reach the optimal solutions, three iterations were suffcient as illustrated in the figure.

In the following section, BBS is used to find the gate location in a box geometry that will fill the mold in minimum time.

7.4.3   Case study: minimum fill time

During an injection system design, filling the mold as quickly as possible is an important goal as it will reduce the cycle time and increase the profitability of the process. Also for quick-curing resins, it lowers the possibility of resin gelling and hardening before the mold is completely filled. The mold of interest and its mesh are illustrated in Fig. 7.13(a). It is a two chamber, 65 × 30 × 25 cm³ hollow box, and the dividing wall has a circular hole in it. The mesh of the mold has 3378 nodes and 3316 elements.

An injection system with a single gate is designed. The resin is injected through the gate under the constant pressure of 0.2 MPa. The resin viscosity is assumed to be 0.1 Pa s. The preform is assumed to be isotropic with a

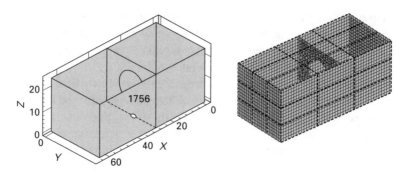

7.13 *Left:* box mold. *Right:* the mesh of the box mold and partitioning scheme.

permeability of 10^{-6} m^2. The gate is desired at a location on the mold geometry that will minimize the fill time. Hence, the cost function for this problem is as follows:

$$f = f(x): \text{the fill time} \qquad\qquad 7.6$$

where x is the location of the single gate. The vents are placed in the last regions to fill or the mold can be assumed to be under perfect vacuum, so the possibility of having dry spots is non-existent. For a given set S_i, pseudo-average $\bar{f}_{S_i} = f(S_i)$ is estimated by equating the injection pressures of each node in the set to the preselected constant pressure value, $\forall x \in S_i$, $P(x) = P_{\text{injection}}$ and then simulating mold filling using LIMS.

Initially, the problem is solved using the exhaustive search (ES). After 3378 simulations, ES showed that the optimal gate location is at Node 1756 ($x = 30$ cm, $y = 16.5$ cm, $z = 0$ cm) with a fill time of 1000 seconds. Node 1756, which is located on a hidden line, is marked with an ellipse in the left-hand figure in Fig. 7.13. During BBS, a $3 \times 3 \times 3$ partitioning is applied throughout the optimization procedure. After a total of 3 iterations and 57 simulations, BBS arrived at the same result. Overall, BBS was 98% more efficient computationally than ES, which illustrates the advantages of optimization algorithms. More details and examples can be found in Gokce [5] and Gokce and Advani [44].

7.5   Disturbances in the mold filling process

During the mold filling stage of LCM processes, the textile sheets must be completely saturated with resin. The gates and the vents must be placed appropriately to ensure complete saturation. Usually, the gate locations are chosen first to achieve certain objectives such as minimum fill time, and minimum pressure requirements. Next, the vents are placed at locations where resin arrives last.[1] Such design decisions are made based on the process model and subsequent simulations. If the actual values of the process parameters are significantly different from the predicted values, then the subsequent designs may fail to satisfy the intended objectives.

The constitutive relationship between the resin velocity field and the pressure field in the mold cavity is a function of the resin viscosity and the permeability field as stated by eqn 7.1. Variations in these two parameters may cause the mold filling stage to go astray, resulting in a defective part. In the mold filling process, the fill time and the flow pattern are important output variables. The fill time is the total duration of the mold filling process and is a function of several parameters such the resin viscosity, permeability field and the injection conditions. The flow pattern can be defined as the filling sequence of the mold cavity by the resin. The flow pattern is a function of the normalized form of the permeability field in the mold cavity. The fill time and the flow pattern are independent of each other.

The variations in the resin viscosity and the permeability field cause the fill time and the flow pattern to vary. Variations in the fill time are tolerable. A very small fill time implies fast filling of the mold, which may cause micro-bubbles in the fiber tows, an issue not addressed here. If the mold filling stage takes too much time, resin may start gelling, which will complicate and possibly fail the mold filling operation. On the other hand, variations in the flow pattern may cause significant defects in the composite part. Since gates and vents are placed into the mold based on a predicted flow pattern, a moderate alteration in the flow pattern due to disturbances may cause the gate and vent design to fail. Hence, variations in the flow pattern are more critical than the variations in the fill time. Some disturbances that may influence these two outputs are summarized below.

- *Resin viscosity*: Variations in the resin viscosity η, may cause the resin to flow slower or faster and affect the fill time. Under certain conditions, the isothermal-Newtonian flow assumption may fail. If the resin begins to gel before the entire mold is filled owing to faulty chemical composition, its viscosity will increase by several orders. Consequently, the resin flow will slow down in certain parts of the mold cavity and alter the flow pattern.
- *Defects in the preform*: Manufacturing defects or impurities in the textile preform may cause local variations in the permeability field. As a result, resin may flow faster or slower at certain parts of the mold, changing the flow pattern. Note that major preform defects can be detected easily before

---

[1]Note that the term *last* is not used in the chronological sense. It means the last region to fill in a contiguous dry area, after which the dry area disappears, regardless of the time scale.

the preforms are placed into the mold and minor defects do not cause significant disturbances. Hence, this is a minor issue.

- *Textile preform compaction during mold closing*: In LCM processes, the textile sheets in the mold cavity are under pressure. (This pressure is not related to the injection pressure applied at the gates.) In vacuum-assisted processes with single-sided molds, atmospheric pressure is applied on the textile sheets through the plastic covering. In matched die molds, the pressure on the preform can be much higher, depending on the number of preform layers and the preform architecture. Depending on the pressure level, the fiber architecture may change in various ways, affecting the shape of the flow channels, which determine the permeability of the porous medium, as illustrated in Fig. 7.14(a). In general, compaction affects the preform permeability globally, hence it may affect the fill time but not the flow pattern.
- *Interaction of textile layers during mold closing*: Since architecture of most textile preforms consists of repeated patterns, there are regions with

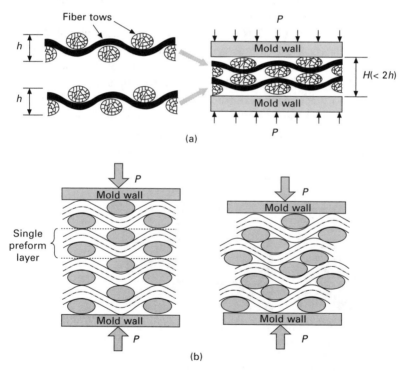

7.14 Interaction of preform layers in a closed mold: (a) *compaction*: the preform layers become thinner under pressure; (b) *nesting*: preform layers mesh with each other under pressure. There is no nesting on the left figure, whereas there is a significant level of nesting on the right figure.

high and low fiber fractions. Under pressure, the textile preform layers protrude into each other as illustrated in Fig. 7.14(b), a phenomenon called the *nesting effect*. The level of nesting effect is a function of the relative position of textile sheets with respect to each other and the applied pressure on the textile preform. The nesting effect influences the structure of flow channels in the mold cavity, hence the permeability field of the porous medium. Its effect on the permeability field is global, hence it does not affect the flow pattern. In this case the nesting effect is related to the compaction of textile sheets.

- *Accidental flow paths in the mold cavity*: Especially in matched die molds, the mold cavity is rigid. Achieving an exact fit between the textile preform and the mold cavity is a difficult challenge. When there are gaps between the inner mold surface and the textile preforms, channels that offer low resistance to resin flow may be created unintentionally in the mold cavity. Resin flows faster in these channels, altering the flow pattern significantly, a phenomenon known as *racetracking*. Figure 7.15 shows an example of altered flow front due to a gap along an edge. These channels, called *racetracking channels*, are more likely to form at certain locations in the mold as shown in Fig. 7.16. Several researchers have attempted to model these channels and their effects on the flow pattern [47–53]. Traditionally, the textile preform is assumed to fit the mold cavity exactly. However, because of process variations during preform cutting/placement and preform compaction mechanisms during mold closing, it is diffcult to achieve a perfect contact between the bulk preform and the inner mold surface. Consequently, racetracking emerges as the most common disturbance in the mold filling stage of the certain LCM processes.

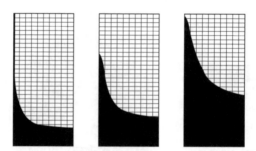

7.15 Experimental snapshots of racetracking along the left edge of a rectangular mold cavity [45, 46]. The resin enters the mold cavity from a line injection gate along the upper edge. Horizontal flow fronts would be developed if there was no racetracking.

The flow pattern, the key measure of successful mold filling, is a function of the racetracking conditions, which depend on the state of the contact between the textile preform and the inner mold surface. Since the contact

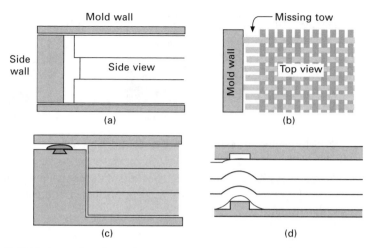

7.16 Various situations that cause racetracking: (a) poor dimensional control during perform cutting; (b) low volume fraction at the edges, possibly due to missing fiber tows; (c) insufficient mold closing; (d) poor mold design.

state is unrepeatable from one experiment to the other, racetracking is a major cause of process unrepeatability in the mold filling process. Because of its frequent occurrence and significant adverse effects, racetracking is the only disturbance studied in this chapter. However, the tools and methodologies developed for racetracking can be applied to other disturbances.

7.5.1   Modeling a racetracking channel

A racetracking channel is created because the resistance to flow along a certain path is affected by the type of fiber architecture, precision of a cut and placement of the textile preform. Resin flow through a racetracking channel usually meets a lower resistance than the textile preform, hence permeability along a racetracking channel is higher. The resin flow in the channel can be modeled as slow, fully developed, steady-state viscous duct flow and one can develop an equivalent permeability expression for it in terms of channel dimensions as [49, 52, 54]:

$$K_{rt} = \frac{a^2}{12}\left[1 - \frac{192a}{\pi^5 b}\sum_{i=1,3,5..}^{\infty}\frac{\tanh\left(i\pi b/2a\right)}{i^5}\right] \qquad 7.7$$

where K_{rt} is the equivalent permeability of the racetracking channel, b is the distance between the preform and the mold wall and a is the height of the mold cavity (Fig. 7.17). Note that the wider the gap between the preform and the mold wall, the stronger will be the racetracking.

The equivalent permeability of the racetracking channel in eqn 7.7 is not

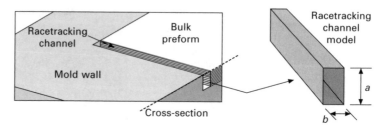

7.17 Modeling of resin flow in the racetracking channel as duct flow.

suffcient to quantify racetracking alone. The ratio of the equivalent permeability K_{rt} to the textile preform permeability $K_{preform}$ is chosen as a measure of the racetracking strength;

$$\rho = \frac{K_{rt}}{K_{preform}}$$ 7.8

where ρ is the racetracking strength. Note that $K_{preform}$ can vary by several orders of magnitude for various preforms. Table 7.1 shows racetracking strength for five different preforms with an identical racetracking channel of 0.5 mm wide and 6.35 mm high. In the simulation environment, these five preforms are placed in a rectangular mold successively such that a 0.5×6.35 mm^2 racetracking channel is along the top edge. The injection gate is placed on the left edge and mold filling is simulated under constant flow rate injection for all five cases: $\rho = \{1, 4.1, 52.5, 124.5, 562.5\}$. Note that $\rho = 1$ means there is no racetracking in the mold. The flow front positions at the same time step for all five cases are superimposed in Fig. 7.18. The higher the racetracking strength ρ, the more resin flow deviates from the nominal case ($\rho = 1$). Figure 7.18 shows that even if the racetracking dimensions are identical in all the cases, racetracking takes place at varying levels as it is a function of the permeability of the textile preform as well.

A racetracking channel is not likely to have a perfect duct shape as in Fig. 7.17. The dimensions of the channel, a and b, in eqn 7.7 may vary along the channel; $a = a(s)$ and $b = b(s)$, where s is the coordinate along the length of

Table 7.1 Racetacking strengths for various performas with identical racetracking channels

Type of preform [55]	Preform permeability (m²)	Racetracking strength, ρ
Random glass fiber	4.88×10^{-9}	4.1
3D woven glass fibers	3.77×10^{-10}	52.5
Unidirectional (warp) woven glass fibers	1.59×10^{-10}	124.5
Non-woven stitched glass fiber	3.53×10^{-11}	562.5

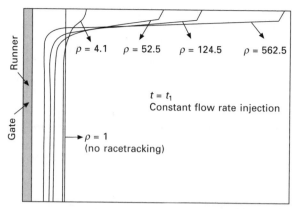

7.18 Flow front locations at the same time step for five different racetracking strengths.

the channel. Hence, $K_{rt} = K_{rt}(s)$ and $\rho = \rho(s)$. Since racetracking is unrepeatable and may appear in many different forms and strengths, it is not straightforward to model $\rho(s)$ in the simulation environment. Thus, researchers have assumed that racetracking is uniform along a racetracking channel. Gokce and Advani justified this assumption [44].

7.5.2   Experimental measurement of racetracking

A rectangular mold with an impermeable L-shaped insert, called the L-mold, has been designed [54]. The flow patterns in the mold were studied in three stages: (i) the simulation stage, (ii) the experiment stage and (iii) the racetracking analysis stage.

Simulation stage

Five potential racetracking channels were identified in the mold as illustrated in Fig. 7.19. During a moldfilling experiment, the racetracking strength along a channel can be of any value. Since it is not possible to include all ρ values in the study, a finite set of ρ values, called a *racetracking set*, is selected to represent the entire ρ range of a channel. During an experiment, the actual racetracking strength in a channel will be approximated by the closest value from its racetracking set. Thus, the continuous ρ range for the racetracking channel is discretized into a finite number of ρ values. The more elements there are in the racetracking set, the better the actual racetracking strength can be approximated.

The racetracking conditions in channels I, II, III and IV were modeled by the racetracking set $\mathscr{R} = \{3, 7, 14, 32, 70\}$, and channel V was modeled by the racetracking set $\mathscr{R} = \{2, 6\}$. By taking one value from the racetracking

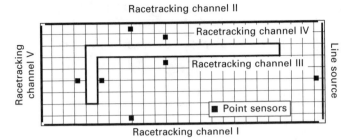

7.19 Potential racetracking channels and sensor locations in the L-mold.

set of each channel, 1250 ($5 \times 5 \times 5 \times 5 \times 2$) racetracking scenarios were created. The collection of all scenarios is called a *scenario space S.*

The mold filling was simulated for all 1250 racetracking scenarios.[2] Seven point sensor locations were identified in the mold for racetracking measurements (Fig. 7.19). A table that includes resin arrival times at the sensors for all scenarios was created from the simulation results. The table, called the *detection table,* would be used to compare simulation results with the experimental arrival times to identify the racetracking scenario that is closest to the mold filling state in the mold.

Experiment stage

A rectangular mold and an L-shaped insert were manufactured and assembled to create an L-mold. Point sensors and a CCD camera were used as the sensors (Fig. 7.20). Fifty mold filling experiments were conducted using the same mold and materials under the same processing conditions. During each experiment, resin arrival times at the point sensors were recorded to identify the closest scenario from the scenario space *S*. Note that actual racetracking strength in a racetracking channel will probably not be equal to any of the values in the racetracking set associated with the channel. Consequently, an actual flow pattern will not exactly match any of the 1250 racetracking scenarios. However, this is a source of error common to all discretization operations. If the discrete case is suffciently close to the actual case, then the results returned by the discrete case will be within the range.

---

[2]To make the text more readable, the term *scenario* will be used to represent a *mold (filling operation) with specific racetracking conditions.* Therefore, the phrases a *scenario is filled/drained/saved* must be understood accordingly.

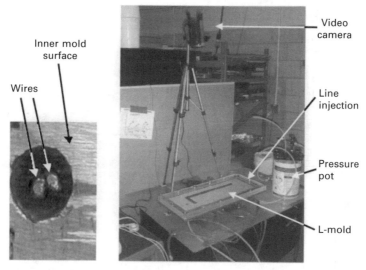

Inner mold surface

Wires

Video camera

Line injection

Pressure pot

L-mold

7.20 (a) A point sensor; (b) Experimental set-up [56].

Racetracking analysis

The flow pattern of each experiment was approximated by a racetracking scenario using the detection table. Thus, 50 ρ measurements were obtained for each of the five racetracking channels in the mold. The frequency of the elements in set $\mathcal{R}_i$ of racetracking channel i is displayed and plotted in Fig. 7.21. The data show that racetracking occurred frequently at varying levels in all channels, which illustrates the significance of racetracking.

7.5.3   Successful mold filling despite racetracking

Traditionally, in RTM processes the gate locations are selected first and then vents are placed at the last-filled regions. The case without any disturbances is called the *nominal case* and the corresponding vent configuration is called the *nominal vent configuration*. Racetracking even at moderate levels can alter the flow pattern significantly and cause the resin to arrive at the vents earlier, blocking the air drainage. The nominal vent configuration, which was designed for the nominal flow pattern, cannot drain the air successfully for an altered flow pattern. There are two approaches to address the disagreement between the altered flow pattern and the nominal vent configuration:

1. *Active control*: One way is to counteract the influence of racetracking on the flow pattern using feedback control. The resin flow is steered by modifying the injection conditions at the gates towards the existing (nominal) vent configuration so that all air is drained from the mold cavity.

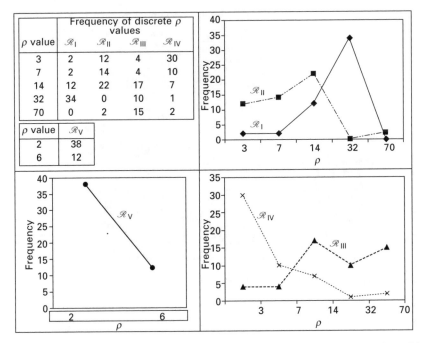

ρ value	Frequency of discrete ρ values			
	$\mathscr{R}_\mathrm{I}$	$\mathscr{R}_\mathrm{II}$	$\mathscr{R}_\mathrm{III}$	$\mathscr{R}_\mathrm{IV}$
3	2	12	4	30
7	2	14	4	10
14	12	22	17	7
32	34	0	10	1
70	0	2	15	2

ρ value	$\mathscr{R}_\mathrm{V}$
2	38
6	12

7.21 Experimental outcomes for racetracking channels in the L-mold.

2. *Passive control*: The vents act as constraints on the mold filling process by dictating the last points resin must arrive for a successful mold filling. Instead of driving the resin towards a nominal vent configuration as in feedback control, one might take the inverse approach and augment the nominal vent configuration by adding more vents, hence relaxing the constraint on the mold filling process. A suffcient number of vents can be placed optimally in the mold such that there will always be a vent in the vicinity of a region where resin arrives last.

7.6   Active control

The central idea in feedback control is to manipulate the resin flow in the mold cavity by modifying the injection conditions at the gates. A feedback control system has three major components:

1. *Data acquisition hardware*: This includes the sensory system and signal processing circuitry. The sensory system provides data from the closed mold, which enables one to track resin flow in the mold cavity. The sensors used to monitor mold filling can be direct current (DC), dielectric, optic, ultrasound or microwave-based, and can be classified in various ways depending on the method used, such as point-sensing versus

continuous-sensing sensors, or intrusive versus non-intrusive sensors. The variables that are measured by the sensors can be conductivity, impedance, electric current and reflection properties of light/laser beams, ultrasound waves or microwaves. Signal processing cards usually come with the sensory system, otherwise these are available off-the-shelf.

2. *Controller*: This is a computer program that can evaluate the incoming process data and generate input signals for the gates that will manipulate the resin flow in the desired direction.

3. *Actuators*: These are the controllable process inputs that drive the physical system. In the mold filling process, the only actuators are the injection gates in the mold. One can manipulate the resin flow by varying the injection conditions at the gates.

The control methods that were used for the mold filling process can be divided into two as *set point-based control* and *scenario-based control*. In set point-based control, the injection system with two or more gates and *a* certain number of vents is designed at first. A target flow pattern – usually the nominal flow pattern – is designated. During the mold filling operation, the control is conducted periodically: during a *sampling time* Δt, the flow front position is determined through the sensors and is compared with the target flow front position at the current time step. The error is estimated and, based on the *traction error*, the controller generates the optimal input signals for the injection gates to minimize the error between the actual and target flow front position. Neural networks, adaptive control and model-based control are some of the set point-based control methods [57–61].

There is an offine stage and an online stage in the scenario-based control. In the offline step, the injection system is designed for the nominal case. A scenario space that aims to model the disturbances is created. The sensory system is designed and detection tables are created to identify scenarios in the closed mold. Auxiliary injection gates are placed into the mold. Using various algorithms, control actions that fill each scenario successfully are determined, concluding the offline stage. The online stage can divided into two: the *recognition step* and the *control step*. During the recognition step, the sensory system tracks the resin flow and associates it with a scenario from the scenario space. Once the closest scenario has been found, the associated control actions are undertaken in an open-loop manner. Decision trees, genetic algorithms (GA) and ES are used for scenario-based control [7, 62, 63].

7.6.1   Simulation-based liquid injection control (SLIC)

A scenario-based learning control system has been developed at the University of Delaware [63]. The system, named Intelligent RTM, consists of an offline stage called simulation-based liquid injection control (SLIC) and an online stage called AutoRTM, which is a LabView program. Intelligent RTM has eight modules:

- *Level 0*: Given the mesh of the part, nominal material parameters and the disturbance ranges, a scenario space is created.
- *Level 1*: The injection system for the disturbance-free case is designed. Optimal gate and vent locations are determined simultaneously using GA.
- *Level 2* (Only for LCM processes that utilize distribution media): The distribution media for the part is designed for the disturbance-free case.
- *Level 3:* The locations of the auxiliary gates that will undertake control actions and the locations of the point sensors that will activate the gates are determined optimally using GA. The auxiliary system is tested in the scenario space.
- *Level 4:* A sufficient number of vents are placed optimally in the mold using GA and detection tables are created to choose the scenario from the scenario space that is closest to the actual mold filling state.
- *Level 5:* The data from the offline stage are converted into a form (Lab View format) that is usable by the online components of the system. This step completes the offline stage of the control system. The mold filling operation can be initiated and controlled using the data from the offline stage.
- *Level 6:* When a racetracking scenario that does not belong to the scenario space takes place during mold filling, it is added to the detection table and control actions that will fill the scenario successfully using the existing hardware are sought offline. This *learning* ability gives flexibility to SLIC.
- *Level 7* (under development): Performance statistics reports that include the composite quality, cycle time and cost are prepared automatically.

7.6.2   Case study: rectangular mold with triangular insert

SLIC is demonstrated on a rectangular part with a triangular insert as shown in Fig. 7.22. The top and bottom edges, and three edges of the triangle are identified as potential racetracking channels. In all channels, the racetracking strength is modeled with the racetracking set $\mathscr{R} = \{20, 100\}$, which yields a scenario space with $2^5 = 32$ scenarios. The resin is injected from two injection lines on the left and right edges of the geometry. Auxiliary gates are placed optimally into the mold and the control actions that fill each scenario successfully are determined in level 3. In level 4, the scenario detection sensors are placed into the mold and a detection table is created. The snapshots from an experiment are illustrated in Fig. 7.23 along with the flow pattern of the associated scenario, scenario 29 with $\rho = 100$ in channels 1, 2 and 5 and $\rho = 20$ in channels 3 and 4. After the detection stage, the control actions

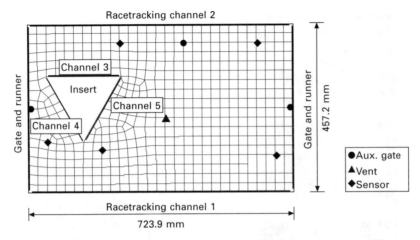

7.22 The mesh of the part used for the SLIC demonstration. The detection sensors and the auxiliary gate locations are also shown in the figure.

associated with scenario 29 – close all injection gates and open the auxiliary gate at the top – were taken to complete the mold filling successfully, which would otherwise fail.

7.6.3   Gate effectiveness

The mold filling process is irreversible: the wetted region always expands, it does not shrink. It is impractical, if not impossible, to reverse the process by sucking the resin back out of the mold. Mathematically, feedback control allows one to control the pressure at a certain number of points in the mold cavity, thus modifying the pressure gradients that drive the resin flow. However, owing to the process physics, the effects of such control on the pressure gradients are highly diffusive and lack any directional preference. Hence, as the resin flow front moves away from the gates, the ability of the gate to manipulate the resin flow, called the *gate effectiveness*, is diminished.[3] There is only a limited time window within which a gate can significantly modify

---

[3] From the *control theory* perspective, the loss of gate influence in steering the flow front indicates loss of *controllability*, which is defined as the ability to drive a dynamic system from an initial state to a target state in finite time [64]. The dynamic systems that are governed by partial differential equations such as mold filling process are called *distributed parameter systems* (or infinite dimensional systems) [65–67]. Theoretical operator, Hilbert uniqueness and moment are some of the methods used to investigate controllability of distributed parameter systems [68–70].

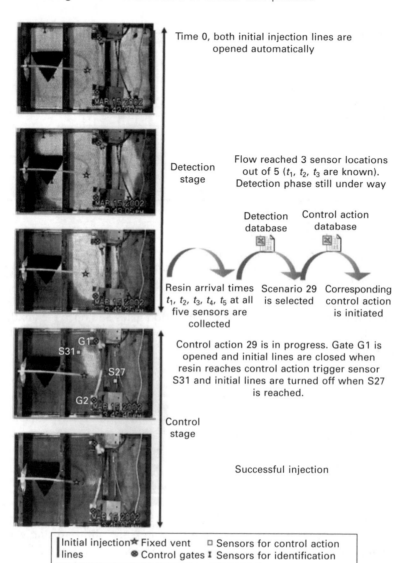

7.23 Snapshots from a mold-filling experiment. The flow pattern was identified as scenario 29 from the scenario space and the associated control actions were taken to achieve complete filling.

the resin flow pattern. This phenomenon is illustrated in the simulation environment with a case study below.

Mold III, a long rectangular mold with a line resin source on the far left and an injection gate on the bottom edge, is designed as illustrated in Fig. 7.24(a). Resin proceeds in straight lines from left to right. A coordinate system is attached to the mold such that its origin coincides with the point

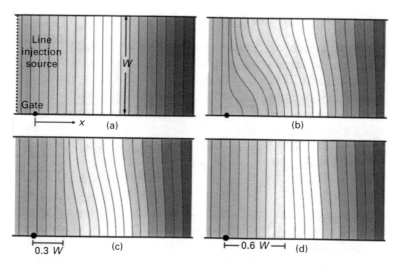

7.24 (a) Mold III, a rectangular mold with a line injection at the left and an injection gate at the bottom. (b) The gate at $x = 0$ is switched on as soon as the flow reaches it. (c) The gate at $x = 0$ is switched on after the flow reaches $x = 0.3W$, where W is the width of the mold. (d) The gate at $x = 0$ is switched on after the flow reaches $x = 0.6W$. In all cases, the flow proceeds from left to right from a constant flow rate line injection source on the left. When the gate is switched on, the line source is switched off.

gate location and its x axis is aligned with the flow direction. Principal directions of the permeability tensor are aligned with x and y directions and the preform is isotropic: $K_{xx} = K_{11} = K_{yy} = K_{22}$. The gate is initially closed. In the first case, the gate is switched on as soon as the flow front arrives there (Fig. 7.24b). In the second case, flow is allowed to proceed until $x = 0.3W$, where W is the width of the mold (Fig. 7.24c), before the point gate is switched on. In the final case, the gate is turned on when the flow is at $x = 0.6W$ (Fig. 7.24d). In all cases, the line source at the left edge is turned off when the gate is turned on. The simulations show that the effect of the gate on the flow front strongly depends on the position of the gate location with respect to the flow front position.

Since the flow front is continuously steered through a set of injection gates in set point-based control, the gate effectiveness is more crucial in this case. The gate effectiveness has been addressed as a secondary topic in a few studies [58, 60]. Gokce and Advani studied gate effectiveness in one-dimensional flow and used the results to derive guidelines for the gate effectiveness in higher-dimensional flows [71]. They investigated the effectiveness of the gate in Mold III through a statistical study by varying the vertical location of the gate and the flow front location separately.

7.7   Passive control

In passive control, instead of steering the resin towards the nominal vent locations, additional vents are placed into the mold at locations where resin is likely to arrive last. Depending on the racetracking situation, last-filled regions may take place at various locations of the mold cavity, which may require an impractical number of vents. Therefore, it is necessary to determine which regions of the mold cavity are more likely to receive resin last and place vents at these locations only. Hence, the racetracking conditions and their probability must firstly be predicted.

Although racetracking conditions are not repeatable, they depend on various parameters such as preform cutting/placement tolerances and variations in manufacturing conditions. Figure 7.21 in section 7.5.2 displays the racetracking measurements from 50 experiments. Assuming the manufacturing conditions are stable and sufficient number of experiments are conducted, the distribution of ρ values in the figure can be used as a racetracking forecast for future experiments. Note that racetracking sets and scenario spaces were created in sections 7.5.2 and 7.6.2. Below, a more general and comprehensive approach to globally model racetracking conditions in the mold cavity is presented [44] and each step is illustrated on the simplified version of a window pane mold, which will be used in a case study in section 7.7.4.

- Possible racetracking channels in the mold cavity are identified. Figure 7.25 shows the potential racetracking channels in the window pane mold.

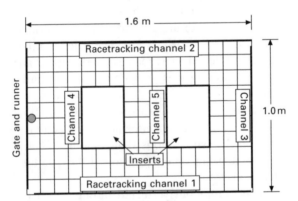

7.25 Potential racetracking channels in the window pane mold geometry.

- While the racetracking strength in a channel (ρ) is ultimately unpredictable during an experiment, it is possible to develop a probabilistic measure such as, 'On a selected side of the mold, it is 20% likely that there will be negligible racetracking, 50% likely that there will be low racetracking,

20% likely that there will be medium racetracking and 10% likely that there will be strong racetracking'. Density functions from the probability theory are utilized as probabilistic measures of the racetracking conditions in the channels. The probability that the racetracking strength in a channel will occur within a certain range, $\rho_1 \geq \rho \geq \rho_2$, can be estimated from a density function. The density function that characterizes racetracking behavior in a racetracking channel is called as the *racetracking forecast*. The ideal way to obtain the density function is to use process statistics from the manufacturing floor. The tolerances of preform cutting and the equipment used to place the fabrics in the mold, plus the experience of the operators, will also influence the density function. Density functions can be derived only under steady manufacturing conditions, and the degree of agreement between the density function and the actual racetracking conditions in the mold will influence the efficiency of the optimized results on the manufacturing floor. Variations of Weibull density function (Fig. 7.26a) are used to model racetracking conditions in the window pane mold.

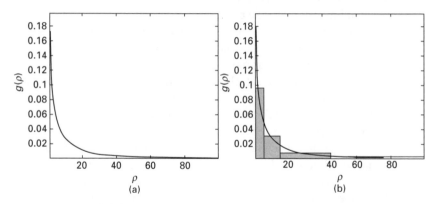

7.26 (a) Weibull probability density function. (b) Continuous probability density function is discretized into a racetracking set with five racetracking values.

- The density functions are usually continuous. Since a continuous function is not usable in optimization studies, it is discretized into a racetracking set, $\mathscr{R}_i = \{\rho_1, \ldots, \rho_m\}$, where i is the index of the racetracking channel and m is the number of elements in the racetracking set. Each element in the racetracking set represents a ρ interval from the ρ range. A probability value is calculated for each element in $\mathscr{R}$ from the density function. The likelihood that an actual racetracking value will be in a certain ρ interval is given by the probability value of the $\mathscr{R}$ element which represents that interval. The sum of the probability values is equal to 1. All five density functions used for the window pane mold are discretized into four ρ values as shown in Table 7.2 and corresponding probability values are estimated from the Weibull density function (Fig. 7.26b).

Table 7.2 Racetracking sets of the channels in the window pane mold and associated probability values

	Racetracking set	Associated probability values
Channel 1	$\mathscr{C}_1$ = {1, 10, 30, 100}	P_1 = {0.44, 0.31, 0.20, 0.05}
Channel 2	$\mathscr{C}_2$ = {1, 10, 30, 100}	P_2 = {0.44, 0.31, 0.20, 0.05}
Channel 3	$\mathscr{C}_3$ = {1, 10, 25, 80}	P_3 = {0.39, 0.35, 0.19, 0.07}
Channel 4	$\mathscr{C}_4$ = {1, 10, 25, 80}	P_4 = {0.39, 0.35, 0.19, 0.07}
Channel 5	$\mathscr{C}_5$ = {1, 10, 25, 80}	P_5 = {0.39, 0.35, 0.19, 0.07}

- After all channels are modeled with appropriate racetracking sets, a scenario space S is created by taking all possible racetracking arrangements: $\mathscr{C}_1 \times \mathscr{C}_2 \times \ldots \times \mathscr{C}_n$, where $\times$ is the Cartesian product[4] operator and n is the number of racetracking channels in the mold. Each scenario represents a process window that may take place during the actual mold filling process and has a certain probability of occurrence. Assuming the racetracking forecast of each channel is independent of the other racetracking channels, the probability of each scenario is estimated by multiplying the probability values of the racetracking set elements that make up the scenario. The probability of a scenario determines its weight in the scenario space, which has a total probability of 1. With the addition of the probability dimension, the scenario space represents mold filling conditions, including preform cutting and placement operations. Practically, it predicts what might happen during mold filling on the manufacturing floor. Note that scenario space is an approximation to the racetracking forecast. As for the window pane mold, the scenario space is created by taking the Cartesian product of the racetracking sets, $\mathscr{C}_1 \times \mathscr{C}_2 \times \mathscr{C}_3 \times \mathscr{C}_4 \times \mathscr{C}_5$, which yields 1024 scenarios. The runner location on the left edge was known beforehand, which simplified the racetracking modeling. Otherwise, the scenario space is independent of the gate locations and injection conditions.

7.7.1   Optimization problem

The locations of the vents do not have an effect on the mold filling process until resin arrival. *Assuming they are closed shortly after resin arrivals*, vent

---

[4] The Cartesian product in set theory resembles *outer multiplication operation* in *tensor notation*, except that the operands are not actually multiplied in the Cartesian product but are grouped together.

locations act as a constraint on the mold filling process, rather than a process variable.

> **Constraint I.** At any given time during mold filling $t = t_i$, each contiguous air bubble in the mold cavity must have at least one vent within its volume; $p(\mathscr{V}) : \mathscr{D}_i \cap \mathscr{V} \neq 0$, where p is the constraint, $\mathscr{D}_i$ is the set of nodes in the contiguous air bubble i and $\mathscr{V}$ is the vent configuration. A vent configuration is defined as the set of all nodes that possess a vent.

When an air bubble is no longer connected with any vent ($\mathscr{D}_i \cap \mathscr{V} = 0$) then Constraint I is violated and a dry spot forms[5]. Therefore, given the racetracking forecast, the vents should be placed optimally in the mold to minimize the likelihood of dry spot formation. Consequently, the optimization problem can be stated as follows:

> **Problem statement:** Given a racetracking forecast (approximated by scenario space S), a non-empty, closed set $\mathscr{A} \subset R$ {: integer} and a function[6] $f : \mathscr{A}^n \rightarrow R$, find at least one vent configuration $\mathscr{V}^* \in \mathscr{A}^n$ satisfying $f(\mathscr{V}^*) \geq f(\mathscr{V})$ for all $\mathscr{V} \subset \mathscr{A}^n$, where n is the number of vents and $\mathscr{A}^n = \mathscr{A} \times \ldots \times \mathscr{A}$ (n times Cartesian product).

In LCM, $\mathscr{A}$ is the set of nodes in the finite element mesh of the mold where a vent can be placed. Function f is the *performance index* (cost function) that needs to be maximized. The performance index is chosen as the *mold filling success rate*, which is defined as the cumulative probability of the scenarios that are successfully drained by a given vent configuration $\mathscr{V}$:

$$\text{Performance index of } \mathscr{V} : f(\mathscr{V}) = P(\mathscr{B} = \{ \text{ scenarios drained by } \mathscr{V} \})$$

7.9

where event $\mathscr{B}$ is a subset of scenario space $\mathscr{S}$. For example, in a given problem with a certain racetracking forecast, if scenarios ξ_2, ξ_7 and ξ_{19} are drained successfully by vent configuration $\mathscr{V}_1$, then $\mathscr{B} = \{\xi_2, \xi_7, \xi_{19}\}$ and $f(\mathscr{V}_1) = P(\xi_2 + (P(\xi_2) + P(\xi_7) + P(\xi_{19})$ according to eqn 7.9. Note that function p in Constraint I is integrated into f in the optimization problem statement above. Since an analytical expression for f is not available, mold filling simulation output from LIMS is used to estimate the value of $f(\mathscr{V})$.

---

[5] In this chapter, as soon as a dry spot forms, it is recorded and the dynamics of the air bubble motion in the closed mold are ignored.

[6] f maps elements of domain $\mathscr{A}^n$ to real numbers R.

7.7.2   Last-filled region and vent fitness maps

There is a close relationship between the last-filled regions and the optimal vent locations in the mold. Once the scenario space has been created and the injection system designed, the mold filling is simulated for the entire scenario space. In the window pane mold, the resin is injected through a gate and number on the left edge. Mold filling was simulated for 1024 scenarios in the scenario space. Figure 7.27 illustrates the flow patterns and the last-filled regions from six randomly selected scenarios. The racetracking values (ρ) of the channels in each scenario are shown below the flow pattern: the first value belongs to channel 1, the second value belongs to channel 2, etc. Next the last-filled regions from all scenarios are superimposed – taking scenario probabilities into account – to determine where last-filled regions concentrate over the mold geometry. The distribution of the last-filled regions over the window pane mold is illustrated in Fig. 728(a), which is called a *last-filled region map*. The resin is more likely to arrive at darker mold regions in the figure.

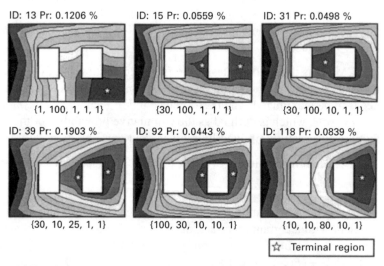

7.27 The flow patterns of six sample scenarios from the scenario space of the window pane mold. The resin runner lies along the left edge. ID is the index of the scenario in the scenario space and Pr is the (percentage) probability of the scenario. Five values under the figures show the racetracking strengths along the five channels in the mold for the given scenario.

Ideally, a *successful mold filling* implies that there are no unfilled nodes remaining when the resin arrives at the last vent. However, this is a very restrictive definition for the optimization problem under study, because it will be a challenge to find a feasibly small vent configuration with a significant

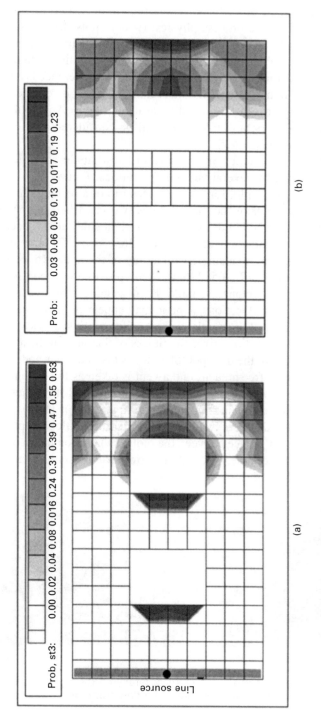

Line source

Prob: 0.03 0.06 0.09 0.13 0.017 0.19 0.23

(a)

(b)

7.28 (a) Last-filled region distribution in window pane mold. (b) Corresponding vent distribution plot that shows the merit of nodes to become vent locations.

performance index. In order to relax this requirement, two tolerances are introduced, based on the process physics: (i) dry spots that include fewer nodes than a preset value at the time of formation are ignored, and (ii) in order to drain air bubbles in the vicinity of a vent, a certain amount of resin is allowed to bleed after it arrives at the vent. The first tolerance is quantified in terms of nodes in a dry spot and the second tolerance is represented as a dry spot-proof region of certain radius around a vent. Any mold filling simulation that satisfies both tolerances will be considered as *successfully drained* (or *successfully filled*).

Owing to the introduction of the tolerances, a single scenario can be successfully drained by several vent configurations. Figure 7.29 illustrates the mold filling contour plot for a certain scenario of a rectangular mold. A single gate at the center of the mold creates two last filled regions on the right and left sides of the mold. The air on the right side can be successfully drained by a single vent at any of the nodes {265, 276, 277}. The air on the left side can be drained by a single vent at any of the nodes {5, 6, 16, 17}. Hence, nodes {5, 6, 16, 17, 265, 276, 277} are possible vent configurations in this scenario of the rectangular mold. Each scenario will give a different set of nodes as the vent locations. A node may appear in many scenarios as a possible vent location. The cumulative probability of the scenarios it appears in as a possible vent location, gives the fitness of that node to be a vent. For the window pane mold, the dry spot tolerance is taken as four nodes and the radius of the dry spot-free area around a vent is taken as the distance nodes between two. The vents are found for all scenarios and the fitness of the nodes to be vents are estimated and plotted in Fig. 7.28(b). The two contour plots in Fig. 7.28 are different, yet display similar morphological properties. The reason for the difference between the plots is the introduction of the dry spot tolerances. For example, Fig. 7.28(a) shows that dry spots are frequently

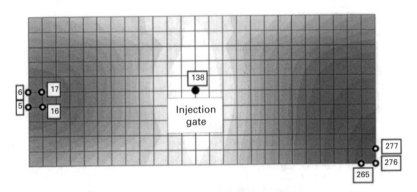

7.29 Nodes that can drain air successfully on the right and left sides of the mold. Contour plot shows the flow front locations at several time steps.

captured to the left of the two inserts in the mold. However, Fig. 7.28(b) does not suggest any vent in those two regions, because the sizes of the dry spots are smaller than the dry spot tolerance. The vent distribution plot shows the merit of nodes becoming vents, hence it can be used as an elementary vent optimization tool.

Although vent distribution plots display the merit of nodes becoming vent locations, they are not sufficient to choose optimal vent locations, as they exclude an important pieces of information: (i) each scenario is independent of other scenarios and (ii) each scenario requires all necessary vents. Even if a single required vent is missing, a scenario will not be drained successfully. For example, for the scenario in Fig. 7.29, the possible vent locations are {5, 6, 16, 17, 265, 276, 277}. However, it must be stated that there must be at least two vents in the mold to drain this scenario: one vent from set {265, 276, 277} and one vent from set {5, 6, 16, 17}. If either vent is missing, the mold filling will not be successful in the process window represented by this scenario. Vent distribution plots indicate the merit of individual nodes to be vents, but exclude the fact that certain nodes should act together in a vent configuration to be useful. Hence, a combinatorial search algorithm that works on a scenario basis is developed in the following section.

7.7.3   Combinatorial search (CS)

It has been shown by Gokce and Advani [44] that a vent configuration that is created by merging two or more vent configurations includes all the scenarios from the individual vent configurations. Thus, in combinatorial search, superior small vent configurations are found first and the optimal vent configuration is sought by merging the small vent configurations as explained below:

- *Step I.* For each scenario, the vent configurations with the smallest number of vents that save the scenario are found. Note that a scenario is saved by several vent configurations owing to the dry spot tolerances. These small vent configurations are called the *elementary vent configurations* (EVC) and are used as the building blocks of large vent configurations (i.e. vent configurations with a large number of vents) during the optimization procedure. The collection of all EVCs is called the EVC set E.
- *Step II.* The data from the previous step are rearranged to find the scenarios saved by each EVC. The cumulative probability of the scenarios saved by an EVC gives its success rate (performance index).
- *Step III.* EVCs with low success rates are eliminated from set ε. The remaining EVCs are grouped with respect to the number of vents they include. $^{i}\varepsilon$ is used to represent the group of elementary vent configurations with i vents.
- *Step IV.* Note that there is a constraint on the number of vents, n. There are

various ways to combine elementary vent configurations from ε that satisfy this constraint. For example, if four vents are allowed in a mold and the set ε has vent configurations with one, two or three vents ($\varepsilon = {}^1\varepsilon \cup {}^2\varepsilon \cup {}^3\varepsilon$), then all vent configurations that include four vents can be created as follows:

$$ {}^4\mathscr{W} = \{{}^1\varepsilon \times {}^1\varepsilon \times {}^1\varepsilon \times {}^1\varepsilon\} \cup |{}^1\varepsilon \times {}^1\varepsilon \times {}^2\varepsilon\} \cup \{{}^2\varepsilon \times {}^2\varepsilon\} \cup \{{}^1\varepsilon \times {}^3\varepsilon\} $$

7.10

- *Step V.* Instead of creating ${}^4\mathscr{W}$ at once, one may use the following incremental procedure, which is based on the underlying theory of the combinatorial search:

  $i = 1$: ${}^1\mathscr{W} = {}^1\varepsilon$

  $i = 2$: ${}^2\mathscr{W} = \{{}^1\mathscr{W} \times {}^1\varepsilon\} \cup \{{}^2\varepsilon\}$

  $i = 3$: ${}^3\mathscr{W} = \{{}^2\mathscr{W} \times {}^1\varepsilon\} \cup \{{}^1\mathscr{W} \times {}^2\varepsilon\} \cup \{{}^3\varepsilon\}$

  $i = n = 4$: ${}^4\mathscr{W} = \{{}^3\mathscr{W} \times {}^1\varepsilon\} \cup |{}^2\mathscr{W} \times {}^2\varepsilon\} \cup \{{}^1\mathscr{W} \times {}^3\varepsilon\}$

At each step during the procedure, only superior vent configurations with high-performance indices (success rate) in ${}^i\mathscr{W}$ are retained for the next step, reducing the computational cost significantly. In, this way, it is possible to introduce *minimum required performance* as a second optimization stop criterion, which will yield fewer vents with suffcient performance. Note that optimal vent configurations that include one, two or three vents are also determined as a by-product of the procedure above.

7.7.4   Case study: the window pane mold

The combinatorial search is demonstrated on the window pane, a 1.6×1.0 m^2 rectangular mold containing two inserts (Fig. 7.25). The mesh of the geometry includes 175 nodes, 136 2D elements (quads) and 80 1D elements (bars). 1D elements are used to model racetracking channels. Top, bottom and right edges, and the perimeters of the inserts are identified as five racetracking channels. The racetracking conditions in each channel are forecast using variations of the Weibull density function. The racetracking sets and associated probability values are shown in Table 7.2. A scenario space with 1024 scenarios is created to model the racetracking condition in the mold. The injection scheme is specified as constant flow rate through node 70 ($x = 0$, $y = 0.5$ m) on the left edge, which is connected to a runner along the left edge. The last-filled region map is illustrated in Fig. 7.28(a).

The number of vents is selected as four. Dry spots with three or fewer nodes are tolerated and an area of 2-node radius around a vent is considered dry spot-proof. Set ε is generated with 465 EVCs, which include one or two vents, i.e. $\varepsilon = {}^1\varepsilon \cup {}^2\varepsilon$. Using the incremental procedure at step V above, the optimal vent configurations with one, two, three and four vents are found consecutively. The optimal vent configurations are also investigated using an exhaustive search. A vent candidate pool $\mathscr{P}$ is created from the nodes that appeared in a last filled region, at least in one scenario. Vent configurations are created by taking Cartesian products of $\mathscr{P}$. When the computational cost becomes prohibitive for large vent configurations, the candidate pool is sampled evenly (geometrically) to reduce the pool size. The ES conducted using a downsized candidate pool is called *partial exhaustive search* (partial ES). While ES provides the optimum solution, a partial ES may not provide the optimum solution, but will indicate near-optimal solutions. The results of CS and ES are compared for accuracy and efficiency. The computational cost is reported in terms of the duration of the optimization process. Gateway 2000 computers with Pentium 4 1600 MHz processors and 256MB RAM were used for this study. The results for CS and (partial) ES are listed in Table 7.3. The results for the 3-vent and 4-vent cases are illustrated in Fig. 7.30. In each case, all nodes that appeared as vent locations are shown on the figure. The performance index of the CS result is shown with a pie chart on each figure. The computational cost comparison between the CS and ES (or partial ES) is illustrated with a bar plot. CS results are as accurate as ES results and better than partial ES results. Computational savings are 57%, 91%, 99% and 97% for 1-vent, 2-vent, 3-vent and 4-vent optimization, respectively.

Table 7.3 CS and ES optimization results for window pane mold. Performance index is the cumulative probability of the scenarios that are drained by $\mathscr{V}^*$

Optimization method	Optimal vent configuration[a], $\mathscr{V}^*$ (node numbers)	Performance index, $f(\mathscr{V}^*)$	Computational cost (s)
Creation of set ε	–	–	69
CS	118	27.6%	69
ES	118	27.6%	162
CS	107 117	48.2%	209
ES	107 117	48.2%	2,439
CS	57 107 117	69.0%	301
ES	57 107 117	69.0%	44,570
CS	57 106 118 168	87.9%	1,364
Partial ES	57 107 117 169	87.3%	51,165

[a]x and y coordinates of the nodes are as follows (in m): 57 (1.4, 0.0), 106 (1.3, 0.5), 107 (1.3, 0.6), 117 (1.6, 0.4), 118 (1.6, 0.5), 168 (1.4, 0.9), 169 (1.4, 1.0).

As a benchmark, an intuitive vent design, illustrated in Fig. 7.31, was tested in the scenario space. The success rate was only 28% as compared to 69% with the optimal 3-vent vent configuration.

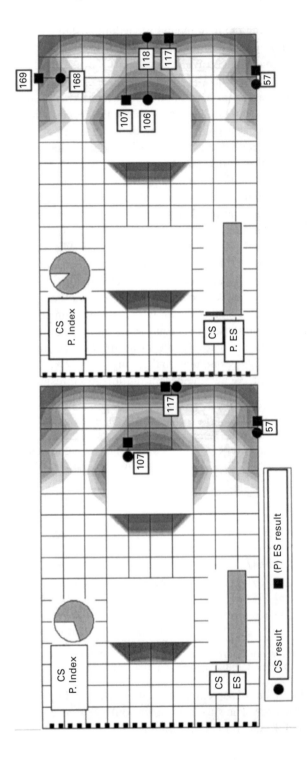

7.30 Optimal 3-vent (*left*) and 4-vent (*right*) vent configurations in the window pane mold as found by combinatorial search (CS) and exhaustive search (ES). Since each optimization method returned different results, vent locations are shown by different markers, as illustrated in the legend.

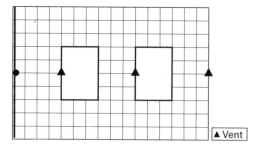

7.31 The vent configuration used as a benchmark to evaluate the performance of the optimization algorithm Combinatorial Search.

Mold symmetry and scenario space

The window pane mold has many symmetric features: the geometry, injection conditions, density function and the racetracking sets of symmetric channels. Consequently, last-filled region and vent fitness maps are also symmetric. Therefore:

- Two vent configurations that are symmetries of each other will yield the same results.
- It is erroneous to investigate a section of the geometry only, since several scenarios in the scenario space are not symmetric owing to different racetracking conditions along the symmetric channels. Yet, several scenarios are symmetric with respect to each other. Two scenarios are called *symmetric* if the resin flow pattern of one is a symmetry of the flow pattern in the other across an axis. The size of the scenario space can be reduced by eliminating one of the symmetric scenario groups. Later, the optimal vent configuration for the reduced scenario space can be mirrored across symmetry axes to create the optimal vent configuration for the entire scenario space.

7.8   Conclusion

Resin flow through textile preforms in the mold cavity was modeled in section 7.2 using Darcy's law. Liquid injection molding simulation (LIMS), which was used to simulate resin flow in the mold cavity, was introduced in section 7.3. Its features and utilization in design, optimization and control studies were presented. The branch and bound search was adapted to gate location optimization in section 7.4 and its application was demonstrated with a case study. Section 7.5 introduced the disturbances associated with the mold filling process. Racetracking was identified as the key disturbance and it was shown with a case study that it is not repeatable. Active and passive control methods that enable successful mold filling in the presence of racetracking were proposed in sections 7.6 and 7.7, respectively. In active control, which was exemplified by simulation-based liquid injection control (SLIC), the resin flow is redirected by varying the injection conditions at the injection gates. In passive control, racetracking conditions were modeled probabilistically and a sufficient number of vents are placed at locations where resin is more likely to arrive last. The approaches and procedures to create these strategies were presented with pedagogic examples.

7.9   Outlook

The knowledge base about LCM processes is growing every day. Several tools and methodologies that address modeling, design, optimization and control issues are available or under development. Ultimately, the aim is to develop the LCM technology to the extent that simple, high-volume parts will be manufactured quickly, and large, complex and low volume parts will be manufactured successfully. A maximum level of automation is desirable, especially for the former group. Since automation makes the manufacturing steps repeatable, the average quality of the parts will increase. Automation also cuts down the labor costs and decreases the health hazards by decreasing the exposure of the workers to the resin fumes and preform soot. Reducing the reject rate is a prime goal in all cases.

Further research and development is necessary to bring LCM processes to their full potential. The work presented in this chapter spans several important issues in mold filling. Some of the future research directions are presented below.

7.9.1   Multiple gate optimization

This topic is rarely addressed by the researchers, possibly because of the extensive computational effort required. There is a permutational relationship between the number of the gates and the problem size: in a mold with N nodes, there are N possibilities for the first gate, $N(N-1)$ possibilities for the second gate, etc. An optimally designed multi-injection system can reduce the fill time significantly and hence the need to explore optimization methods that can address this issue effectively.

7.9.2   Optimal placement of flow enhancement media

In LCM processes such as VARTM (or vacuum infusion), a flow enhancement medium is placed on the entire surface of the preform to distribute the resin quickly. However, this method is not applicable when there are mold regions that are inaccessible for the flow enhancement medium or the vacuum line as shown in Fig. 7.32. Also, if the mold geometry is complex and includes

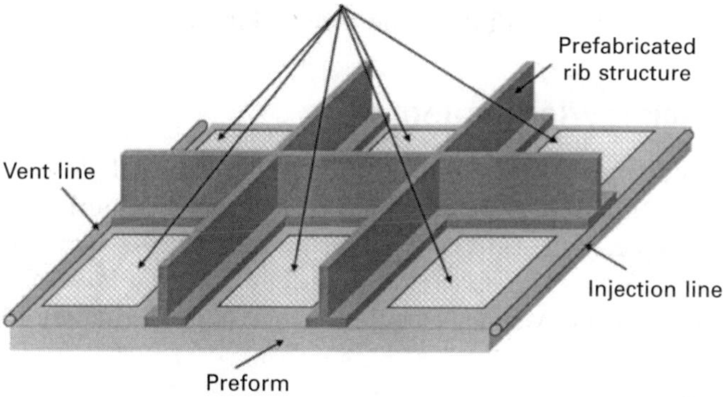

Flow enhancement media (number of plies = 2, 1 or 0)

Prefabricated
rib structure

Vent line

Injection line

Preform

7.32 A composite part geometry that does not allow the placement of flow enhancement media over the entire geometry.

inserts or ribs, such a design may fail to fill the mold successfully. Therefore, the flow enhancement medium needs to be designed optimally in these cases in order to increase the efficiency of the mold filling operation.

7.9.3 Use of manufacturing data for racetracking forecast

In section 7.7, a generic density function was used to forecast racetracking in the racetracking channels for illustration purposes. In actual applications, the racetracking forecast must capture the manufacturing conditions, which are assumed to be stable. Statistics from the manufacturing floor and equipment, material and processing tolerances must be incorporated into the racetracking forecast. Even if it is not possible to capture the actual case exactly, a sufficiently accurate racetracking forecast will be very useful to test various designs in the simulation environment.

7.9.4 Feasible target flow pattern selection in set point-based active control

Set point-based feedback controllers generate control inputs based on the tracking error between the target flow pattern and the actual flow pattern. A flow pattern strongly depends on the permeability field in the mold cavity. Given a permeability field and a gate configuration, there is a finite range of flow patterns that can be created by modifying the injection conditions at the gates, which is called the *reachable set*. For a controller to be successful in minimizing the tracking error, the target flow pattern must belong to this set. Therefore, it is crucial to predict the reachable set in the presence of disturbances and select a feasible target flow pattern. Multiple target flow patterns can be selected as needed depending on the racetracking forecast.

7.10   Acknowledgements

The financial support, provided by the Office of Naval Research (ONR) under Grant N00014-97-C-0415 for the Advanced Materials Intelligent Processing Center at the University of Delaware, made the work presented in this chapter possible. The invaluable contributions of Dr Pavel Simacek, Dr Kuang-Ting Hsiao and Mathieu Devillard to this chapter are kindly appreciated.

7.11   References

1. Advani S.G. and Sozer M.E., *Process Modeling in Composites Manufacturing.* Marcel Dekker, New York, 2003.
2. Advani S.G., Resin transfer molding in *Flow and Rheology in Polymeric Composites Manufacturing*, pages 465–516. Elsevier Publishers, Amsterdam, 1994.
3. Rosato D.V. and Lee (ed.) S.M., Polymeric Matrix Composites in *International Encyclopedia of Composites*, volume 3, pages 148–181. VCH Publishers, Inc., New York, 1990.
4. Schwartz M.M., *Composite Materials Handbook.* McGraw Hill, New York, 1992.
5. Gokce A., Modeling of flow continuum and optimal design of control-oriented injection systems in liquid composite molding processes. PhD thesis, University of Delaware, 2003.
6. Bickerton S., Modeling and control of flow during impregnation of heterogeneous porous media, with application to composite mold filling processes. PhD thesis, University of Delaware, Spring 1999.
7. Stadtfeld H.C., Experimental validation of impregnation schemes with control strategies in liquid composite molding processes. Master's thesis, University of Delaware, 2000.
8. Kueh S.R.M., Fiber optic flow monitoring system and methodology for sensor placement to identify flow disturbances in liquid composite molding. PhD thesis, University of Delaware, Summer 2001.
9. Kozeny J., *Sitzungsberichte Abt IIa.* Wiener Akademie der Wissenschaft, 1927.
10. Carman P.C., *Fluid Flow Through Granular Beds.* Transactions 15, Institution of Chemical Engineers, 150–166, 1937.
11. Dave R., Kardos J.L. and Duducovic M.P., A model for resin flow during composite processing. part 2: Numerical analysis for unidirectional graphite/epoxy laminates. *Polymer Composites*, 8(2), 1987.
12. Lam R.C. and Kardos J.L., The permeability and compressibility of aligned and cross-plied carbon fiber beds during processing of composites. In *Proceedings of ANTEC, SPE*, New York, May 1989.
13. Bruschke M.V., A predictive model for permeability and non-isothermal flow of viscous and shear-thinning fluids in anisotropic fibrous media. PhD thesis, University of Delaware, 1992.
14. Chiemlewski C., Petty C.A. and Jeyaraman K., Crossflow permeation of viscous and viscoelastic liquids in cylinder arrays. In *Proceedings of the American Society for Composites Fifth Technical Conference*, East Lansing, June 1990.

15. Simacek P. and Advani S.G., Permeability model for woven fabrics. *Polymer Composites*, 17(6): 887–899, December 1996.

16. Chang W. and Kikuchi N., Analysis of non-isothermal mold filling process in resin transfer molding (RTM) and structural reaction injection molding (SRIM). *Computational Mechanics*, 16(1): 22–35, 1995.

17. Lee C.K., Sun C.C. and Mei C.C., Computation of permeability and dispersivities of solute or heat in periodic porous media. *International Journal of Heat and Mass Transfer*, 39(4): 661–676, 1996.

18. Dimitrovova Z. and Faria L., Finite element modeling of the resin transfer molding process based on homogenization techniques. *Computers and Structures*, 76(1): 379–397, 2000.

19. Koponen A., Kataja M. and Timonen J., Statistical physics, plasmas, fluids, and related interdisciplinary topics. *Physical Review E*, 56(3): 3319–3320, 1997.

20. Spaid M. and Phelan F., Modeling void formation dynamics in fibrous porous media with the lattice boltzman method. *Composites Part A*, 29 (7): 749–755, 1998.

21. Gokce A. and Advani S.G., Permeability estimation with the method of cells. *Journal of Composite Materials*, 35(8): 713–728, 2001.

22. Nedanov P., Permeability characterization of fibrous porous media. Master's thesis, University of Delaware, 2002.

23. Gebart B.R., Permeability of unidirectional reinforcements for RTM. *Journal of Composite Materials*, 26(8): 1100–1133, 1992.

24. Verheus A.S. and Peeters J.H., A. The role of reinforcement permeability in resin transfer molding. *Composites Manufacturing*, 4(1): 33–38, 1993.

25. Parnas R.S. and Salem A.J. A comparison of the unidirectional and radial in-plane flow of fluids through woven composite reinforcements. *Polymer Composites*, 14(5): 383–394, 1993.

26. Ferland P., Guittard D. and Trochu F., Concurrent methods for permeability measurement in resin transfer molding. *Polymer Composites*, 17(1):149–158, 1996.

27. Wang J.T., Wu C.H. and Lee L.J., In-plane permeability measurement and analysis in liquid composite molding. *Polymer Composites*, 15(4): 278–288, 1994.

28. Weitzenbock J.R., Shenoi R. A. and Wilson P.A., Measurement of three-dimensional permeability. *Composites Part A*, 29(1–2): 159–169, 1998.

29. Bruschke M.V. and Advani S.G., A finite element/control volume approach to mold filling in anisotropic porous media. *Polymer Composites*, 11: 398–405, 1990.

30. Bruschke M.V. and Advani S.G., A numerical approach to model non-isothermal viscous flow through fibrous media with free surfaces. *International Journal for Numerical Methods in Fluids*, 19: 575–603, 1994.

31. Lee L.J., Young W.B. and Lin R.J., Mold filling and cure modeling of RTM and SRIM processes. *Composite Structures*, 27(1–2): 109–120, 1994.

32. Tredoux L. and Westhuizen J., Van der. Development of a numerical code that simulates combined heat transfer, resin flow and compaction during composite processing. *Composites Manufacturing*, 6(2): 85–92, 1995.

33. Fracchia C.A., Castro J. and Tucker III C.L., A finite element/control volume simulation of resin transfer molding. In *Proceedings of the ASC 4th Annual Technical Conference*, Santa Monica, October 1995.

34. Young W.-B., Consolidation and cure simulation for laminated composites. *Polymer Composites*, 17(1): 142–148, 1996.

35. Liu X.-L., Isothermal flow simulation of liquid composite molding. *Composites Part A: Applied Science and Manufacturing*, 31(12): 1295–1302, Dec 2000.

36. Simacek P., Sozer E.M. and Advani S.G., *User Manual for DRAPE 1.1 and LIMS 4.0 Liquid Injection Molding Simulation, Technical Report UD-CCM 98-01.* Center for Composite Materials, University of Delaware, Newark, DE, 1998.

37. The Mathworks, Inc., http://www.mathworks.com. *Learning MATLAB.*

38. Mathur R., Advani S.G. and Fink B.K., Use of genetic algorithms to optimize gate and vent locations for the resin transfer molding process. *Polymer Composites*, 20: 167–178, 1999.

39. Gokce A., Hsiao K.-T. and Advani S.G., Branch and bound search to optimize injection gate locations in liquid composite molding processes. *Composites: Part A*, 33: 1263–1272, 2002.

40. Papalambros P.Y. and Wilde D.E., *Principles of Optimal Design: Modeling and Computation.* Cambridge University Press, Cambridge, 1988.

41. Rao S.S., *Optimization: Theory and Applications.* Wiley Eastern Limited, New Delhi, 1979.

42. Parker R.G. and Rardin R.L., *Discrete Optimization.* Academic Press Inc., San Diego, 1988.

43. Horst R. and Tuy H., *Global Optimization: Deterministic Approaches.* Springer-Verlag, Berlin, 3rd edition, 1996.

44. Gokce A. and Advani S.G., Combinatorial search to optimize vent locations in the presence of disturbances in liquid composite molding processes. *Materials and Manufacturing Processes*, 18(2): 261–285, 2003.

45. Bickerton S. and Advani S.G., Characterization of corner and edge permeabilities during mold filling in resin transfer molding. In *Proceedings of the ASME AMD-MD Summer Meeting*, volume 56, pages 143–150, Los Angeles, CA, June 1995.

46. Bickerton S. and Advani S.G., Characterization of racetracking in liquid composite molding processes. *Composites Science and Technology*, 59: 2215–2229, 1999.

47. Han K., Wu C.H. and Lee L.J., Characterization and simulation of resin transfer molding – racetracking and dry spot formation. In *Proceedings of Ninth Annual ASM-ESD Advanced Composite Conference,* Michigan, 1993.

48. Leek R., Carpenter G., Donnellan T. and Phelan F., Simulation of edge flow effects in resin transfer molding. In *Proceedings of the 25th International SAMPE Technical Conference*, pages 233–245, Philadelphia, Oct 1993.

49. Ni J., Zhao Y., Lee L.J. and Nakamura S., Analysis of two regional flow in liquid composite molding. *Polymer Composites*, 18(2): 254–269, 1997.

50. Sheard J., Senft V., Mantell S.C. and Vogel J.H., Determination of corner and edge permeability in resin transfer molding. *Polymer Composites*, 19(1): 96–105, 1998.

51. Hammami A., Gauvin R. and Trochu F., Modelling the edge effect in liquid composite molding. *Composites Part A: Applied Science and Manufacturing*, 29(5–6): 603–609, 1998.

52. Bickerton S. and Advani S.G., Characterization of racetracking in liquid composite molding processes. *Composites Science and Technology*, 59: 2215–2229, 1999.

53. Chaneske J., Jayaraman K., Norman D. and Robertson R., Effects of preform architecture on racetracking in liquid molding. In *Proceedings of International SAMPE Symposium and Exhibition*, pages 954–966, Long Beach, CA, May 2000.

54. Tadmor Z. and Costas G.G., *Principles of Polymer Processing.* John Wiley and Sons, New York, 1979.

55. Data compiled and evaluated by Parnas R.S. and Flynn K.M., *NIST Data Base on Reinforcement Permeability Values, Version 1.0.* NIST Std. Ref. Database 63.

56. Devillard M., Hsiao K.-T., Gokce A. and Advani S.G., On-line characterization of

bulk permeability and race-tracking during the filling stage in resin transfer molding process. In *34th International SAMPE Technical Conference*, Baltimore, November 2002.

57. Demirci H.H. and Coulter J.P., Neural network based control of molding processes. *Journal of Materials Processing and Manufacturing Science*, 2: 335–354, 1994.

58. Demirci H.H. and Coulter J.P., Control of flow progression during molding processes. *Journal of Materials Processing and Manufacturing Science*, 3(4): 409–425, 1995.

59. Demirci H.H., Coulter J.P. and Guceri S.I., Numerical and experimental investigation of neural network-based intelligent control of molding processes. *Journal of Manufacturing Science and Engineering, Transactions of the ASME*, 119(1): 88–94, 1997.

60. Berker B., Barooah P., Yoon M.K. and Sun J.Q., Sensor based modeling and control of fluid flow in resin transfer molding. *Journal of Materials Processing and Manufacturing Science*, 7(2): 195–214, 1998.

61. Nielsen D. and Pitchumani R., Real time model-predictive control of preform permeation in liquid composite molding processes. In *Proceedings of the ASME National Heat Conference*, Pittsburgh, August 2000.

62. Sozer E.M., Bickerton S. and Advani S.G., Modeling and control of liquid composite mold filling process. In *Proceedings of Flow Processes in Composite Materials (FPCM)*, Plymouth, UK, pages 109–124, July 1999.

63. Hsiao K.-T., Devillard M. and Advani S.G., Streamlined intelligent rtm processing: From design to automation. In *Proceedings of 47th International SAMPE Symposium and Exhibition*, volume 47, pages 454–465, Long Beach, CA, May 2002.

64. Westphal L.C., *Sourcebook of Control Systems Engineering*. Chapman & Hall, London, 1992.

65. Kazimierz M., Zbigniew N. and Malgorzata P., *Modelling and Optimization of Distributed Parameter Systems*. Chapman & Hall, London, 1996.

66. Tzou H.S. and Bergman L.A., *Dynamics and Control of Distributed Systems*. Cambridge University Press, Cambridge, 1998.

67. Tzafestas S.G., *Distributed Parameter Control Systems*. Pergamon Press, Oxford, 1982.

68. Avdonin S.A. and Ivanov S.A., *The Method of Moments in Controllability Problems, for Distributed Parameter Systems*. Cambridge University Press, Cambridge, 1995.

69. Grinberg A.S., Lototskii V.A. and Shklyar B., Sh. Controllability and observability of dynamic systems. *Automation and Remote Control* (English translation of *Avtomatika i Telemekhanika*), 52(1): 1–16, June 1991.

70. Cao Y., Gunzburger M. and Turner J., Controllability of systems governed by parabolic differential equations. *Journal of Mathematical Analysis and Applications*, 215(1): 174–189, Nov. 1997.

71. Gokce A. and Advani S.G., Gate effectiveness in controlling resin advance in liquid composite molding processes. *Journal of Manufacturing Science and Engineering*, 125(3): 548–555, 2003.

8

Mechanical properties of textile composites

I A J O N E S, University of Nottingham, UK and
A K P I C K E T T, Cranfield University, UK

8.1   Introduction

Textile composites are used typically because of their high strength-to-weight and stiffness-to-weight ratios. In order to exploit these properties it is essential to have a good understanding of their behaviour under load, beginning with the capability to model their stiffness and later consider their strength via suitable failure criteria. It is normal to assume that composites (especially reinforced thermosets) are linear elastic up to the point of failure, although there is some evidence that even thermoset matrix materials can yield in shear. Moreover, there is now increasing interest not just in the modelling of damage onset but also the degradation in properties of a structure as damage progresses. However, in order to be able to reach that stage, an accurate understanding of the structure's elastic behaviour must be established. Hence this chapter starts with a description of elastic behaviour, from simple approaches based on micromechanics and classical laminate theory to more sophisticated analyses developed specifically for textile composites. This is followed by a description of failure behaviour, with particular emphasis on impact loading.

8.2   Elastic behaviour

The application of elasticity to composite materials (even idealised ones, let alone those with the complex fibre architecture of textile-reinforced composites) is an extremely complex area, and any summary of this field in a few pages needs to be accompanied by some strong notes of caution. The first of these is that the ensuing description is only a flavour of the basic theory, and the reader should have no difficulty in locating a number of texts covering this basic material (and its underlying assumptions and limitations) in more detail. For a more in-depth treatment of some aspects, the reader is referred to the very detailed text by Bogdanovitch and Pastore[1], and is encouraged to review the latest literature in the field, some of which is referenced in the present chapter.

The second note of caution is that there are many variants on the notation and definitions used within the theory, with differences ranging from the superficial (the choice of symbols used to represent the various quantities) to the more fundamental (the use of matrices and contracted notation, cf. tensorial notation; mathematical, cf. engineering definition of shear strains). It is therefore vital to ensure that the correct notation and conventions are used before making use of any formula or computer program. The notation used here is essentially that of of Hull and Clyne[2] and of Jones[3], with some minor variations. A summary of the notation, and a more extended presentation of the material in section 8.2.1 is available in Appendix A of Owen et al[4].

8.2.1   Fundamentals: orthotropic solids

In order to understand the behaviour of textile reinforced composites it is necessary first to study the concepts of anisotropic and orthotropic behaviour, and to introduce relevant concepts from laminate analysis. Viewed at a microscopic level, textile composites (irrespective of their fibre architecture) are assembled from bundles of parallel fibres embedded in (and assumed to be rigidly bonded to) a matrix material that is usually polymeric and has a much lower modulus than the fibres, and is usually assumed to be homogeneous and isotropic (i.e. properties constant throughout irrespective of position or orientation). This structure will clearly have stiffness and strength properties which are much greater in the direction of the fibre than in any perpendicular direction. Figure 8.1 shows an idealisation of this structure, where the symmetry of the structure is apparent, and enables us to simplify our treatment of the elastic behaviour from the general case of true anisotropy (material properties different in all directions with no symmetry).

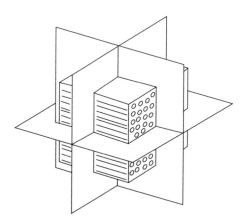

8.1 Schematic diagram of fibre-reinforced composite as an orthotropic solid showing three planes of symmetry.

The elastic behaviour (i.e. the relationship between stresses and strains in the three principal directions) of a material possessing three such planes of symmetry is characterised by the orthotropic constitutive model. This is fully defined by three elastic moduli E_1, E_2 and E_3, three independent Poisson's ratios v_{12}, v_{23} and v_{31} and three shear moduli G_{12}, G_{23} and G_{31}. Three further Poisson's ratios are related to the aforementioned moduli and Poisson's ratios:

$$v_{21} = v_{12} \frac{E_2}{E_1}; \quad v_{32} = v_{23} \frac{E_3}{E_2}; \quad v_{13} = v_{31} \frac{E_1}{E_3} \qquad 8.1$$

The elastic behaviour may be expressed via Hooke's law for orthotropic materials:

$$\varepsilon_1 = \frac{1}{E_1}(\sigma_1 - v_{12}\sigma_2 - v_{13}\sigma_3)$$

$$\varepsilon_2 = \frac{1}{E_2}(\sigma_2 - v_{21}\sigma_1 - v_{23}\sigma_3)$$

$$\varepsilon_3 = \frac{1}{E_3}(\sigma_3 - v_{31}\sigma_1 - v_{32}\sigma_2)$$

$$\gamma_{12} = \frac{\tau_{12}}{G_{12}} \quad \gamma_{23} = \frac{\tau_{23}}{G_{23}} \quad \gamma_{13} = \frac{\tau_{13}}{G_{13}} \qquad 8.2$$

This is a generalisation of the isotropic three-dimensional (3D) expression of Hooke's law, one of the first mathematical relationships to which students of solid mechanics are exposed. In practice we are generally interested in a thin layer or lamina of material (or, eventually, an assembly of them); in this case it can usually be assumed that through-thickness stress (defined to be σ_3) is negligible and the through-thickness strain (ε_3) is unimportant. Equation 8.2 can therefore be simplified to express the strains within the 1–2 plane in terms of the stresses in that plane:

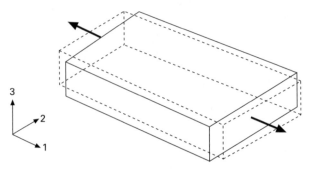

3

2

1

8.2 Orthotropic solid loaded along its axis displays no tension-shear coupling.

$$\varepsilon_1 = \frac{1}{E_1}(\sigma_1 - \nu_{12}\sigma_2)$$

$$\varepsilon_2 = \frac{1}{E_2}(\sigma_2 - \nu_{21}\sigma_1)$$

$$\gamma_{12} = \frac{\tau_{12}}{G_{12}} \qquad\qquad 8.3$$

These relationships may be expressed more conveniently in matrix form:

$$\begin{Bmatrix} \varepsilon_1 \\ \varepsilon_2 \\ \gamma_{12} \end{Bmatrix} = \begin{bmatrix} \dfrac{1}{E_1} & -\dfrac{\nu_{12}}{E_1} & 0 \\ -\dfrac{\nu_{21}}{E_2} & \dfrac{1}{E_2} & 0 \\ 0 & 0 & \dfrac{1}{\tau_{12}} \end{bmatrix} \begin{Bmatrix} \sigma_1 \\ \sigma_2 \\ \sigma_{12} \end{Bmatrix} = \begin{bmatrix} S_{11} & S_{12} & 0 \\ S_{12} & S_{22} & 0 \\ 0 & 0 & S_{66} \end{bmatrix} \begin{Bmatrix} \sigma_1 \\ \sigma_2 \\ \tau_{12} \end{Bmatrix}$$

or $\qquad \{\varepsilon\} = [S]\{\sigma\} \qquad\qquad 8.4$

where S_{11}, etc., are elements of the on-axis compliance matrix $[S]$ of a lamina. For elastic behaviour the compliance matrix must be symmetric, so that $\nu_{12}/E_1 = \nu_{21}/E_2$ which follows on from eqn 8.1. Therefore only four independent constants (E_1, E_2, G_{12} and either of the Poisson's ratios) are needed to define the in-plane elastic behaviour of a lamina. Note the presence of the zeros in the matrices in eqn 8.4; these mean that the material does not distort in shear when stretched in the on-axis direction (Fig. 8.2).

This behaviour is a direct consequence of the symmetry noted in Fig. 8.1. Sometimes it is preferable instead to express stresses in terms of strains:

$$\begin{Bmatrix} \sigma_1 \\ \sigma_2 \\ \tau_{12} \end{Bmatrix} = \begin{bmatrix} \dfrac{E_1}{1-\nu_{12}\nu_{21}} & \dfrac{\nu_{12}E_2}{1-\nu_{12}\nu_{21}} & 0 \\ \dfrac{\nu_{21}E_1}{1-\nu_{12}\nu_{21}} & \dfrac{E_1}{1-\nu_{12}\nu_{21}} & 0 \\ 0 & 0 & G_{12} \end{bmatrix} \begin{Bmatrix} \varepsilon_1 \\ \varepsilon_2 \\ \gamma_{12} \end{Bmatrix} = \begin{bmatrix} Q_{11} & Q_{12} & 0 \\ Q_{12} & Q_{22} & 0 \\ 0 & 0 & Q_{66} \end{bmatrix} \begin{Bmatrix} \varepsilon_1 \\ \varepsilon_2 \\ \gamma_{12} \end{Bmatrix}$$

or $\qquad \{\sigma\} = [Q]\{\varepsilon\} \qquad\qquad 8.5$

where Q_{11}, etc., are elements of the on-axis 'reduced' (plane stress) stiffness matrix $[Q]$ of the lamina, and $[Q] \equiv [S]^{-1}$. Note that these reduced stiffness terms are *not* equal to the corresponding terms in the 3D stiffness matrix for the material. In section 8.2.3 we shall return to the manipulation of these elastic stiffnesses and compliances, but it is useful at the present stage to consider how values for the above moduli and Poisson's ratios can be obtained.

8.2.2   Obtaining the unidirectional properties: testing and micromechanics

The above analysis assumed the availability of the unidirectional bulk properties of the composite material, and later sections will assume knowledge of the local properties of the impregnated fibre bundles (tows) which (along with the regions of pure resin) constitute textile reinforced composite materials. However, obtaining these properties is problematic. It must be realised that composite materials differ in one vital respect from conventional engineering materials such as metals: they do not generally exist as an entity until the manufacture of the component from which they are made. In other words, they are not (with some exceptions) produced by melting or machining stock material but are created from resin and fibre (or a mixture of the fibre and an uncured resin or thermoplastic matrix) at the time of moulding, and their properties depend heavily upon the manufacturing processing conditions. There are two main approaches: experimental measurement using standard testing procedures, and theoretical prediction from the properties of the constituent materials (fibre and matrix). By far the best way of obtaining material properties for design purposes is to perform tests on statistically large samples of materials. Such testing is time consuming and the resulting properties often constitute closely guarded proprietary data. It is important that such tests are conducted to recognised standards.

Measurement of unidirectional and laminate properties: testing methods

Experimental testing may be applied both to the purely unidirectional material and to the finished laminate. A detailed discussion of testing procedures is inappropriate here; instead, reference is made to the various standards in use which define various methods of obtaining the required properties. In particular, BS2782-10 covers various aspects of testing of reinforced plastics (notably method 1003[5], which covers tensile testing, and 1005[6], which covers flexural testing). Standardised test methods for unidirectional materials are also available (for example, ISO 527-5:1997[7], which covers the testing of unidirectional reinforced plastics). Experimental determination of shear properties is problematic; one widely used method is the Iosipescu test embodied in ASTM standard D5379[8], which involves the testing of small, vee-notched specimens loaded within a special frame that subjects them to pure shear within the gauge region. Shear strength properties are easily obtained from the maximum load measured; elastic properties may be determined via the use of a ±45° strain gauge bonded onto the gauge region of the specimen.

Prediction of unidirectional properties: micromechanics

Instead of physically constructing a unidirectional material or a laminate and performing tests, an alternative approach is to predict the unidirectional behaviour of the composite from the properties of its constituent materials, namely its fibre and matrix. This approach is clearly better than nothing, but has some important limitations:

- The spatial distribution of the fibres within the matrix is unknown.
- Wet-out and interfacial bonding are normally assumed to be perfect but are unlikely to be so in practice (there may, for instance, be dry regions within each tow of fibres).
- While the on-axis elastic behaviour of composites is fibre-dominated and relatively easy to predict, the transverse and shear behaviours result from complex stress distributions within the composite and are much less easy to predict theoretically.

The prediction of composite properties from those of the fibre and matrix is one example of the discipline known as micromechanics, and a number of theoretical and semi-empirical models exist. The simplest is known as the 'rule of mixtures', and probably the most commonly used is a refinement of that method known as the Halpin–Tsai equations, although more sophisticated approaches exist such as the composite cylinder assemblage approach, which can take account of anisotropic properties in the fibres themselves.

Rule of mixtures

The rule-of-mixtures approach[3] is applicable particularly to the determination of the longitudinal modulus of a unidirectional composite and of its major Poisson's ratio; its application to transverse and shear moduli gives less accurate results. In the case of the longitudinal modulus, the assumption is made that the strain along a two-dimensional (3D) block of material (Fig. 8.3) is uniform (the iso-strain assumption explored in more detail in section 8.2.3) and that no transverse stresses are present. For a given value of strain,

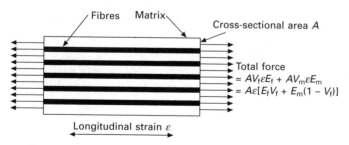

8.3 Rule of mixtures approach to determine longitudinal modulus.

the loads carried by the fibre and matrix components will be proportional both to their moduli and to the cross-sectional area of the fibre and matrix, with the respective areas being proportional to the fibre and matrix volume fractions; in other words, the fibres and matrix are assumed to be elastic bodies (springs) acting in parallel. This leads to the following expression for longitudinal modulus E_1:

$$E_1 = E_f V_f + E_m V_m = E_f V_f + E_m (1 - V_f) \qquad 8.6$$

where it is assumed that there is no void content in the composite, and where E_f and E_m are respectively the elastic moduli of the fibre and matrix, and V_f and $V_m \equiv 1 - V_f$ are respectively the volume fractions of the fibre and matrix. A slightly less trivial derivation leads to a very similar expression for the major Poisson's ratio v_{12} (which in a UD material can be assumed to be identical to v_{13}):

$$v_{12} = v_f V_f + v_m (1 - V_f) \qquad 8.7$$

Both these expressions are found to give answers of good accuracy. By contrast, the application of the rule of mixtures to the transverse elastic properties requires the assumption that the fibres and matrix are elastic bodies in series (requiring the iso-stress assumption, again explained in section 8.2.3); this is very much a simplification since the true arrangement is three-dimensional and is a combination of series and parallel behaviour. The resulting expression for E_2 ($\equiv E_3$) is rather inaccurate but is presented below for completeness:

$$\frac{1}{E_2} = \frac{1}{E_f} V_f + \frac{1}{E_m} (1 - V_f) \qquad 8.8$$

Similar arguments apply to the shear moduli (for which both the series and parallel assumptions give poor results), but these will not be pursued further here.

The Halpin–Tsai equations

The Halpin–Tsai equations[9, 10] are effectively a refinement and generalisation of the rule of mixtures. Indeed, for the longitudinal modulus E_1 and major Poisson's ratio v_{12} the rule of mixtures is used directly via eqns 8.6 and 8.7. The basis of the approach for the transverse and shear moduli is to assume intermediate situations between the extremes of the series and parallel models to which the rule of mixtures is restricted. For these moduli, the form of the equation is as follows:

$$\frac{M}{M_m} = \frac{1 + \xi \eta V_f}{1 - \eta V_f} \qquad 8.9a$$

where: $\eta = \dfrac{(M_f/M_m) - 1}{(M_f/M_m) + \xi}$ 8.9b

in which:

$\quad\quad M = E_2, G_{12}$ and G_{23} for composite
$\quad\quad M_f$ = fibre property E_f or G_f
$\quad\quad M_m$ = matrix property E_m or G_m
$\quad\quad V_f$ = fibre volume fraction
$\quad\quad$ (proportion of composite by volume)

ξ is a constant relating to the way the load is shared between fibre and matrix within the model; its value must lie between the extremes of $\xi = 0$ (in which case eqn 8.9(a) simplifies to the series model similar to eqn 8.8) and $\xi = \infty$ (when eqn 8.9(a) simplifies to the parallel model similar to eqn 8.6). For low fibre volume fractions (up to 65%) ξ may be assumed to be 2 and 1 respectively for the calculation of E_2 (and E_3) and G_{12} (and G_{13}). For higher volume fractions (as may be encountered within densely packed fibre tows), correction factors may be applied. For the calculation of G_{23} a simple value of ξ is insufficient so a formula is required in any case. The three formulae for ξ are therefore[9]:

$\quad\quad \xi \cong 2 + 40V_f^{10}$ for calculation of E_2 and E_3

$\quad\quad \xi \cong 1 + 40V_f^{10}$ for calculation of G_{12} and G_{13}

$\quad\quad \xi \cong \dfrac{1}{4 - 3v_m}$ for calculation of G_{23} 8.10

Halpin[10] states that the Poisson's ratio v_{23} in the transverse plane may be calculated approximately from the relevant direct and shear moduli using the following relationship:

$\quad\quad v_{23} \cong 1 - \dfrac{E_2}{G_{23}}$ 8.11

Hashin's composite cylinder assemblage model

Although the above formulae are widely quoted and used, no published results have been found that evaluate their accuracy for use with fibres (such as carbon or aramid) whose properties are themselves anisotropic. One model which does take account of fibre anisotropy is Hashin's composite cylinder assemblage (CCA) model[11, 12]. This gives closed form expressions for E_1, G_{12}, the bulk modulus and v_{12}, and upper and lower bounds on the transverse moduli. The expressions are rather lengthy, and the interested reader is advised to refer to Hashin[11] or the appendix of Naik and Shembekar[13].

8.2.3   Off-axis behaviour and laminates

The analysis in section 8.2.1, and the properties discussed in section 8.2.2, relate purely to material where the fibres all lie with their axes in the same direction, defined to be the 1-direction of the material's local coordinate system. Practical textile reinforced composites can of course have very complex fibre architectures, so this discussion will concentrate next upon one of the simplest of such structures, a planar plate constructed using layers of non-crimp textile reinforcement. The analysis which follows may be found in almost any text on composite materials, such as those by Hull and Clyne[2] and Jones[3]. A global Cartesian coordinate system provides a frame of reference for definition of stresses and strains in each layer, and for the overall loads on and deformations of the plate: the x and y axes are coplanar with the plate and the z axis is perpendicular to the plate and measured from its mid-surface. Consider a thin lamina within such a structure, with its fibre axes orientated at angle θ to the global x direction. The stresses present in that lamina may be expressed in terms of the in-plane strains which that lamina undergoes:

$$\begin{Bmatrix} \sigma_x \\ \sigma_y \\ \tau_{xy} \end{Bmatrix} = \begin{bmatrix} \bar{Q}_{11} & \bar{Q}_{12} & \bar{Q}_{16} \\ \bar{Q}_{12} & \bar{Q}_{22} & \bar{Q}_{26} \\ \bar{Q}_{16} & \bar{Q}_{26} & \bar{Q}_{66} \end{bmatrix} \begin{Bmatrix} \varepsilon_x \\ \varepsilon_y \\ \gamma_{xy} \end{Bmatrix} \qquad\qquad 8.12$$

where $\bar{Q}_{11}$ etc. are the elements in the off-axis reduced stiffness matrix $[\bar{Q}]$ of the lamina. Note that in general there are no zero terms in the $[\bar{Q}]$ matrix: this denotes that stretch along either the x or y axis will result in shearing of the lamina (Fig. 8.4).

$[\bar{Q}]$ is related to $[Q]$ by the following fourth rank tensor transformations:

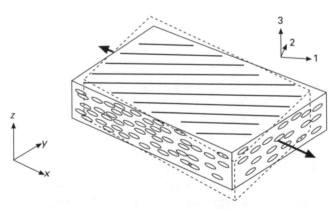

8.4 Example of tension-shearing coupling in an off-axis composite.

$$\overline{Q}_{11} = Q_{11}m^4 + Q_{22}n^4 + 2(Q_{12} + 2Q_{66})m^2n^2$$

$$\overline{Q}_{22} = Q_{11}n^4 + Q_{22}m^4 + 2(Q_{12} + 2Q_{66})m^2n^2$$

$$\overline{Q}_{12} = (Q_{11} + Q_{22} - 4Q_{66})m^2n^2 + Q_{12}(m^4 + n^4) \qquad 8.13$$

$$\overline{Q}_{66} = (Q_{11} + Q_{22} - 2Q_{12})m^2n^2 + Q_{66}(m^4 + n^4 - 2m^2n^2)$$

$$\overline{Q}_{16} = (Q_{11} - Q_{12} - 2Q_{66})m^3n - (Q_{22} - Q_{12} - 2Q_{66})mn^3$$

$$\overline{Q}_{26} = (Q_{11} - Q_{12} - 2Q_{66})mn^3 - (Q_{22} - Q_{12} - 2Q_{66})m^3n$$

in which $m = \cos\theta$ and $n = \sin\theta$.

In practice a composite laminate will be made up, for instance, of layers of non-crimp textile reinforcement with each ith layer of fibres within this architecture having its own ply angle θ_i and hence its own off-axis stiffness matrix $[\overline{Q}]_i$. It is then standard bookwork to use *classical laminated plate theory* or *classical laminate theory* (CLT) to assemble the three stiffness matrices $[A]$, $[B]$ and $[D]$ which relate the forces $\{F\}$ and moments $\{M\}$ per unit of plate length (the 'unit forces and moments') to the mid-plane strains $\{\varepsilon^0\}$ and curvatures/twists $\{\kappa\}$ with which the laminate deforms:

$$\begin{Bmatrix} N_x \\ N_y \\ N_{xy} \end{Bmatrix} = \begin{bmatrix} A_{11} & A_{12} & A_{16} \\ A_{12} & A_{22} & A_{26} \\ A_{16} & A_{22} & A_{66} \end{bmatrix} \begin{Bmatrix} \varepsilon_x^0 \\ \varepsilon_y^0 \\ \gamma_{xy}^0 \end{Bmatrix} + \begin{bmatrix} B_{11} & B_{12} & B_{16} \\ B_{12} & B_{22} & B_{26} \\ B_{16} & B_{26} & B_{66} \end{bmatrix} \begin{Bmatrix} \kappa_x \\ \kappa_y \\ \kappa_{xy} \end{Bmatrix} \qquad 8.14$$

and:

$$\begin{Bmatrix} M_x \\ M_y \\ M_{xy} \end{Bmatrix} = \begin{bmatrix} B_{11} & B_{12} & B_{16} \\ B_{12} & B_{22} & B_{26} \\ B_{16} & B_{26} & B_{66} \end{bmatrix} \begin{Bmatrix} \varepsilon_x^0 \\ \varepsilon_y^0 \\ \gamma_{xy}^0 \end{Bmatrix} + \begin{bmatrix} D_{11} & D_{12} & D_{16} \\ D_{12} & D_{22} & D_{26} \\ D_{16} & D_{26} & D_{66} \end{bmatrix} \begin{Bmatrix} \kappa_x \\ \kappa_y \\ \kappa_{xy} \end{Bmatrix} \qquad 8.15$$

where:

$$A_{ij} = \sum_{k=1}^{n} (\overline{Q}_{ij})_k \cdot (z_k - z_{k-1})$$

$$B_{ij} = \frac{1}{2} \sum_{k=1}^{n} (\overline{Q}_{ij})_k \cdot (z_k^2 - z_{k-1}^2) \qquad 8.16$$

$$D_{ij} = \frac{1}{3} \sum_{k=1}^{n} (\overline{Q}_{ij})_k \cdot (z_k^3 - z_{k-1}^3)$$

in which z_k and z_{k-1} are the through-thickness coordinates of top and bottom surfaces respectively of the kth layer. While CLT is a commonly used approximation to the behaviour of composite laminates, it is just that: an approximation, based upon a number of simplifying assumptions generally known as the Kirchhoff–Love assumptions, which include the assumption that initially straight section lines passing perpendicularly through the laminate (sometimes known as normals) remain perpendicular and straight during deformation (Fig. 8.5).

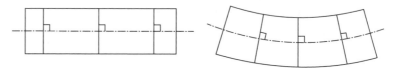

8.5 Section through shear-rigid plate illustrating the assumption of non-deformable normals.

This implies a constraint against transverse shearing effects, and is a 2D extension of the well-known assumption in engineering beam theory that initially plane perpendicular sections remain plane and perpendicular. In fact the transverse shear stiffness of practical laminates, especially those made using carbon fibre reinforcements, can be very low relative to their in-plane stiffness especially if they are of significant thickness. In order to model the behaviour of such a laminate accurately, it is necessary to relax the assumption of non-deformable normals by using a shear-deformable analogue of CLT. Such theories can assume, with progressively increasing levels of complexity, that the normals remain straight but not normal (e.g. Fig. 8.6), or that they deform from straightness following a polynomial function, or that they follow a more complex mode of deformation still, e.g. a piecewise function possibly involving exponentials. The interested reader is referred to the book of Reddy[14] and to papers by Soldatos and coworkers[15, 16], which generalise a range of approaches and develop advanced versions of such theories whose results closely approximate the exact results obtainable from 3D elasticity. Theoretical discussion of such approaches is beyond the scope of the present chapter, but the concepts of 'thin' (shear-rigid) and 'thick' (shear-flexible) plate and shell assumptions will be further discussed in the context of finite elements.

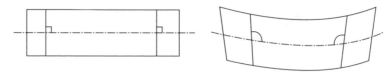

8.6 Section through plate illustrating the assumption of uniform transverse shear deformation.

8.2.4   Analysis of practical textile-reinforced composites

Analysis of homogeneous composite materials, whether they be orthotropic, anisotropic or quasi-isotropic, can be handled via the kind of mathematical techniques described earlier. Textile composites, especially those made from woven textiles, are, however, inhomogeneous and typically have properties that vary continuously throughout the structure. In the latter case there are no easy and accurate answers, although some useful approximations can be obtained. Useful overviews of some of the possible approaches are provided by Hofstee and van Keulen[17] and Tan et al.[18]; a selection of the most important ones are outlined here.

Analysis of non-crimp textile composites

For non-crimped (stitch-bonded) textile composites, laminate theory can be used directly since the fibres remain substantially straight without serious localised waviness, and each group of fibres retains a distinct position within the laminate structure which may quite properly be discretised into plies. However, the following factors may need to be taken into account:

- Such textiles are often used where significant forming or draping is required, and this can greatly affect the content of fibres lying in particular directions. The effects of this behaviour on elastic properties have been modelled recently by Crookston et al.[19], but this is by no means standard practice.
- Since there is only fairly loose constraint on the positions of the tows especially within non-crimped fabrics, statistical variations in fibre direction, content, etc., may be significant and may in turn have a significant effect upon the mechanical properties. This issue is under investigation at the University of Nottingham, and an overview of the subject is presented by Long et al.[20]
- In some textile reinforcements (e.g. those that are predominantly unidirectional, having 95% of fibres in the 0° direction and the remaining 5% in the 90° direction) the tows clearly retain their distinctive shape and the true structure is not truly made up of uniform layers of differing thickness. Similarly, the true shape of the tows means that in some cases 'nesting' between successive layers can occur. As far as can be ascertained, these issues have not yet been fully investigated although the effects of tow shape and related issues are again under investigation at Nottingham[21].

Analysis of woven, braided and knitted textiles

By contrast, woven, braided and knitted textiles have more complex structures and their composites require a more detailed treatment. Leaving aside knitted textiles, a range of models of increasing complexity can be used to obtain the elastic properties of composites with crimped tows. There is a clear trade-off between the manual and computational effort involved and the accuracy obtained, although even the most complex approaches still result in errors of several per cent compared with experimental values. The simplest and least accurate approach is to ignore the true fibre architecture of the textile reinforcement and simply to treat each set of tows as being a different layer with a given orientation within a uniform laminate. Such an approach ignores the reduction in strength and stiffness due to fibre waviness or crimp. In modelling approaches that attempt to model the fibre architecture, a so-called representative (or repeating) volume element (RVE), sometimes called a unit cell, is considered, and suitable boundary conditions at its edges are assumed so that the behaviour of this element can be extrapolated to that of a continuous sheet of the composite. One extension of the laminate approach, which takes account of the fibre crimp, is that due to Hofstee and van Keulen[17], and will be considered later. An alternative extension to this approach, sometimes termed the mosaic approach, treats the textile composite as a laminate with a laminate sequence which varies continuously over the RVE, and Naik and Shembekar's model[13] follows this approach again taking account of fibre crimp. Such approaches are essentially 2D simplifications of a highly complex 3D structure and stress system.

An alternative approach discretises the volume of the RVE into small sub-cells or voxels; there are several variants on this approach including those described by Bogdanovitch and Pastore[1] and the approach of Bigaud and Hamelin[22] described later. More rigorous still is the use of 3D models of the RVE which separately model the true or idealised geometry of the fibre tows and the regions of matrix between them using solid finite elements. Such an approach has been used by Chapman and Whitcomb[23] and many others, and this body of work is explored below. While such an approach may be regarded as a true simulation of the physical situation, the meshes required are large and complex, and the data preparation and run times are both considerable. With the automation of the mesh generation process and the increasing power of computers, such an approach is becoming more attractive; however, it is more appropriate in the present discussion to review some examples of simpler approaches that enable the engineer to approximate the behaviour of a textile composite without undertaking a major analysis task.

Iso-stress and iso-strain assumptions

Of the analytical approaches that take account of the variations of properties over the textile composite, there are two main sets of assumptions used in averaging the varying effective properties over a representative length, volume or area. These are known as iso-stress and iso-strain assumptions, or alternatively as the series and parallel models respectively. In its 1D form, the iso-stress assumptions regard elemental lengths of the composite as being subjected to equal stress and effectively acting as elastic elements or springs in series. The iso-strain assumption regards elemental lengths as being subjected to equal strain; this is normally equivalent to placing their stiffnesses in parallel. In practice, a combination of both approaches is often used, with the local elastic behaviour being averaged over the length, area or volume of an RVE or unit cell of the composite.

Approaches based upon classical laminate theory

Hofstee and van Keulen[17] use a simplified geometry of a crimped tow[24] within a plain weave composite, consisting of straight and curved (circular arc) segments, to obtain equivalent properties for the crimped tow in each direction (Fig. 8.7). Either iso-stress or iso-strain assumptions may be made,

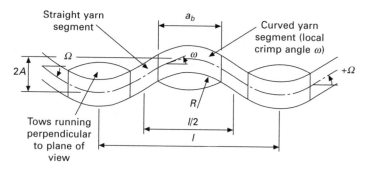

8.7 Crimped tow geometry assumed by Hofstee (adapted from references 17 and 24).

involving integration of the locally transformed compliances or stiffnesses of the tow as the fibre crimp angle changes. The resulting volume averaged stiffnesses $\langle C \rangle$ are (for the iso-strain assumption):

$$\langle C \rangle^\varepsilon = \frac{a_b}{l\Omega} \int_{-\Omega}^{\Omega} C(\omega)d\omega + \left(\frac{1}{2} - \frac{a_b}{l} \right)[C(-\Omega) + C(\Omega)] \qquad 8.17$$

and (for the iso-stress assumption):

$$\langle C \rangle^\sigma = \left\{ \frac{a_b}{l\Omega} \int_{-\Omega}^{\Omega} S(\omega)d\omega + \left(\frac{1}{2} - \frac{a_b}{l} \right)[S(-\Omega) + S(\Omega)] \right\}^{-1} \qquad 8.18$$

where $C(\omega)$ and $S(\omega)$ are respectively the transformed stiffness and compliance matrices at the local crimp angle ω of the curved portion of the fibre, and Ω is the maximum crimp angle. The resulting crimped tow properties are used within a simplified model of each textile layer within the laminate, consisting of separate sub-layers representing the warp, weft (fill) and resin components of the layer.

This model is then used as the basis of a parametric study of the variation on modulus with shear angle (to represent deformation during draping), along with a study of the effect of fabric stretching (resulting in straightening of one set of tows at the expense of increasing the crimp in the other set). It is found that the iso-strain version of the model yields results that are almost identical to those of the model when crimp is neglected, while the iso-stress version yields moduli in the fibre directions which are significantly lower (around 10–30% lower) than those obtained using the iso-strain version. The results of the two versions of the model are compared with experimental data[25] and it is found that in the direction of the fibres, the effects of crimp become significant and the iso-stress version of the model yields results in closer agreement with experiment than the iso-strain version.

The iso-stress and iso-strain concepts are combined with CLT in the approach by Naik and Shembekar[13] for the analysis of unsheared woven composites. Two main variants of the theory are presented, both of which involve discretising a representative volume element of the composite along and perpendicular to the direction of loading relevant to each property. In contrast to the simple application of iso-stress and iso-strain assumptions to the averaging of tensile stiffnesses, the approach of Naik and Shembekar[13] involves the averaging of direct, bending and coupling stiffnesses over the representative volume element of the composite. In the parallel–series (PS) model, the components $A_{ij}(x, y)$, etc., of the local laminate stiffness matrices $[A]$, etc. at position (x, y) are averaged across the width perpendicular to the loading (x) direction following an iso-strain assumption:

$$\overline{A}_{ij}^{P}(x), \overline{B}_{ij}^{P}(x), \overline{D}_{ij}^{P}(x)$$

$$= \frac{1}{a_f + g_f} \int_0^{a_f + g_f} A_{ij}(x, y), B_{ij}(x, y), D_{ij}(x, y) \, dy \qquad 8.19$$

where a_f and g_f are respectively the width of the fill tows and of the gaps between them. These averaged stiffnesses $\overline{A}_{ij}^{P}(x)$, etc., are then inverted and the resulting compliances $\overline{a}_{ij}^{P}(x)$, etc., are averaged along the loading direction x according to an iso-stress assumption:

$$\overline{a}_{ij}^{ps}, \overline{b}_{ij}^{ps}, \overline{d}_{ij}^{ps} = \frac{1}{a_w + g_w} \int_0^{a_w + g_w} \overline{a}_{ij}^{P}(x), \overline{b}_{ij}^{P}(x), \overline{d}_{ij}^{P}(x) \, dx \qquad 8.20$$

where a_w and g_w are respectively the width of the warp tows and of the gaps between them. These averaged compliances are finally used to calculate the equivalent moduli and Poisson's ratio. In the SP model, an iso-strain approach is used to average the laminate compliances along the loading direction. These averaged compliances are inverted and used in turn with an iso-stress assumption by averaging across the width perpendicular to the loading direction. Finally, the resulting overall laminate stiffnesses are inverted to give the overall compliances and hence the equivalent moduli and Poisson's ratio. The authors compare the results of these two methods with 1D parallel and series models and with experimental results and conclude that the PS approach gives results in best agreement with experiment.

Approaches based upon sub-cells or voxels

Another approach combining the iso-stress and iso-strain assumptions is that of Bigaud and Hamelin[22]. While they initially explored the use of rigorous finite element (FE) models of unit cells, they turned away from this approach owing to the difficulties in automatic meshing and the long run times involved. Instead, they divide the unit cell into voxels or sub-cells for which the properties of each sub-cell are scaled according to the number of sub-cell vertices lying within a given yarn or tow. The sub-cell properties are then averaged over the unit cell using different combinations of iso-stress and iso-strain assumptions as appropriate. For example, for calculating the direct modulus in a given direction (say the y direction), the stiffnesses are averaged over each slice in the x–z plane according to an iso-strain assumption before inverting the results and averaging the resulting compliances in the y direction according to an iso-stress assumption and using the overall compliance to calculate the modulus. A broadly similar approach is used for calculating the Poisson's ratios. By contrast, the shear moduli are calculated from global shear stiffnesses which are purely the averages of the relevant sub-cell stiffness terms. The authors compare the results from their analytical model with those from a numerical approach, and demonstrate that the various results agree within a few per cent for the moduli but their results for the Poisson's ratios (especially in the plane perpendicular to the fibre direction) are in less good agreement with the alternative solutions.

Approaches based upon 3D FE modelling of the unit cell

Reference has already been made to the difficulties of this approach, although it is potentially the most accurate method for predicting elastic and failure behaviour of textile composites. In its most rigorous form, it involves constructing an FE model of an RVE of the textile using 3D continuum elements (see section 8.2.5) that replicates the idealised fibre architecture of the textile, with separate regions to represent the directional (orthotropic) tows and the isotropic matrix material. Two of the earliest examples of this approach appear to be those of Whitcomb[26] and of Guedes and Kikuchi[27]. This theme has been continued in more recent work, for example by Chapman and Whitcomb[23] and by Ng et al.[28]. Dasgupta et al.[29] similarly predicted the elastic moduli and coefficients of thermal expansion using unit cell FE models. Several later authors, e.g. Zako et al.[30], have used this approach within a non-linear analysis to obtain predictions of the development of damage within the composite material, while other authors, e.g. Aitharaju and Averill[31], have used 3D FE unit cell models as a method of validating simpler and more computationally efficient approaches to obtaining material properties. There are two main difficulties associated with the use of full 3D models of the tow/matrix architecture:

1. The geometry, especially of the matrix region, involves very complex shapes with features that are very difficult to mesh. For example, there are thin sections between converging surfaces of tows as they cross, leading to the potential for very distorted elements. This makes the construction of the model non-trivial, making the use of links with solid modelling techniques[32, 33] and/or fibre architecture models[34] highly desirable. Improved availability of solid modelling systems and the increasing integration of solid modelling with FE can help with this issue.
2. The number of elements involved in such a 3D mesh can become very large especially as mesh convergence studies are undertaken. While ongoing improvements in computing mean that this problem will become less significant with time, the problem nonetheless remains that complete convergence of calculated stresses may not be practical.

It must be concluded that at present the use of 3D FE models provides a valuable research tool for predicting tow-level behaviour within a textile composite, at the expense of considerable data preparation and run time. An example analysis produced using this approach is shown in Fig. 8.8. This represents uniaxial loading of a 2 × 2 twill weave fabric, where the mesh was generated automatically using the University of Nottingham 'TexGen' textile modelling software[34].

8.2.5   Finite element modelling of textile composite components

For all components except the most simple, it is likely that structural analysis will be undertaken using the finite element method (FEM), in which the structure is broken down into small regions whose constitutive properties are easily evaluated, prior to assembling a mathematical model of the global behaviour of the system from the properties of all the elements. It is not appropriate here to provide a detailed overview of the FEM, although numerous comprehensive texts exist (for example the standard text by Zienkiewicz and Taylor[35]) and a brief overview of the FEM in the context of composite materials is given by Owen et al.[4] However, it is useful to explore the kinds of elements that are available in the analysis of textile composites.

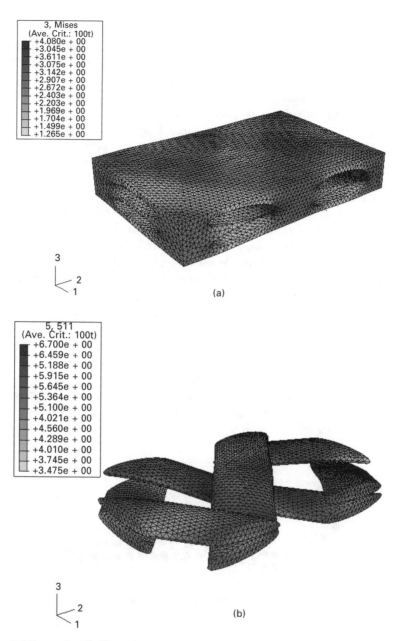

8.8 Example 3D FE analysis for a 2 × 2 twill weave glass/polyester composite. (a) Von Mises stress distribution in the matrix; (b) stress in the fibre direction for the impregnated tows.

Elements for analysing thin laminates

These are usually extensions of thin shell isotropic elements. Four independent elastic constants (normally E_1, E_2, v_{12} and G_{12}), along with the orientation of the material's principal directions (relative to a global or local coordinate system), are required to define each layer. A layer is assumed here to consist of a ply or group of plies having a given set of orthotropic material properties orientated at a single angle. In practice, it may consist of one of the sets of tows in a non-crimp fabric ply or even (as a crude approximation, or with the properties modified as in the method of Hofstee and van Keulen[17]) one of the sets of tows in a woven fabric. Material directions will normally lie parallel to the laminate mid-surface, and care must be taken that the material directions are correctly specified so that this is indeed the case. Thin shell elements typically have five or six degrees of freedom (DOFs) at each node: three displacement DOFs, along with two or three rotational DOFs to enforce continuity of slope between elements. A range of different element topologies are available: these include three-noded elements (which have rather poor accuracy, resulting in a need for many elements in order to obtain a converged solution), four- or eight-noded triangular elements (which use higher orders of interpolation to overcome the problems of the three-noded elements) and eight-noded quadrilateral elements (see Fig. 8.9). These shell elements are applicable to thin shells where transverse shear strains have little effect, and the Kirchhoff–Love assumption of zero transverse shear strain may be enforced mathematically (i.e. analytically, within the formulation) over the whole element or numerically at certain points on the element (but permitting slight shearing elsewhere).

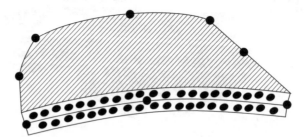

8.9 Eight-noded, doubly curved laminated quadrilateral shell element.

Elements for analysing moderately thick laminates

Thin shell elements are adequate for reasonably thin metallic plates and shells, but composites are very flexible in transverse shear relative to their overall (direct and bending) stiffness. It is therefore necessary to take account of transverse shearing for moderately thick laminates or poor results will be obtained. So-called 'thick shell' or shear-deformable plate and shell elements are therefore used. Instead of the Kirchhoff–Love assumptions these use the Reissner–Mindlin assumption, which may be stated as: 'straight sections remain straight but do not necessarily remain normal'.

G_{23} and G_{31} must now be considered in order to take account of transverse shear stiffness. Nodal degrees of freedom still usually consist of three displacements and three rotations but the rotations now represent the combined effects of mid-surface rotation of the shell due to bending, and rotation of shell normals (relative to the mid-surface) due to shearing. It should be noted that when transverse (interlaminar) shear stresses are calculated (either from thin shell or thick shell elements), they are usually reconstructed using equilibrium considerations, not calculated directly from the strain field assumed above.

Although 'thick shell' elements are able to model transverse shearing effects, they do have a pitfall. If they were to be used for modelling of thin laminates where shearing deformations are insignificant, a spurious mode of energy storage occurs within the element leading to so-called 'shear locking', a stiffening of the element that tends to rigid behaviour as the shell thickness tends to zero. Some care is therefore needed when deciding whether to use 'thin' or 'thick' shell elements (or even 3D elements), although it is very difficult to give clear guidance on this issue and there is no substitute for experience and benchmarking.

As implied by the lists of engineering properties required for each type of element, it is normally necessary to specify equivalent elastic properties of each ply of textile-reinforced composite. An alternative form of the input data is the set of laminate stiffnesses (i.e. [A], [B], [D] and the transverse shear stiffnesses). Some of the approaches for predicting these elastic constants for textile composites were presented in section 8.2.4.

3D orthotropic elements

These are of limited usefulness for modelling conventional (2D) textile composite laminates, although they are useful where solid pieces of composite are to be modelled. They could also be used in the modelling of composites manufactured from thick (3D) textiles such as 3D woven or braided reinforcements. Moreover, they are essential in the creation of 3D models of representative volume elements or unit cells of textile composites. Typical 3D elements are shown in Fig. 8.10.

In most FE systems, all the usual 3D elements are available with orthotropic properties, e.g. 4- and 10-noded tetrahedral, and 8- and 20-noded 'bricks', along with other geometries including wedge elements, and orthotropic plane stress, plane strain and generalised plane strain 2D elements.

Some FE systems allow layered bricks and other layered 3D elements,

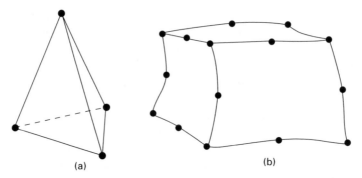

(a)   (b)

8.10 Conventional 3D solid elements with the opportunity to use orthotropic properties: (a) four-noded tetrahedron and (b) twenty-noded hexahedron ('brick').

which form a useful halfway house between explicitly modelling each layer separately with its own 3D elements and using shell elements. However, once again, care must be taken to ensure that the fibre directions are correctly specified for each layer.

8.2.6   Conclusions

This section has examined the elastic behaviour of composite materials in general, with particular emphasis on the difficulties of extending this kind of analysis to cope with the complex fibre architectures found in composites made with woven and braided textiles. It may be concluded that, as with any form of engineering analysis, there is a trade-off between the accuracy obtained using a given analysis method (ranging from simplistic treatment of different tow sets as different plies, though to rigorous 3D finite element modelling of representative volume elements) versus the complexity of the analysis and the computational and data preparation times involved. Furthermore, only elastic properties have been considered thus far, whereas the onset of failure and the progression of damage will be examined in the following section.

8.3   Failure and impact behaviour

Impact and failure analysis of textile composites is appreciably more difficult than for metals owing to the complexity and multitude of failure modes that can occur. Possible failure modes include, among others, tensile fibre failure, compressive fibre buckling, matrix crushing, transverse matrix cracking, fibre–matrix debonding and delamination between adjacent plies in the case of a laminated textile composite. The interaction of these failure modes further complicates the problem, making a predictive numerical simulation difficult, but not impossible. This section illustrates some of the numerical methods that are often employed for such simulations. For the purpose of this discussion the methods outlined are typical of those currently available in some commercial FE analysis codes.

Most textile composites comprise one or more plies laid over each other to form a laminate. The plies may be unidirectional or bidirectional, or more intricate weaves and knits which are bound together with resin that is either pre-impregnated, or post-impregnated using an injection process. Regardless of the manufacturing route, the basic laminate comprises plies and resin interfaces, each of which may independently, or concurrently, fail during excessive loading. Other types of textile composites could have a 2.5D or 3D architecture in which yarns are intertwined through the thickness direction; this will eliminate delamination failure.

A large body of experimental and analytical work on the failure of composites and textile composites is available in the literature; see for example Hinton et al.[36] and Cox and Flanagan[37]. However, for practical analysis and failure prediction of large-scale composite structures only the FE technique is suitable and is, therefore, the only method to be discussed here. Two popular FE methods are available; namely, the implicit and explicit techniques, each of which has its advantages and disadvantages[38]. The two methods are briefly introduced, followed by some composite failure modelling approaches that are equally applicable to both analysis techniques.

8.3.1   Solution strategies for FE failure analyses

Usually commercial FE codes use either an implicit or an explicit solution to perform an analysis. The two solution algorithms render each method more, or less, appropriate to different classes of problems. Generally, implicit methods are most suitable for straightforward failure analysis, particularly if the loading is 'static', while explicit methods are more appropriate for dynamic, highly non-linear problems which would include composites impact (and crash) analysis.

The implicit method is most commonly used in general-purpose FE codes for linear and non-linear problems. Essentially implicit FE analysis requires the composite structure to be discretised using (modelled as) an assembly of appropriate finite elements. Each element has a stiffness matrix dependent on the material constitutive relationship and the nature of the chosen element. Each element stiffness matrix is computed and assembled to give the global structure stiffness matrix $[K]$. This matrix provides the relationship between nodal displacements $\{u\}$ and applied nodal forces $\{P\}$ of the structure:

$$\{P\} = [K]\{u\} \quad \text{or} \quad \{u\} = [K]^{-1}\{P\} \qquad 8.21$$

In contact and non-linear material problems, such as impact analyses for composites, the matrix $[K]$ is not constant and iterative solution schemes are necessary to find the nodal displacements for a given applied loading. Often these solution schemes require repetitive assembly and inversion of the global stiffness matrix, which can be prohibitively expensive and unreliable.

In recent years the explicit method has become popular for highly non-linear dynamic problems, especially if contact between bodies is involved: crashworthiness and impact simulation are two notable examples. The same meshing techniques used in the implicit method are used to discretise the structure. The difference is the solution algorithm and the problem is now formulated as a dynamic problem using the linearised equations of motion, from which a solution in the time domain is obtained:

$$[M]\{\ddot{u}\}_n + [C]\{\dot{u}\}_n + [K]\{u\}_n = \{F_{ext}\}_n \qquad 8.22$$

where $\{u\}$, $\{\dot{u}\}$ and $\{\ddot{u}\}$ are vectors of nodal displacement, velocity and acceleration, n is the cycle number at time position T_n (after n time steps each of length ΔT); $[M]$, $[C]$ and $[K]$ are the mass, damping and stiffness matrices respectively and $\{F_{ext}\}$ is a vector of applied external nodal forces. Material damping can be neglected in short duration dynamic problems and, replacing the term $[K]\{u\}_n$ with the equivalent internal nodal force vector $\{F_{int}\}$, this gives Newton's second law of motion,

$$[M]\{\ddot{u}\}_n = \{F_{ext}\}_n - \{F_{int}\}_n \qquad 8.23$$

If a lumped mass distribution is assumed the mass matrix $[M]$ is diagonal and a solution for nodal accelerations $\{\ddot{u}\}$ is trivial:

$$\{\ddot{u}\}_n = [M]^{-1}(\{F_{ext}\} - \{F_{int}\})_n \qquad 8.24$$

The nodal velocities $\{\dot{u}\}_{n+1/2}$ and nodal displacements $\{u\}_{n+1}$ may then be obtained by integration in the time domain using the central finite difference operators:

$$\{\dot{u}\}_{n+1/2} = \{\dot{u}\}_{n-1/2} + \{\ddot{u}\}_n \Delta T_n$$

$$\{u\}_{n+1} = \{u\}_n + \{\dot{u}\}_{n+1/2}\Delta T_{n+1/2} \qquad 8.25$$

Equations 8.24 and 8.25 are 'conditionally stable' and restrict the allowable integration time step to $\Delta T_{critical}$, which is dependent on the element size and material properties.

An important advantage of the above scheme is that the system of equations are uncoupled, allowing an element-by-element solution; thus the formation and inversion of a large global stiffness matrix are not needed. Furthermore, instabilities due to sheet buckling, or material softening due to damage, are easily treated using the explicit dynamic formulation. Contact is also easily handled by first identifying its occurrence and then imposing temporary resisting 'penalty' forces, as additional external forces, in eqn 8.24. Treating these phenomena using an implicit method, where global stiffness matrices must be assembled, inverted and used in iterative schemes to yield a solution does present formidable computational difficulties; however, linear static analysis and identification of initial failure is more easily handled.

8.3.2   Ply and delamination failure modelling

Delamination is important because it causes ply separation and loss in bending stiffness of the laminate; whereas in-plane ply failure controls in-plane stiffness and ultimate failure of the laminate. Most FE codes have ply constitutive models to treat mechanical stiffness, and they usually have a variety of popular failure models to identify rupture. Usually delamination is not considered, but some codes, including implicit[39] and explicit[40], have now implemented delamination models to characterise mechanical stiffness and failure of the resin-rich ply interface. The following sections outline the principles of ply and delamination models used in FE codes.

Ply constitutive and failure models

Most commercial FE codes allow composites to be represented using 2D shell or 3D solid elements; usually the shells are 'multilayered' so that a laminate with many plies can be treated using a single element. In each case the composite orthotropic material properties are specified in the 'fibre' frame and the relative orientation of this frame, with respect to the 'global' structural axis system, must be given. The 2D and 3D orthotropic constitutive relations are given in many composites textbooks[2, 3]; the plane stress relations for an orthotropic material are described in section 8.2.1.

Many ply failure models have been proposed that can predict failure for a given state of loading. Usually impact and crash analysis codes tend to prefer a 'damage mechanics' approach in which failure is first identified and thereafter the elastic properties are progressively reduced until full rupture occurs. The nature of the failure mode dictates whether sudden rupture and rapid damage will occur (e.g. tension), or whether slow progressive damage and energy absorption are more representative (e.g. compressive crushing).

Finite element codes often allow the user a choice of different failure criteria, or damage mechanics methods. A linear implicit analysis will provide element stresses and strains for a given applied loading. For a selected failure criterion the instance of initial material failure is then found by identifying the critical location, and linearly scaling the applied loads to achieve failure at this location (failure index $f = 1$, see below). This would correspond to 'first-ply failure', which does not necessarily represent the ultimate strength of the laminate or structure. More sophisticated FE techniques can allow this first-ply to be excluded, or certain stress components to be excluded, and further loading applied. Clearly the structure stiffness matrix must be reformulated and iterative solution strategies are required. Alternatively, damage mechanics methods can be used in explicit (time) or implicit (load) integration schemes in which the structure is progressively loaded and the evolution of damage followed.

The following outlines some commonly used ply failure criteria; this is then followed by an example of a typical damage mechanics model. These models were originally developed for unidirectional and woven plies, but could be applied to more advanced textiles. The reader should be aware, however, that there is considerable controversy over composite failure models[41]. Nevertheless, if an appropriate test programme is conducted, in which the failure modes tested and used to define the failure law are comparable to those expected in the actual structure under consideration, then these models can be used with a good level of confidence.

Maximum stress and maximum strain

This criterion assumes failure occurs if any stress component in the fibre frame (σ_1, σ_2, τ_{12}) exceeds its corresponding allowable limit (σ_{1u}, σ_{2u}, τ_{12u}). These limits are normally determined from appropriate coupon tests. The criterion for a 2D finite element is:

$$f = \max \left(\left| \frac{\sigma_1}{X_\sigma} \right|, \left| \frac{\sigma_2}{Y_\sigma} \right|, \left| \frac{\tau_{12}}{S_\sigma} \right| \right), \quad \text{for failure } f \geq 1 \qquad 8.26$$

with,

$$X_\sigma = \sigma_{1u}^t \quad \text{if} \quad \sigma_1 \geq 0, \quad Y_\sigma = \sigma_{2u}^t \quad \text{if} \quad \sigma_2 \geq 0$$

$$X_\sigma = \sigma_{1u}^c \quad \text{if} \quad \sigma_1 < 0, \quad Y_\sigma = \sigma_{2u}^c \quad \text{if} \quad \sigma_2 < 0$$

$$S_\sigma = \tau_{12u}$$

Note that different failure limits are possible for in-plane tension (σ_{1u}^t) and compression (σ_{1u}^c). The maximum strain criterion has an identical form except that stress terms are replaced with strains and the failure quantities are now failure strains. A major concern of this model is that failure is not influenced by the interaction of the different stress–strain components.

The Tsai–Wu quadratic failure criterion

The Tsai–Wu criterion is typical of a second class of general failure criteria for anisotropic composites in which a single expression is derived to express failure. The general form of this quadratic failure criterion in stress space is given by:

$$f^2 = \sum_{i=1}^{6} F_i \sigma_i + \sum_{i,j=1}^{6} F_{ij} \sigma_i \sigma_j \qquad \text{8.27}$$

The F_{12} term provides interaction of stress components and the differences between tensile and compressive failure stresses are introduced in the coefficient F_i, F_{ij}. For plane stress the above reduces to:

$$F_1 \sigma_1 + F_2 \sigma_2 + F_{11} \sigma_1^2 + F_{22} \sigma_2^2 + F_{66} \tau_{12}^2 + 2 F_{12} \sigma_1 \sigma_2 = f^2 \qquad \text{8.28}$$

where

$$F_1 = \frac{1}{\sigma_{1u}^t} - \frac{1}{\sigma_{1u}^c}; \qquad F_{11} = \frac{1}{\sigma_{1u}^t \sigma_{1u}^c}$$

$$F_2 = \frac{1}{\sigma_{2u}^t} - \frac{1}{\sigma_{2u}^c}; \qquad F_{22} = \frac{1}{\sigma_{2u}^t \sigma_{2u}^c}$$

$$F_{66} = \frac{1}{\tau_{12u}^p \tau_{12u}^n} \qquad F_{12} = \frac{1}{2} \sqrt{F_{11} F_{22}}$$

Many similar failure criteria are available in commercial codes including, for example, the Modified Puck, Hoffmann and Tsai–Hill criteria. An excellent review of these criteria and other models, compared with test measurements, is given by Hinton et al.[36]

Ply failure modelling using damage mechanics

Continuum damage mechanics provides a simple method to treat ply (and delamination) failure and is often used in advanced FE codes for impact and crash analysis of composite structures. Damage is a permanent weakening of the material due to resin micro-cracking, fibre–matrix debonding, fibre failure, etc., and occurs prior to complete rupture. Its effect can be approximated by reducing (damaging) the material elastic properties via a damage parameter (d). The stress–strain relation is then given by:

$$\sigma = E_d \cdot \varepsilon \quad \text{where} \quad E_d = E_o (1 - d) \qquad \text{8.29}$$

The elastic and damaged moduli are E_o and E_d respectively, and d is a scalar damage parameter that varies from zero, for strains below a damage initiation strain ε_i, and increases linearly to d_u at full material damage strain ε_u. Damage d is usually linked to the state of strain or strain energy release rate. Complete material rupture at strain ε_u is easily imposed by setting the damage $d = 1$. The evolution of damage and the resulting stress–strain curve is shown in Fig. 8.11.

For orthotropic materials several material moduli are present, which may be independently, or concurrently, damaged. The essential requirement is a valid set of coupon tests that determine the mechanical properties and damage

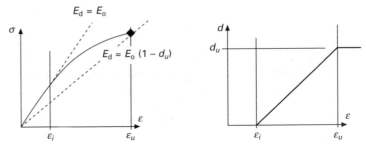

8.11 Illustration of damage mechanics: (a) variation of stress and strain; (b) Variation of damage *d*.

evolution leading to failure for each of the different possible failure modes. The following illustrate the principles and application of continuum damage mechanics to a woven fabric composite[42]; this work is an extension of a previous model originally proposed for unidirectional composites[43]. The plane stress orthotropic constitutive law with damage parameters is given by:

$$
\begin{Bmatrix} \varepsilon_1 \\ \varepsilon_2 \\ \gamma_{12} \end{Bmatrix} = \begin{bmatrix} 1/E_1(1-d_1) & -\upsilon_{12}/E_1 & 0 \\ -\upsilon_{12}/E_2 & 1/E_2(1-d_2) & 0 \\ 0 & 0 & 1/G_{12}(1-d_{12}) \end{bmatrix} \begin{Bmatrix} \sigma_1 \\ \sigma_2 \\ \tau_{12} \end{Bmatrix} = [S]\{\sigma\}
$$

8.30

where the compliance matrix $[S]$ comprises four undamaged elastic constants E_1, E_2, G_{12} and υ_{12} and elastic damage parameters (d_1, d_2) for the two principal fibre directions and damage parameter (d_{12}) for inelastic shear response.

Figure 8.12 shows typical force versus displacement curves for composite coupons loaded in the 0°/90° and ±45° directions (test curves are shown

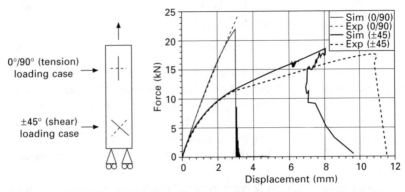

8.12 Test and simulation curves for a typical balanced glass fabric–epoxy composite under axial 0°/90° and shear ±45° loading.

dashed). The material is a woven R-glass fabric with Fibredux 924 epoxy resin. Usually at least these tests are required to determine the damage parameters d_1, d_2 and d_{12}. However, some models account for the effect of interaction of stress components on damage and additional tests are necessary; this effect will be more important for textile composites than simple UD plies. The Ladevèze model[42, 43], for example, clearly defines the test program needed and the methods to derive and specify the damage evolution parameters. Following these procedures reasonable agreement between test and simulation curves, Fig. 8.12, has been obtained using the crashworthiness FE code PAM-CRASH[40].

Delamination modelling

The detection of delamination and the ability to model its growth can have important consequences on the laminate stiffness and sequence of ply failures leading to ultimate laminate failure. Delamination also absorbs a significant proportion of impact energy and should be properly modelled in any numerical model of impact failure.

The following approach to treat delamination is gaining favour and is available in at least two commercial FE codes[39, 40]. The method involves modelling the laminate as a stack of discrete finite elements, in which each element represents one ply or a subset of plies. These elements are tied together using interface constraints that characterise the mechanical stiffness, strength and delamination energy absorption of the resin-rich interface. The interface algorithm performs two key operations; first, at the start of the calculation, matching 'pairs' of nodes and adjacent elements are identified and, second, during the calculation these matching pairs are constrained to move relative to each other dependent on a linear elastic with damage stress-displacement law. Figure 8.13 shows a schematic view of one element of the interface.

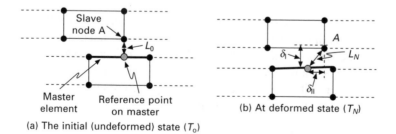

(a) The initial (undeformed) state (T_0)

(b) At deformed state (T_N)

8.13 Identification of the 'slave' node and the attached 'master' surface and the mode I (normal) and mode II (shear) separation distances.

The main features of the interface mechanical law are shown in Fig. 8.14 for normal (mode I) loading. A simple linear elastic law is assumed up to a failure stress σ_{Ic}. Thereafter, linear damaging is activated such that at final separation δ_{Imax}, the fracture energy of the fibre–resin system has been absorbed. That is, the area under the curve corresponds to the fracture energy G_{Ic} of the resin–fibre interface. Thus σ_{Imax}, δ_{Ic} and δ_{Imax} are selected so that the elastic-damage curve fulfils the required criteria. Identical arguments are used to define the normal (mode I) and shear (mode II) curves which have different parameters; typically crack growth under mode II loading will absorb about ten times the energy of mode I crack growth.

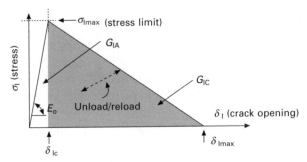

8.14 Diagram of the stress-crack opening curve for mode I loading.

The following formulae describe mode I failure; identical equations are used for mode II by interchanging E_0, σ_{Imax}, δ_I and G_{Ic} with G_0, σ_{IImax}, δ_{II} and G_{IIc}. The resin through-thickness normal modulus E_0 is used for the elastic range of the stress-displacement curve. When a maximum stress σ_{Imax} is reached, the stress displacement law follows a linear damage equation of the type:

$$\sigma_I = (1 - d_I) \cdot E_0 \cdot \varepsilon_I = (1 - d_I) \cdot E_0 \cdot \frac{\delta_I}{L_0} \qquad 8.31$$

where d_I is the damage parameter, varying between the values of 0 (undamaged) and 1 (fully damaged), E_0 is the modulus, L_0 the normal distance between the original position of the slave node and the master element and δ_I the deformed normal separation distance of the slave node and master element.

The area under the elastic range is the fracture energy required to start damage of the interface G_{IA}, and is given by:

$$G_{IA} = \frac{1}{2} \frac{\sigma_{Imax}^2}{E_0} L_0 \qquad 8.32$$

The total area under the load–displacement curve is the critical fracture energy required for failure of the interface (including the elastic and damaging zones) and is given by:

$$G_{Ic} = \frac{1}{2}\, \sigma_{Imax}\, \delta_{Imax} \qquad\qquad 8.33$$

The values of G_{Ic} and G_{IIc} can be obtained via the standard double cantilever beam (DCB) and end notched flexure (ENF) tests[44, 45] respectively, from which the necessary parameters for the model can be found.

Figure 8.15 illustrates mode I loading of a simple single element test case. Path (a_1) shows the case of tensile loading. Damage initiates when σ_{Imax} is reached and path (a_2) is then followed. Path (b_1) represents unloading using the new partially damaged material modulus. Note that if unloading causes the gap to fully close then the original undamaged modulus is used, path (b_2). Reloading follows paths (b_1) and (b_2) to the point where unloading initiated, after which path (a_3) is followed until full failure occurs.

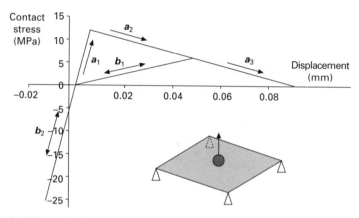

8.15 Example loading–unloading for mode I of a single element test case.

Invariably, in practice, there is a strong coupling between normal and shear loading at the interface. Indeed a specific test procedure has been devised to investigate and quantify this type of mixed mode loading using a mixed mode bending (MMB) test[46]. This work has shown that, for many materials, a linear coupling model can reasonably represent the mixed-mode failure process, thus:

$$\frac{G_I}{G_{In}} + \frac{G_{II}}{G_{IIn}} = 1 \qquad\qquad 8.34$$

where $n = A$ for the onset of fracture and $n = C$ for fracture; G_I and G_{II} are the instantaneous values and G_{In} and G_{IIn} corresponding to test values as measured by the DCB and ENF tests. Example solutions using this approach, including validation against DCB, ENF and MMB tests, have been presented by Lourenco[47]. Figure 8.16 shows the experimental set-up for a DCB test

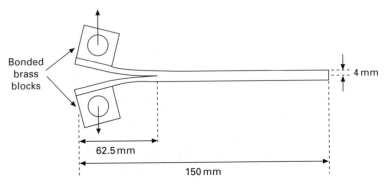

8.16 Experimental test of the double cantilever beam to determine the mode I fracture energy for a non-crimp fabric-based composite.

where the Mode I delamination crack growth in a non-crimp fabric (NCF) reinforced composite is evaluated. Figure 8.17 shows the comparison between test and simulation for this NCF laminate in which the simulation model uses the experimentally determined value for G_{IC}.

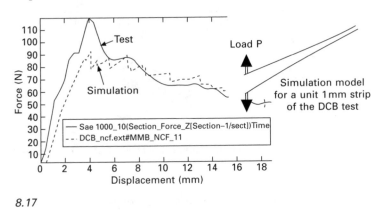

8.17

8.3.3   Examples of impact simulations of textile composites

The following examples illustrate the state-of-the-art simulation of textile composite structures under impact loading. In each case the fabric model[42] and delamination model[48] have been used in the crashworthiness simulation code PAM-CRASH.

Impact simulation of a laminated carbon fabric composite plate

The first example considers a 50 mm diameter rigid ball of mass 21 kg striking a flat, simply supported composite plate at 6.28 m/s; Fig. 8.18. The material is a woven fabric with T300 carbon fibres and Fibredux 924 epoxy resin. The plate is 300 mm square and has 16 plies of quasi-isotropic lay-up $[(0,90)/(+45, -45)]_{4s}$, giving a thickness of 4.6 mm. A comparison of a conventional multilayered shell (no delamination) and the modelling method with delamination between plies is presented and shows the improved agreement between test and simulation results.

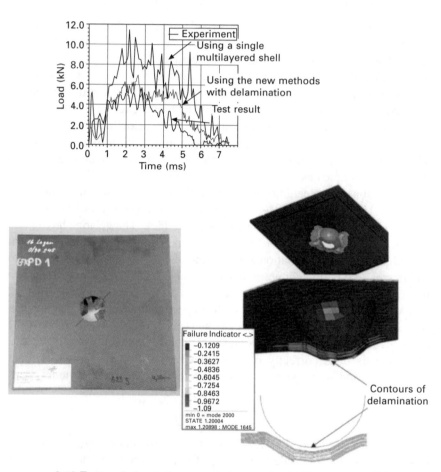

8.18 Test and simulation results for a spherical impactor striking a flat CFRP plate.

Delamination occurs initially under the impact point at the centre of the laminate, where the shear is a maximum, causing it to separate into two sublaminates. These sublaminates then delaminate at their mid-plane as the ball intrudes further into the plate creating further sublaminates and delamination until, possibly, all plies are delaminated. The prediction of extensive delamination is valuable information that agrees well with test Cscan results. At this velocity the simulation correctly calculates penetration of the plate by the impactor; for a lower velocity of 2.33 m/s (not shown) the simulation and test both find that the punch rebounds, albeit with significant ply and delamination damage.

Impact simulation of a laminated glass fabric aircraft leading edge

A second example concerns the 'bird-strike' impact of a demonstrator aircraft leading edge constructed from woven R-glass fabric and Fibredux 924 epoxy resin. The fluid-like behaviour of the bird at high velocity is usually approximated in tests using a gelatine projectile. Smoothed particle hydrodynamics is a specialised FE technique used here for the projectile, since conventional finite elements cannot handle the excessive deformations that occur in this material. The gelatine, at high velocities, behaves like a fluid and is treated as a nearly incompressible liquid with zero shear response. The laminate is modelled using the same stacked shell with delamination interface techniques as used in the previous flat plate.

The demonstrator leading edge is approximately 200 mm long with a nose radius of 15 mm and consists of eight quasi-isotropic layers $[(0,90)/(+45, -45)]_{2s}$ giving a laminate approximately 2 mm thick. The gelatine has a mass of 30 g and impact velocity of 132 m/s. Test and simulation results are shown in Fig. 8.19. The delamination zone for the test part is marked on the photo and is seen to be non-symmetric in shape and extends over a large area. The numerical model predicts a comparable total area of delamination which is, as may be expected, symmetric about the impact point. The general agreement of the global area of delamination damage is encouraging. Most probably the difference in the areas affected is due to off-centre impact loading in the test. The analysis does correctly compute deflection of the gelatine, without penetration, and shows similar ply damage to that observed in the test part.

Failure simulation of stitched composite 'T' joint

Within a German-funded BMBF project[49, 50] new technologies have been developed for the 'one sided' stitching of textile composite preforms. As a part of that work, the following study demonstrates the potential for simulation of stitched jointing systems. Figure 8.20 shows some details of the manufacturing process for this preform.

The previous fabric models are used to represent the 'T' and outer composite panel and the delamination model is used to represent the resin part of the interface between the 'T' and the outer panel. In addition, beam elements are used to represent the stitches; these elements have a tensile failure criterion (maximum strain) and stiffness to represent the stitching properly. Figure 8.21 shows details of the geometry and the finite element representation. Only

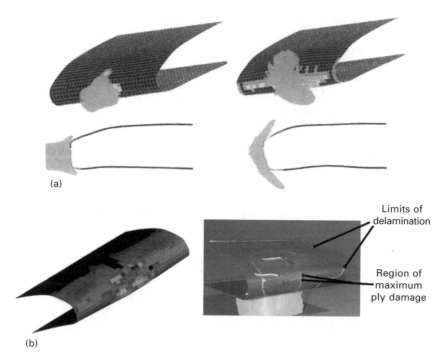

(a)

(b)

Limits of
delamination

Region of
maximum
ply damage

8.19 Test and simulation results for the impact of a demonstrator
leading edge. (a) Early and intermediate deformation states showing
the spread of the gelatine, deformation of the leading edge and the
evolution of delamination damage. (b) Contours of ply damage for
the centre ply (simulation and test).

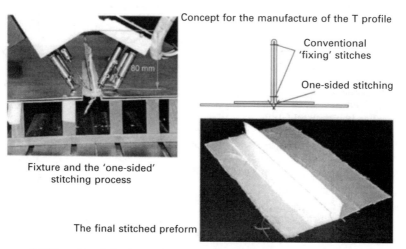

Concept for the manufacture of the T profile

Conventional
'fixing' stitches

One-sided stitching

Fixture and the 'one-sided'
stitching process

The final stitched preform

8.20 Details of the 'one-sided' stitching process to manufacture a 'T'
joint preform (Courtesy: ITA Aachen).

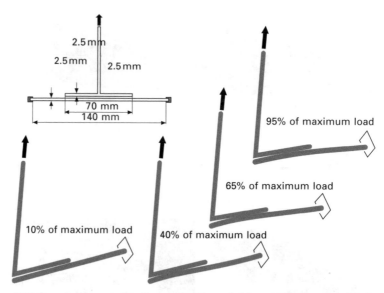

8.21 Geometry and details of the FE model for the composite 'T' joint. The composite is a quasi-isotropic lay-up of non-crimp fabric with RTM6 (Hexcel) epoxy resin.

one symmetric half of a unit strip of the 'T' joint is analysed. In this figure a double chain stitch is used at distances 10 and 30 mm from the symmetry line. Loading is in tension at the free end of the 'T' profile and contours of delamination growth are shown as a function of the maximum failure load.

A simple parametric study, Fig. 8.22, shows the importance of the location

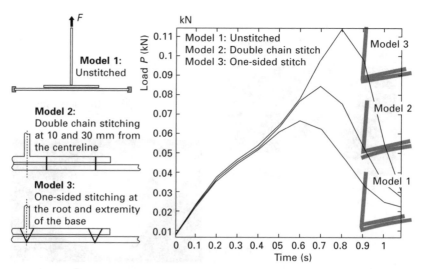

8.22 Example parametric study to optimise stitch location for loading.

of the stitching on the maximum failure load. A 40% increase in failure load can be achieved using optimally located stitching compared with the unstitched case.

8.4   References

1. Bogdanovich A. and Pastore C., *Mechanics of Textile and Laminated Composites*, Chapman and Hall, London, 1996.
2. Hull D. and Clyne T.W., *An Introduction to Composite Materials,* 2nd edn, Cambridge University Press, Cambridge, 1996.
3. Jones R.M., *Mechanics of Composite Materials*, 2nd edn, Taylor and Francis, Washington, DC, 1996.
4. Owen M.J., Middleton V. and Jones I.A. (eds), *Integrated Design and Manufacture using Fibre-reinforced Composites*, Woodhead Publishing Ltd, Cambridge, 2000.
5. *BS2782-10 Method 1003 EN61 Methods of Testing Plastics – Glass reinforced plastics – Determination of tensile properties.* British Standards Institute, UK, 1977.
6. *BS2782 Part 10 Method 1005 EN61 Methods of Testing Plastics – Glass reinforced plastics – Determination of flexural properties. Three point method.* British Standards Institute, UK, 1977.
7. *BS EN ISO 527-5:1997 BS 2782-3: Method 326G:1997 Plastics – Determination of tensile properties – Part 5: Test conditions for unidirectional fibre-reinforced plastic composites.* British Standards Institute, UK, 1997.
8. ASTM D5379/D5379M, Test method for shear properties of composite materials by the V-notched beam method, *Annual Book of ASTM Standards*, ASTM, Philadelphia, USA, 2000, Vol 15.03.
9. Halpin J.C. and Kardos J.L., The Halpin–Tsai equations: a review, *Polymer Engineering and Science*, 1976 **16**(5) 344–352.
10. Halpin J.C., *Primer on Composite Materials Analysis*, 2nd edn., Technomic, Lancaster, PA, 1992.
11. Hashin Z., *Theory of Fibre Reinforced Materials*, CR-1974, NASA, 1972.
12. Hashin Z., Analysis of composite materials – a survey, *J. Appl. Mech.*, 1983 **50** 481–505.
13. Naik N.K. and Shembekar P.S., Elastic behavior of woven fabric composites: I – Lamina analysis, *J. Composite Materials*, 1992 **26**(15) 2196–2225.
14. Reddy J.N., *Mechanics of Laminated Composite Plates: Theory and Analysis*, CRC Press. Boca Raton, FL, 1997.
15. Soldatos K.P., A comparison of some shell theories used for the dynamic analysis of cross-ply laminated circular cylindrical shells, *J. Sound Vib.*, 1984 **97** 305–319.
16. Soldatos K.P. and Timarci T., A unified formulation of laminated composite, shear-deformable, five-degrees-of-freedom cylindrical shell theories, *Composite Structures*, 1993 **25** 165–171.
17. Hofstee J. and van Keulen V., Elastic stiffness analysis of a thermo-formed plain-weave fabric composite. Part II: analytical models, *Composites Sci. Technol.*, 2000 **60** 1249–1261.
18. Tan P., Tong L. and Steven G.P., Predicting the mechanical properties of textile composites – a review, *Composites Part A* 1997 **28A** 903–922.
19. Crookston J.J., Long A.C. and Jones I.A., Modelling effects of reinforcement deformation during manufacturing on elastic properties of textile composites, *Plastics Rubber and Composites* 2002 **32**(2) 58–65.

20 Long A.C., Robitaille F., Rudd C.D. and Jones I.A., 'Modelling strategies for textile composites', *Proc 14th International Conference on Composite Materials (ICCM-14)*, San Diego, CA, 14–18, July 2003.

21. Crookston J.J., Prediction of elastic behaviour and initial failure of textile composites, PhD Thesis, University of Nottingham, 2004.

22. Bigaud D. and Hamelin P., From geometrical description to mechanical prediction – Application to woven fabric composites. *Sci. Eng. Composite Materials*, 1998 **7**(4) 291–298.

23. Chapman C. and Whitcomb J., Effect of assumed tow architecture on predicted moduli and stresses in plain weave composites, *J. Composite Materials*, 1995 **29**(16) 2134–2159.

24. Hofstee J., do Boer H. and van Keulen V., Elastic stiffness analysis of a thermo-formed plain-weave fabric composite. Part I: geometry, *Composites Sci. Technol.*, 2000 **60** 1041–1053.

25. Hofstee J., do Boer H. and van Keulen V., Elastic stiffness analysis of a thermo-formed plain-weave fabric composite – part III: experimental verification, *Composites Sci. Technol.*, 2000 **62** 401–418.

26. Whitcomb J.D., Three-dimensional stress analysis of plain weave composites. *Composite Materials: Fatigue and Stresses* (3rd volume), ASTM STP 1100, ed. T.K., O'Brien. ASTM, Philadelphia, 417–438.

27. Guedes J.M. and Kikuchi N., Preprocessing and postprocessing for materials based on the homogenization method with adaptive finite element methods, *Comp. Meth. App. Mech. Eng.*, 1990 **83**, 143–198.

28. Ng S.-P., Tse P.-C. and Lau K.-J., Numerical and experimental determination of in-plane elastic properties of 2/2 twill weave fabric composites, *Composites Part B*, 1998 **29B** 735–744.

29. Dasgupta A., Agarwal R.K. and Bhandarkar S.M., Three-dimensional modeling of woven-fabric composites for effective thermo-mechanical and thermal properties, *Composites Sci. and Technol.*, 1996 **56** 209–223.

30. Zako M., Uetsuji Y. and Kurashiki T., Finite element analysis of damaged woven composite materials, *Composites Sci. Technol.*, 2003 **63**(3–4) 507–516.

31. Aitharaju V.R. and Averill R.C., Three-dimensional properties of woven-fabric composites, *Composites Sci. Technol.*, 1999 **59** 1901–1911.

32. Brown D., Morgan M. and McIlhagger R., A system for the automatic generation of solid models of woven structures. *Composites Part A*, 2003 **34** 511–515.

33. Sun W., Lin F. and Hu X., Computer-aided design and modeling of composite unit cells. *Composites Sci. Technol.*, 2001 **61** 289–299.

34. Robitaille F., Long A.C., Jones, I.A. and Rudd C.D., Automatically generated geometric descriptions of textile and composite unit cells. *Composites Part A*, 2003 **34** 303–312.

35. Zienkiewicz O.C. and Taylor R.L., *The Finite Element Method* 5th edn, Butterworth-Heinemann, Oxford, 2000.

36. Hinton M.J., Soden P.D. and Kaddour A.S., *Failure Criteria in Fibre-Reinforced-Polymer Composites*, Elsevier, Amsterdam, 2004.

37. Cox B.N. and Flanagan G., *Handbook of Analytical Methods for Textile Composites*, NASA Contract Report 4750, March 1997.

38. Cook R.D., Malkus D.S. and Plesha M.E., *Concepts and Applications of Finite Element Analysis*, 3rd edn, Wiley, New York, 1989.

39. The LUSAS FE software code, www.lusas.co.uk.

40. PAM-CRASH™ FE code, www.esi-group.com.
41. Hart-Smith L.J., The role of biaxial stresses in discriminating between meaningful and illusory composite failure theories, *Composite Structures*, 1993 **25**(1–4) 3–20.
42. Johnson A.F., Pickett A.K. and Rozycki P., 'Computational methods for predicting impact damage in composite structures', *Workshop: Recent Advances in Continuum Damage Mechanics for Composites*, LMT-Cachan, 20–22 Sept. 2000.
43. Ladevèze P. and Le Dantec E., Damage modelling of the elementary ply for laminated composites, *Composites Sci. Technol.*, 1992 **43** 257–267.
44. ISO DIS15024 *Fibre-reinforced plastic composites – Determination of Mode I interlaminar fracture toughness, GIC, for unidirectionally reinforced materials*, International Organization for Standardization, Geneva.
45. ASTM draft standard D30.06: Protocol for Interlaminar Fracture Testing, End-Notched Flexure (ENF), revised April 24, ASTM, Philadelphia, USA.
46. Reeder J.R. and Crews J.H., Mixed-mode bending method for delamination testing, *AIAA J.*, 1989 **28**(7) 1270–1276.
47. Lourenco N.S.F., Predictive finite element method for axial crush of composite tubes, PhD Thesis, University of Nottingham, 2002.
48. Warrior N.A., Pickett A.K. and Lourenco N.S.F., Mixed mode delamination – experimental and numerical studies, *Strain*, 2003 **39** 153–159.
49. BMBF project, 'Textile Integrationstechniken zur herstellung vorkonfektionierter verstärkungsstrukturen für FVK', 1999–2002.
50. Gries T., Laourine E. and Pickett A.K., Potentiale nähtechnischer Fügeverfahren für Faserverbundwerkstoffe (Potential application of sewing technologies as a joining process for FRP materials), *Aachener Textil Tagung*, Nov 2001.

9

Flammability and fire resistance of composites

A R H O R R O C K S and B K K A N D O L A,
University of Bolton, UK

9.1   Introduction

All fibre-reinforced composites comprise matrix materials that are flammable to varying degrees and, compared with metals such as aluminium or steel, can burn vigorously, often with evolution of smoke. While organic fibre reinforcement such as polyester, aramid and even carbon may add fuel to the burning composite, even if inorganic fibres such as E-glass are the reinforcing structures, the general composite fire resistance and performance (including smoke generation) will be determined by that of the organic matrix. The overall physical behaviour under heat and flame conditions will be influenced by the thermal properties of the fibres present since resins are often thermoplastic or deformable except in the highest performance examples such as phenolics and polyimides. Table 9.1 illustrates the thermal properties of typical fibre reinforcement for composites. It is seen that while glass fibres are probably the most commonly used fibres, while they are non-flammable, their relatively low melting point compared with typical flame temperatures of 1000 °C or so will ensure that under fire exposure conditions, glass fibre reinforced composites will start to deform when temperatures reach 500 °C and above. Thus whether or not a composite maintains both a heat and flame barrier to an advancing fire depends on the combined flammable behaviour of the fibres and resins present, coupled with their abilities to withstand the physical aspects of applied heat.

Fire resistance and smoke generation properties of composite materials are major issues these days because, depending on applications, they must pass some type of regulatory fire test in order to ensure public safety. Thus, it is important to understand how individual components of the end-products burn and how best to modify materials to make them flame-resistant without compromising their uniquely valuable low weight to high mechanical property ratios.

This chapter complements our earlier reviews [1, 3], to which the reader is directed for a greater background understanding and which have provided

Table 9.1 Physical and mechanical properties of fibres [1, 2]

Fibre	Diameter (µm)	Tensile strength (GPa)	Initial modulus (GPa)	Density (kg m⁻³)	Second order, T_g or softening temp. (°C)	Max. service temp. (°C)	Limiting oxygen index, LOI (%)
E-Glass	3–20	2–6	50–100	2400–2600	>700	250	–
S-Glass	3–20	3.5	87	2500	>700	250	–
Carbon*	4–10	1.5–7.0	150–800	1500–2000	–	400–450	55–60
Para-aramid	10	2–4	70–150	1410–1450	340	200	30
Boron	100–200	2–4	370–430	2500–2700	–	350	–
UHMWPE	10–30	1.5	70	990–1020	–50	100	18–19
Alumina	10–20	0.5	310	3800–4000	–	1000	–

*Carbon and graphite fibres.

general overviews of composite fire resistance. This chapter, while presenting a brief overview of composite fire behaviour, will focus on recent research that has concentrated on understanding resin thermal behaviour and the resins used and means of enhancing fire and smoke performance using flame retardants that do not use the environmentally and toxicologically questioned antimony–bromine-based flame retardants [4, 5].

9.2   Constituents – their physical, chemical, mechanical and flammability properties

The structures and physical and mechanical characteristics of textile reinforced composites are described elsewhere in this text, but adhesion between two dissimilar phases is necessary to allow uniform load distribution between them and the nature of the sustenance of this bond under thermal conditions is an essential determinant of the physical stability of a composite in a fire. The thermochemical and thermophysical properties of the fibres and matrix will also be significant fire performance-determining issues. Fire properties also depend upon the methods used to combine these components into one material and whether flame-retardant additives or other systems are included. Generally, there are three methods of conferring fire resistance on composites:

1. Use of inherently flame-resistant resins and/or fibres: the use of different generic resins will be discussed below, as will the thermal properties of available fibres. Modifications of the various resin types by inclusion of flame-retardant comonomers (e.g. brominated variants, particularly polyesters) is beyond the scope of this chapter and the reader is referred elsewhere [6].
2. Incorporation of flame retardant additives. These are included along with some our own recent research using intumescent materials.
3. Use of external coatings and outer protective surfaces. These are usually additional to the fundamental composite and may include paints, coatings and ceramic fibrous structures, often as wet-laid nonwoven structures which are incorporated in the surface(s) that will be fire exposed. These will also be briefly reviewed below with a focus on recent research.

9.2.1   Fibres

The main fibres are chosen from the armoury of conventional and high-performance fibres available to the textile and fibre industries in general and Table 9.1 lists the properties of those most commonly used for composites. In this table, resistance to heat and flame is given in terms of the second order, glass transition temperature (T_g), or softening temperature, the maximum service life temperature and the burning propensity defined in terms of limiting oxygen index, LOI – the percentage of oxygen required to sustain burning of a vertical sample in a downward direction [5]. Thus the fire-resistant fibres have the highest values possible for each of these parameters. For burning, it is generally accepted that if a material has LOI > 30, then in an air atmosphere, it will be deemed to be flame retardant. Such a fibre, e.g. para-aramid, will still burn in a well-ventilated fire.

Fibrous arrays can be in the form of woven or nonwoven cloths or layers. Generally, composites will contain at best a single fibre type and occasionally two, e.g. carbon warp and aramid weft. Discontinuous fibres can also be used, e.g. chopped fibres about 30–50 mm long, distributed in a random manner in a plane and held together with a resin binder. Both tows and cloths can be pre-impregnated with resin, processed and then used as 'prepregs' during composite manufacture. Some of the most used fibres are reviewed briefly below.

Glass fibres

Based on different chemical compositions, various grades of glass are available commercially, e.g. E-, S- R- and C-glass [2]. By pulling swiftly and continuously from the melt, glass can be drawn into very fine filaments. Continuous glass fibres are $3–20 \times 10^{-6}$ m in diameter. The physical properties are given in Table 9.1 [2] and for most textile-reinforced composites E- (electrical resistance) and S- (high strength) glass fibres are preferred because of their combined properties of high strength and modulus. The advantages of glass fibres are their combination of chemical inertness and their high tensile and compressive strengths, low cost, good compatibility and good processibility. The disadvantages are associated with their low modulus and physical thermal stability. When heated, while they are flame resistant in respect of their not supporting combustion, they soften at relatively low temperatures and by 500 °C have lost most of their physical strength and so have limited temperature performance ranges.

Carbon fibres

Carbon fibres are manufactured by controlled pyrolysis and cyclisation of certain organic precursors, e.g. polyacrylonitrile (PAN). Carbon fibres have characteristics of low density, high strength and stiffness. As shown in Table 9.1, their stiffness is high compared with glass fibres. Mechanical characteristics of carbon fibres do not deteriorate with temperature increases up to 450 °C, so they can be used for both polymeric and metal matrices. They are used for manufacturing load-carrying panels of aircraft wings and fuselages, drive shafts of cars and parts operating under intense heating. At temperatures above 450 °C, they will chemically start to oxidise and only become really combustible once temperatures approach 1000 °C.

Aramid fibres

Aramid fibres are based on aromatic polyamides, and where at least 85% of the amide groups are connected directly to an aromatic group, they are generically called aramid fibres. For composites, the para-aramids with their superior tensile properties are preferred and these are typified by the various commercial grades of Kevlar® (Du Pont), and Twaron® (AKZO) and similar fibres available. The general chemical formula for these para-aramids is typified by that for poly(p-phenylene terephthalamide) (PPT)

9.1

While they have excellent tensile properties, their second order transition temperatures are much higher than the majority of organic high-performance fibres (see Table 9.1) although their flammability as measured by limiting oxygen index, for example, is moderate (LOI = 30–31) and comparable with that of the meta-aramids (eg Nomex®, Du Pont) and flame-retardant cotton and wool [5].

Boron fibres

Boron fibres are obtained by high-temperature reduction of boron trichloride vapour on a tungsten or carbon substrate. With rise in temperature, fibres start to degrade in air at 400 °C. In order to prevent their oxidative degradation, they are covered with a refractory silicon or boron carbide coating. They are typically $100–200 \times 10^{-6}$ m in diameter. Because of their large diameter and high stiffness, it is not possible to carry out normal textile processes such as weaving. Hence, these are used in the form of single-thickness, parallel-laid, pre-impregnated sheets or narrow continuous tapes [1]. Their main advantages are high stiffness and compression strength, but they are rather expensive. Table 9.1 shows that their ability to withstand working temperatures as high as 350 °C for their whole expected service lives is a significant factor in their selection for high-temperature applications.

Polyethylene fibres

Fibres from ultra-high molecular weight polyethylene (UHMWPE) may be produced to have similar tensile properties as aramids. While their hydrocarbon chains have chemical inertness, not only do they burn typically like any hydrocarbon material (LOI = 18–19) but also a second major disadvantage is their low melting point, 130–150 °C and hence low maximum service temperatures of only 100 °C or so. Examples of UHMWPE are Spectra (Allied fibres) and Dyneema (DSM).

Alumina fibres

Fibres of polycrystalline alumina can be made by extruding a thickened mixture of fine alumina powder suspended in an alginate binder and then sintering the fibrous mass at high temperature. Alumina fibres are very strong and are resistant to temperatures as high as 900–1000 °C. It may thus be deduced that as fibres, they can offer the greatest fire resistance of all used in composite markets. Examples are Nextel® (3M Corp.) and Saffil® (Saffil Ltd., UK) and they may be used with epoxy, polyimide and maleimide resins.

9.2.2   Matrix polymers

The most common matrix materials for composites (and the only ones discussed here) are polymeric, which can be thermoset or thermoplastic; examples are presented in Table 9.2. Thermoset matrices are fabricated from the respective resin, a curing agent, a catalyst or curing initiator and a solvent sometimes introduced for lowering the viscosity and improving impregnation of reinforcements. In thermosets, solidification from the liquid phase takes place by the action of an irreversible chemical crosslinking reaction which produces a tightly bound three-dimensional 3D network of polymer chains. The molecular units forming the network and the length and density of the crosslinks of the structure will influence the mechanical and any residual thermoplastic properties of the material. The level of crosslinking between resin functional groups and often the degree of non-thermoplasticity is a function of the degree of cure, which usually involves application of heat and pressure. However, some resins cure at room temperature.

The second type of polymers are thermoplastic in nature and have the advantage that they can be formed by physical processes of heating and cooling. Thermoplastics readily flow under stress at elevated temperatures, can be fabricated into required components and become solid and retain their shape when cooled to room temperature. However, the reversibility of this process generates composites having a thermoplastic property and, hence,

Table 9.2 Limiting oxygen index values for polymers and composites at 23 °C [7–9]

Resin	LOI (%)	
	Resin	40% (w/w) resin/ 181 glass cloth
Thermoplastic resins		
Acrylonitrile–butadiene–styrene	34	
Polyaryl sulphone (PAS)	36	
Polyether sulphone (PES)	40	
9,9Bis-(4-hydroxyphenyl) fluorene/ polycarbonate-poly(dimethyl siloxane) (BPFC-DMS)	47	
Polyphenylene sulphide (PFS)	50	
Thermoset resins		
Polyester	20–22	
Vinyl ester	20–23	
Epoxy	23	27
Phenolic	25	57
Polyaromatic melamine	30	42
Bismaleimide	35	60

poor physical resistance to heat. The most widely used matrix materials are discussed below along with their thermal degradative characteristics.

Polyester resins

Polyesters are probably the most commonly used of polymeric resin materials. The advantages of polyester matrices are their ability to cure over a wide range of temperatures under moderate pressures and their low viscosities providing good compatibility with fibres. In addition is their ability to be readily modified by other resins. Essentially they consist of a relatively low molecular weight unsaturated polyester chain dissolved in styrene. Curing occurs by the polymerisation of the styrene, which forms crosslinks across unsaturated sites in the polyester. Curing reactions are highly exothermic, and this can affect processing rates as excessive heat can be generated which can damage the final laminate. The general formula for a typical resin [10] is shown in Fig. 9.2. Among the drawbacks of polyester resins are poor mechanical characteristics, low adhesion, relatively large shrinkage and the presence of toxic components of the styrene type.

Most polyesters start to decompose above 250 °C, whereas the main step of weight loss occurs between 300 and 400 °C [6]. During thermal decomposition, polystyrene crosslinks start to decompose first and styrene is volatilised (Fig. 9.3). The linear polyester portion undergoes scission similar to thermoplastic polyesters (Fig. 9.4), undergoing decarbonylation,

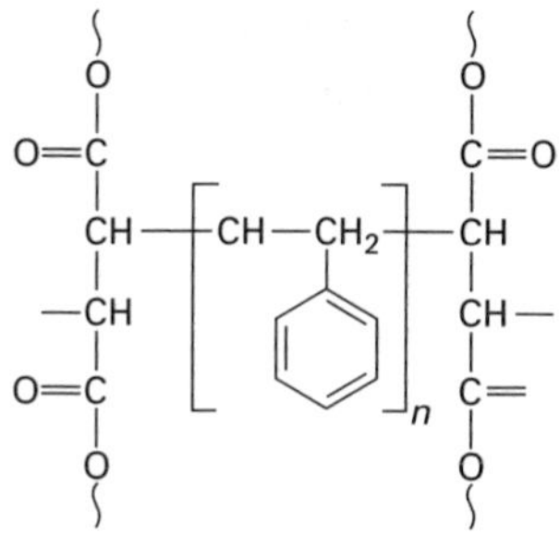

9.2

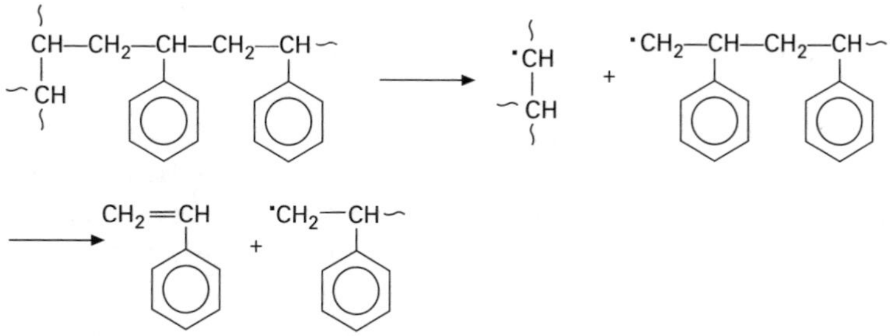

9.3

decarboxylation or splitting off of methylacetylene. Learmonth and Nesbit [11] have shown that during thermal decomposition volatiles are lost up to 400 °C and, above 400 °C, it is solid phase oxidation reactions that predominate with initial attack occurring at crosslinks [12].

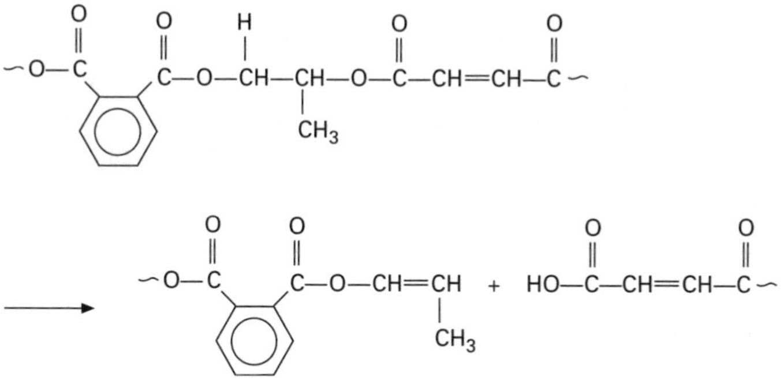

9.4

Because of the ease of formation of these flammable pyrolysis products, polyesters have LOI values of 20–22 and hence, flame readily, and sometimes vigorously, after ignition. Unsaturated polyesters, crosslinked with styrene, burn with heavy sooting. These can be flame retarded by addition of inorganic fillers, addition of organic flame retardants, chemical modification of the acid, alcohol or unsaturated monomer component and the chemical combination of organo-metallic compounds with resins [10].

It is common practice to add inert fillers to polyester resins to reinforce the cured composite, to lower cost and to improve flame retardance. Glass fibre and calcium carbonate often increase the burning rate of the composition [10], but other fillers such as antimony trioxide for halogenated compositions and hydrated alumina are quite effective flame retardants. Modification of the saturated acid component has been by far the most successful commercial method of preparing flame-retardant unsaturated polyesters. Examples are halogenated carboxylic acids, such as chlorendic acid or their anhydrides, tetrachloro- or tetrabromophthalic anhydride [13].

Halogenated alcohols or phenol can also be incorporated into the polymeric chain. Examples are tribromo-neopentyl glycol, tetrabromobisphenol-A and dibromophenol. The crosslinking partner may also be flame-retardant, as in the case of monochloro- or dichlorostyrene and hexachloropentadiene. Examples of halogenated additive compounds are tetrabromo-p-xylene, pentabromobenzyl bromide, pentabromoethyl benzene, pentabromotoluene, tribromocumene, decabromodiphenyl oxide and brominated epoxy resins [13]. The effectiveness of halogenated components is enhanced by simultaneous addition of antimony trioxide.

Phosphorus-containing flame retardants such as phosphonates and dialkyl phosphites can be incorporated into the polyester chain. In addition, allyl or diallyl phosphites may act as crosslinking agents [13].

Vinyl ester resins

Vinyl ester resins like unsaturated polyesters cure by a radical initiated polymerisation. They are mainly derived from reaction of an epoxy resin, e.g. bisphenol A diglycidyl ether, with acrylic or methacrylic acid. Their general formula is shown in Fig. 9.5, where R is any aliphatic or aromatic residue and R' is typically either H or CH_3.

$$R \left[CH_2\overset{\overset{\displaystyle OH}{|}}{C}HCH_2O - \overset{\overset{\displaystyle }{\|}}{\underset{\underset{\displaystyle O}{\|}}{C}} - \overset{\overset{\displaystyle R'}{|}}{C}=CH_2 \right]_n$$

9.5

Like unsaturated polyesters they are copolymerised with diluents such as styrene using similar free radical initiators. They differ from polyesters in that the unsaturation is at the end of the molecule and not along the polymer chain. When methacrylates are used, they offer better chemical resistance than unsaturated polyesters. Their burning behaviour falls between that of polyester and epoxy resins (LOI = 20–23).

Epoxy resins

These resins are extensively used in advanced structural composites particularly in the aerospace industry. They consist of an epoxy resin and a curing agent or hardener. They range from low-viscosity liquids to high melting point solids and can be easily formulated to give suitable products for the manufacture of prepregs by both the solution and hot-melt techniques. They can be easily modified with a variety of different materials. Epoxy resins are manufactured by the reaction of epichlorohydrin with materials such as phenols or aromatic amines. Epoxy resins contain the epoxy or glycidyl group shown in Fig. 9.6, where R is any aliphatic or aromatic residue.

$$R'\left[CH_2-CH-\underset{O}{\overset{}{CH}}-CH_2\right]_n$$

9.6

This group will react typically with phenolic —OH groups and Bisphenol-A type resins are most commonly used for composite structures. Epoxy resins are very reactive, hence both catalytic and reactive curing agents can be used. The general structure of a typical cured epoxy resin is shown in Fig. 9.7, where X can be H and Y depends upon the structure of curing agent.

$$\text{---}\langle O\rangle\text{---}\underset{CH_3}{\overset{CH_3}{C}}\text{---}\langle O\rangle\text{---}O-CH_2-\underset{OX}{\overset{}{CH}}-CH_2-Y$$

9.7

The resin can exist in the uncured state for quite a long time. This property allows the manufacture of prepregs, where the fibres are impregnated with resin and are partially cured [14]. Glass transition temperature of epoxies ranges from 120 to 220 °C [15], hence they can be safely used up to these temperatures. Apart from the simple example above, some of the epoxy resins used in advanced composites are N-glycidyl derivatives of 4,4'-diaminodiphenylmethane and 4-aminophenol, and aromatic di- and polyglycidyl derivatives of Bisphenol A, Bisphenol F, phenol novolacs and tris (4-hydroxyphenyl) methane [15].

Since the catalytic curing agents are not built into the thermoset structure, they do not affect the flammability of the resin. Reactive agents, mostly amines, anhydrides or phenolic resins, on the other hand, strongly affect the crosslinking of these thermosets and hence, their flammabilities [6]. Epoxy resins cured with amines and phenol–formaldehyde resins tend to produce more char than acid or anhydride-cured resin.

During the early stages of the thermal degradation (at lower temperatures) of cured epoxy resins, the reactions are mainly non-chain-scission type, whereas at higher temperatures, chain-scissions occur [16]. The most important non-scission reactions occurring in these resins are the competing dehydration and dehydrogenation reactions associated with secondary alcohol groups in the cured resin structures [16] (Fig. 9.8). The main products are methane, carbon dioxide, formaldehyde and hydrogen. Usually a large amount of methane is liberated before the start of scission reactions, which can be explained because of the reaction in Fig. 9.9.

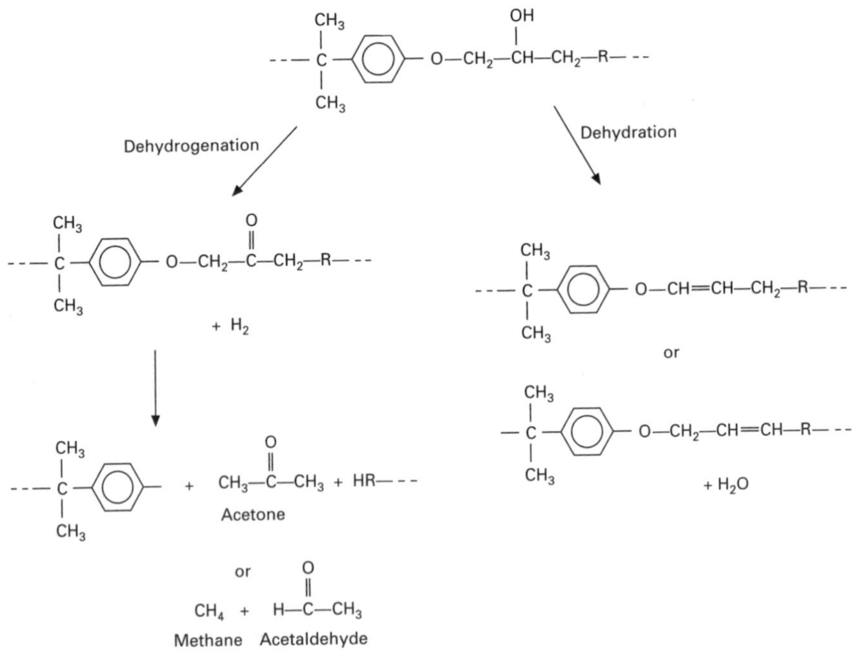

9.8

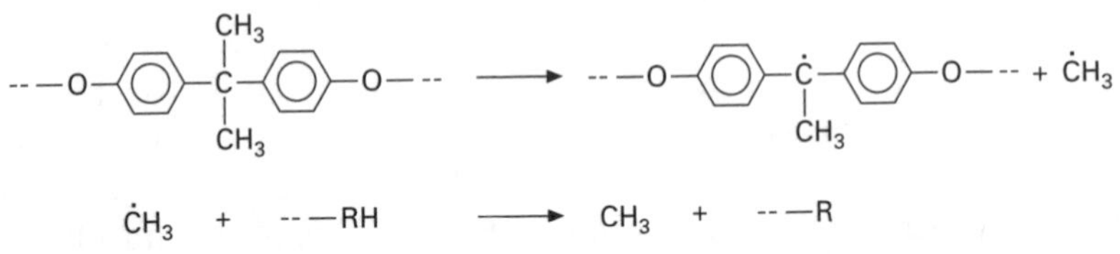

9.9

During chain-scission reactions the aliphatic segments break down into methane and ethylene (and possibly propylene) or acetone, acetaldehyde and methane (and probably carbon monoxide and formaldehyde) all of which are flammable (Fig. 9.10). From the aromatic segments of the polymer, phenol is liberated (Fig. 9.11).

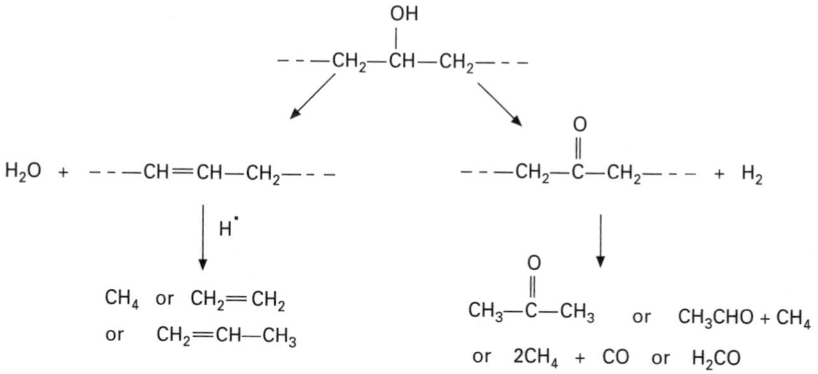

$$\text{OH}$$
$$- - -CH_2-CH-CH_2- - -$$

$$H_2O + - - -CH=CH-CH_2- - - \qquad - - -CH_2-\overset{\overset{\displaystyle O}{\|}}{C}-CH_2- - - + H_2$$

$$\Big\downarrow H^\cdot \qquad\qquad\qquad\qquad \Big\downarrow$$

$$CH_4 \ \text{or}\ CH_2=CH_2 \qquad\qquad CH_3-\overset{\overset{\displaystyle O}{\|}}{C}-CH_3 \quad \text{or} \quad CH_3CHO + CH_4$$
$$\text{or} \quad CH_2=CH-CH_3 \qquad\qquad \text{or}\ \ 2CH_4 + CO \ \text{or}\ H_2CO$$

9.10

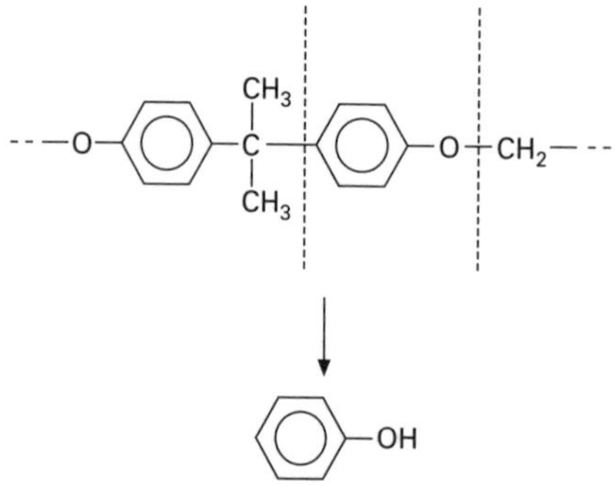

9.11

For phthalic anhydride-cured resins, phthalic anhydride is regenerated together with CO and CO_2. Other degradation products are benzene toluene, o- and p-cresols and higher phenols. In general, these are due to further break down or rearrangement of the aromatic segments of the resins. Phenols and cresols originate from Bisphenol A structural elements, whereas benzene, toluene, etc., originate from aromatic nuclei [16].

Aromatic amine-cured resins give large amounts of water in the temperature range 300–350 °C [17]. Thermal stability of aromatic-amine-cured epoxide resins depends on the aliphatic portion of the network [18]. The linkage present after curing (Fig. 9.12), differs from the glyceryl portion of bisphenol A-based epoxide in that the nitrogen replaces an oxygen atom (Fig. 9.13).

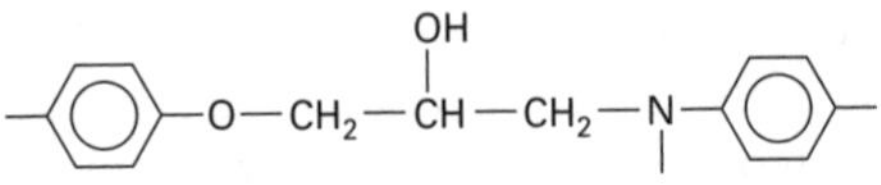

OH
|
—⟨O⟩—O—CH₂—CH—CH₂—N—⟨O⟩—
|

9.12

OH
|
—⟨O⟩—O—CH₂—CH—CH₂—O—⟨O⟩—

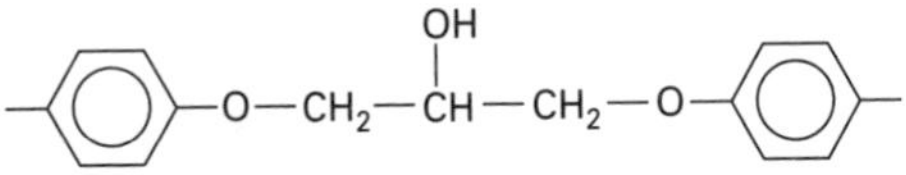

9.13

The flammable volatiles outlined above are, however, produced only in relatively small quantities and this, coupled with their crosslinked and related char-forming character, ensures that epoxy resins are less combustible than polyester resins with higher LOI values in the range 22–23. To confer acceptable levels of flame retardancy requires reactive flame retardants, such as tetrachloro- or tetrabromobisphenol-A and various halogenated epoxides which will act mainly as vapour-phase retardants to raise LOI values easily and significantly. Even the crosslinking agent may be flame retardant, as in the case of chlorendic anhydride, tetrabromo- or tetrachlorophthalic anhydride [19] or possibly phosphorus compounds [20]. Halogenated agents can be supplemented with antimony trioxide [13].

Additive flame retardants such as ammonium polyphosphate, tris (2-chloroethyl) phosphate or other phosphorus-containing plasticisers are also used. Alumina trihydrate used as a filler, is an effective flame retardant for epoxy resins [13].

Phenolic resins

Phenolic resins are manufactured from phenol and formaldehyde. Reaction of phenol with less than equimolar proportions of formaldehyde under acidic conditions gives so-called novolac resins containing aromatic phenol units linked predominantly by methylene bridges. Novolac resins are thermally stable and can be cured by crosslinking with formaldehyde donors such as hexamethylenetetramine. However, the most widely used phenolic resins for composites are resoles manufactured by reacting phenol with a greater than equimolar amount of formaldehyde under alkaline conditions. Resoles are essentially hydroxymethyl functional phenols or polynuclear phenols. Unlike novolacs, they are low-viscosity materials and are easier to process. Phenolic resins can also be prepared from other phenols such as cresols or bisphenols. The general formula is given in Fig. 9.14.

Phenolics are of particular interest in structural applications owing to their inherent fire-resistant properties yielding LOI values of 25 or so, although they tend to increase smoke generation. This high level of inherent flame

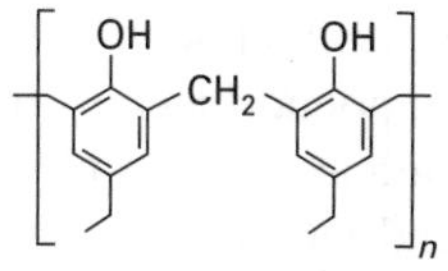

9.14

resistance often means that no further flame retarding is necessary to create composites having required performance levels. However, their main disadvantages are low toughness and a curing reaction that involves the generation of water. The water produced during curing can remain trapped within the composite and during a fire, steam can be generated, which can damage the structure of the material. This evolution is complemented by that generated chemically during the first step of thermal degradation [14], which may be because of phenol-phenol condensation by reactions of the type [10] shown in Fig. 9.15. The released water then helps in the oxidation of methylene groups to carbonyl linkages [6], which then decompose further, releasing CO, CO_2 and other volatile products to yield ultimately char (Fig. 9.16).

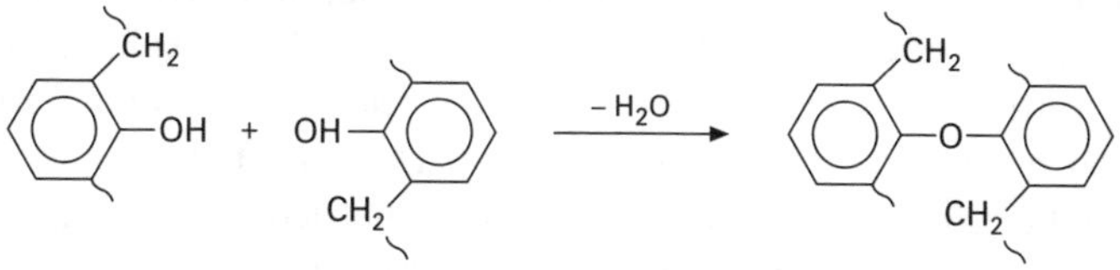

9.15

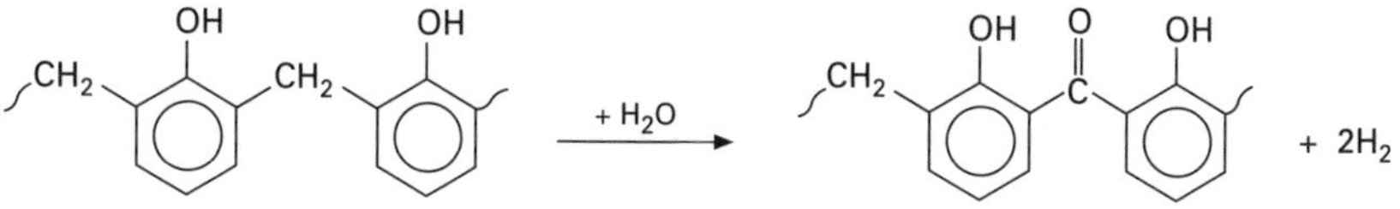

9.16

In the case of highly crosslinked material, water is not released until above 400 °C, and decomposition starts above 500 °C [10]. This was the case for all the phenolic resin samples examined by DTA, by ourselves and published elsewhere [21]. The amount of char depends upon the structure of phenol, initial crosslinks and tendency to crosslink during decomposition [12], and the main decomposition products are methane, acetone, carbon monoxide, propanol and propane.

Where phenolic resins require flame-retardant treatment, additive and reactive flame retardants can be used. Tetrabromobisphenol A, various organic phosphorus compounds, halogenated phenols and aldehydes (e.g. p-bromobenzaldehyde) are some of the reactive flame retardants used for phenolics.

Phosphorus can be introduced by direct reaction of the phenolic resin with phosphorus oxychloride. Likewise inorganic compounds such as boric acid may be incorporated into phenolic resin by chemical reaction [22].

Chlorine compounds (e.g. chloroparaffins) and various thermally stable aromatic bromine compounds may be utilised as additive flame retardant and antimony trioxide is usually added as a synergist. Suitable phosphorus compounds include halogenated phosphoric acid esters such as tris(2-chloroethyl) phosphate, halogenated organic polyphosphates, calcium and ammonium phosphates. Zinc and barium salts of boric acid and aluminium hydroxide also find frequent application [22]. In order to suppress the afterglow of phenolic resins, use is made of compounds such as aluminium chloride, antimony trioxide and organic amides.

Maleimide and polyimide resins

Thermosetting bismaleimide and polyimide resins are used widely in advanced composites. The general formula for polyimide resins is given in Fig. 9.17; their chemistry is often complex [15]. The processing conditions required to manufacture composite components from bismaleimide and other polyimide resins are more severe than used for epoxy systems and the resulting composites are more brittle than those of epoxy matrices. They cure at about 250–350 °C for several hours [15]. However, the glass transition temperature of cured resin is about 100 °C higher than cured epoxy matrices and hence they better retain mechanical properties at higher temperatures.

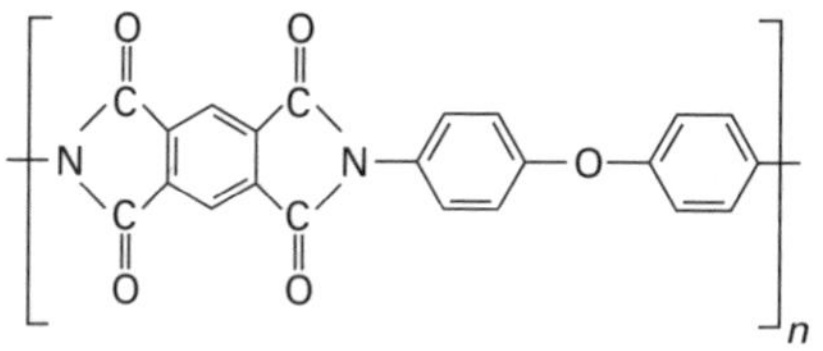

9.17

The aromatic structure of polyimides in particular ensures that they are characterised by high char formation on pyrolysis, low flammability (LOI > 30) and low smoke production when subjected to a flame in a non-vitiated atmosphere. Because of their high cost, they are only used in composites requiring the highest levels of heat and flame resistance.

Thermoplastic resins

Thermoplastic resins are high-molecular weight linear chain molecules with no functional side groups. They are fundamentally different from the thermosets in that they do not undergo irreversible crosslinking reactions but instead melt and flow on application of heat and pressure and resolidify on cooling. However, to give composites with reasonable levels of physical heat resistance, their softening (or glass) transitions must be relatively high, which also influences cost of processing. For this reason, the more common thermoplastics such as polypropylene, polyamides 6 and 6.6 and the poly(alkylene terephthalates) are rarely used when heat and especially fire resistance are required. Some commonly used thermoplastic resins are poly(phenylene sulphide), poly(etheretherketone), poly(etherketone), poly(sulphone), poly(ether imide), poly(phenyl sulphone), poly(ether sulphone), poly(amide imide) and poly(imide). Their glass transition temperatures are 85, 143, 165, 190, 216, 220, 230, 249–288 and 256 °C, respectively [23]. All these resins have aromatic structures and so generally will be inherently flame resistant and have LOI values of at least 30, as shown in Table 9.2.

In a fire, such materials can soften enough to flow under their own weight and drip or run. The extent of dripping depends upon thermal environment, polymer structure, molecular weight, presence of additives, fillers, etc. Dripping can increase or decrease the fire hazard depending upon the fire situation. With small ignition sources, removal of heat and flame by the dripping away of burning polymer can protect the rest of material from spreading of the flame. In other situations, the flaming molten polymer might flow and ignite other materials.

Since thermoplastics are rarely used for rigid composites where the demands of both heat and fire resistance are paramount, the methods to impart flame retardancy are not discussed here. For further details regarding flame retardants for thermoplastic polymers, the reader should consult Kandola and Horrocks [1] and Horrocks [5] and cited references therein.

Resin–matrix interface

The fibre–matrix interface is an important region, which is required to provide adequate chemically and physically stable bonding between the fibres and the matrix. For example, aminosilane is used to bond glass fibre with epoxy matrix systems. Carbon fibre is both surface-treated in order to improve the mechanical properties of the composite, and coated with a sizing agent in order to aid processing of the fibre. Surface treatment creates potentially reactive groups such as hydroxyl and carboxyl groups upon the surface of the fibres, which are capable of reaction with the matrix. Epoxy-based sizing agents are quite common; however, they may not be suitable for the resin matrix [15].

The nature of the interface will affect the burning of the material as well. If the binding material is highly flammable, it will increase the fire hazard of the whole structure. However, if the interface is weak and two phases (fibre and matrix) are pushed apart in case of fire, the matrix will burn more vigorously and inorganic fibres can no longer act as insulators. This situation is typical of layered textiles within a composite where delamination in fires will not only cause increased burning rates but also increase the rate of loss of mechanical properties and hence general product coherence. The use of interlinked reinforcing layers via use of 3D or stitched woven structures, for example, probably yields improved fire behaviour of the resulting composites although no work has been published in this area.

9.3   Flammability of composite structures

As discussed above, composite structures contain two polymeric structures, fibre and resin. Both of these polymeric components in a fire behave differently depending upon their respective thermal stabilities. Composite structures are often layered and thus tend to burn in layers. When heated, the resin of first layer degrades and combustible products formed are ignited. The heat penetrates the adjacent fibre layer and if inorganic fibre is used, it will melt or soften, whereas if organic fibre is used, it will degrade into smaller products depending upon its thermal stability. Heat then penetrates further into the underlying resin, causing its degradation and products formed will then move to the burning zone through the fibrous and, in some cases, resin chars. This will slow the burning front although if the structure is multilayered, it will burn in distinct stages as the heat penetrates subsequent layers and degradation products move to the burning zone through the fibrous layers. In general, the composite thickness of a structure can affect the surface flammability characteristics down to a certain limiting value. At this condition, where it is assumed that the composite has the same temperature through this limiting thickness, the material is said to be 'thermally thin'. However, above this depth, the temperature will be less than at the front face and a temperature gradient will exist where the material is not involved in the early stages of burning; here is said to be 'thermally thick' [24]. The transition from thermally thin to thermally thick is not a constant since it depends on material thermal properties including fibre and resin thermal conductivities.

For a given composite of defined thickness, the condition depends on the intensity of the fire or more correctly, the incident heat flux. While many large scale fire tests involve heat sources or 'simulated fires' having constant and defined fluxes, in real fires, heat fluxes may vary. For example, a domestic room filled with burning furniture at the point of flashover presents a heat flux of about 50 kW/m^2 to the containing wall and door surfaces; larger building fires present fluxes as high as 100 kW/m^2 and hydrocarbon fuel 'pool fires' may exceed 150 kW/m^2. We may examine the heat flux dependence on burning behaviour using calorimetric techniques such as the cone calorimeter [25]. For example, Scudamore [26] has shown by cone calorimetric analysis that the thermally thin to thick effect for glass reinforced polyester, epoxy and phenolic laminates decreases as the external heat flux increases. Generally at heat fluxes of 35 and 50 kW/m^2, thin samples (3 mm) ignited easily compared with thick samples (9.5 mm), but at 75 and 100 kW/m^2 there was not much difference and all samples behaved as if they were 'thermally thin'.

The overall burning behaviour of a composite will be the sum of its component fibres and resin plus any positive (synergistic) or negative (antagonistic) interactive effects. Table 9.1 shows that as measured by limiting oxygen index, most commonly used fibres add little to the fuel content of a composite unless comprising fibres such as UHMW polyethylene or para-aramid. Table 9.2 presents results published in our previous review to illustrate the differences in fundamental component resin burning behaviour [1]. This table also demonstrates that in composites containing a non-flammable fibre reinforcing element such as glass, overall burning performance in terms of LOI reflects that of the resin although clearly, the glass component does have a fire retarding and hence LOI-raising property. From this and a number of studies and reviews [7–9, 27, 28] it may be concluded that ranking of fire resistance of thermoset resin composite components is:

Phenolic > Polyimide > Bismaleimide > Epoxy
> Polyester and vinyl ester

The superior performance of phenolics has been demonstrated above mechanistically in terms of their char-forming ability, which enables composites comprising them to retain mechanical strength for long times under fire conditions [29]. In addition, it is observed that because such composites encapsulate themselves in char, they do not produce much smoke [30]. Epoxy and unsaturated polyesters on the other hand carbonise less than phenolics and as demonstrated above, produce more fuels during pyrolysis and so continue to burn in a fire. Furthermore, those containing aromatic structures such as styrenic moieties produce more smoke. However, while phenolics have inherent flame-retardant properties, their mechanical properties are inferior to other thermoset polymers, such as polyester, vinyl ester and epoxies [29]. Hence, they are less favourable for use in load-bearing structures. Epoxies on the other hand, because of very high mechanical strength, are the more popular choice.

Char formation is the key to achieving low flammability and good fire performance. This is because char is formed at the expense of possible flammable fuel formation (contrast the flammable volatiles formed during polyester and epoxy resin thermal degradation in the previously shown mechanisms with the char-forming tendency of phenolics). In addition, because char 'locks in' the available carbon, less smoke can be formed and the char acts as a barrier to its release should it be formed. Furthermore, the char acts as an insulating layer and protects the underlying composite structure and this also helps to minimise the loss in tensile properties during fire exposure. In other words there is a direct relationship between flammability of a polymer and its char yield as discussed comprehensively by van Krevelen [30]. Gilwee *et al.* [31] and Kourtides [32] have found that a linear relationship exists between limiting oxygen index and char yields for resins and graphite reinforced composites respectively as shown in Fig. 9.18. This shows that composite structures behave similarly to bulk resin polymers, that char formation determines the flammability of the composite and that the presence of inorganic fibre does not improve the flame retardancy of the structure.

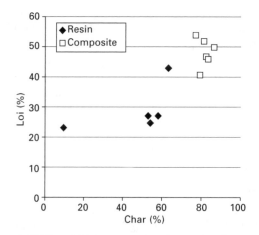

9.18 Plot of LOI versus char formation for a series of resins and graphite fabric (8-harness satin weave) reinforced composites [31, 32].

Brown *et al.* [33] have studied the fire performance of extended-chain polyethylene (ECPE) and aramid fibre-reinforced composites containing epoxy, vinyl ester and phenolic matrix resins by cone calorimetry. Various parameters were determined for ECPE and aramid fabrics only, matrix resins only and their composites and maximum or peak heat release rates (PHRR) only are plotted in (Fig. 9.19). ECPE reduced the flammability of epoxy but increased it for vinyl ester matrix resins. Aramid, on the other hand, had little effect on time to ignition (compared with resin alone) except for the phenolic, but reduced RHR. In general, resin and reinforcement contributions to the composite rate of heat release behaviour as a function of time are discernible and depend on respective flame-retardant mechanisms operating and levels of their transferability and possible synergisms and antagonisms. This indicates that a flame or heat-resistant fibre can be effective for one type of resin but not necessarily for another.

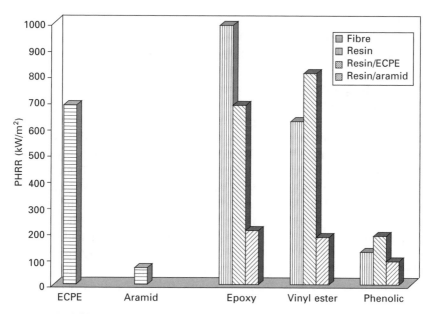

9.19 Maximun or peak, PHRR values for fibre reinforcements, matrix resins and composite materials at 50 kW/m² cone irradiance [33].

9.4   Methods of imparting flame retardancy to composites

As composites continue to replace more conventional materials, their fire performance is increasingly being questioned, especially the poor smoke-generating character of polyester-resinated composites that constitute the majority of the present world market. Unfortunately, imparting flame retardancy and smoke reduction to composites often results in reductions of their mechanical strengths. Therefore, achieving a certain level of flame retardancy while maintaining other such properties is a major challenge. Following our recently published reviews [1, 3] of the work done in this field where we discuss the more basic means of conferring flame retardancy, here we extend these studies to include research work undertaken in our own and other laboratories since 1999.

9.4.1   Use of inherently flame-resistant resins and/or fibres

The reinforcing fibre phase can be rendered flame retardant by appropriate treatment or by the use of high heat and flame-resistant fibres [34], such as aramids or carbon as shown in Table 9.1, although the flame retardancy levels desired should really match those of the matrix if high levels of fire performance are to be realised. Hshieh and Beeson [35] have tested flame-retarded epoxy (brominated epoxy resin) and phenolic composites containing fibre glass, para-aramid (e.g. Kevlar®, Du Pont) and graphite fibre reinforcements using the NASA upward flame test and the controlled atmosphere, cone-calorimeter test. The upward flame propagation test showed that phenolic/graphite had the highest and epoxy/graphite composites had the lowest flame resistance as shown in Fig. 9.20. This is an interesting case that shows that the overall fire performance is not simply the average of the components present. The most flame-resistant graphite or carbon reinforcement has produced the most flammable composite with epoxy possibly because the carbon fibres prevent the liquid decomposition products from the resin from dripping away in the upward flame test – this so-called 'scaffolding effect' is seen in blends of thermoplastic and non-thermoplastic fibres in textiles [5]. Conversely, the presence of the char-forming phenolic will complement the carbon presence in the graphite reinforcement and so present an enhanced carbon shield to the flame. Controlled-atmosphere cone calorimetry also showed that phenolic composites had lower values of time of ignition, peak heat release rate, propensity to flashover and smoke production rate.

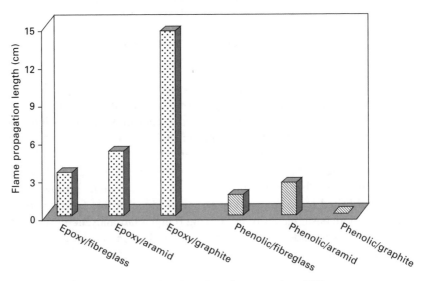

9.20 Flame propagation lengths of composites [35].

9.4.2   Chemical or physical modifications of resin matrix

Conventional flame retardants

Additives such as zinc borate and antimony oxide have been used with halogenated polyester, vinyl ester or epoxy resins [29, 36]. Alumina trihydrate

(ATH) and bromine compounds are other examples [37]. However, many of these resins and additives are ecologically undesirable and in a fire increase the amount of smoke and toxic fumes given off by the burning material. Furthermore, of all methods of improving fire resistance, this usually results in a reduction in the mechanical properties of the composite structure.

Scudamore [26] has studied the effect of flame retardants on the fire performance of glass-reinforced polyester, epoxy and phenolic laminates by cone calorimetry. The polyester laminates examined comprised a brominated resin whereas fibre reinforced (FR) epoxy and phenolic resins contained ATH. ATH was used in the FR phenolic laminate. While generally it was concluded that the fire properties depend on the type of resin and flame retardant, the type of glass reinforcement and for thin laminates, the thickness, more specifically, flame retardants for all resins delay ignition and decrease heat release rates. Again, phenolic laminates showed lower flammability than either FR polyester or epoxy resins and addition of ATH further enhanced flame retardancy by reducing maximum or peak rate of heat release rate values values to less than 100 kW/m^2 at 50 kW/m^2 heat flux.

Morchat and Hiltz [38] and Morchat [39] have studied the effect of the FR additives antimony trioxide, alumina trihydrate and zinc borate on the flammability of flame-retardant polyester, vinyl ester and epoxy resins by thermogravimetric analysis (TGA), smoke production, toxic gas evolution, flame spread and oxygen index methods. Except for epoxy resin, the others contained halogenated materials from which they derived their fire-retardancy characteristics through the vapour phase activity of chlorine and/or bromine. In most cases, with a few exceptions, the additives lowered the rate of flame spread (by 2–70%), increased LOI (by 3–57%) and lowered specific smoke optical density (by 20–85%), depending on the fire retardant and the resin system evaluated. However, for the majority of resins, the addition of antimony trioxide resulted in an increase in smoke production. The best performance was observed upon addition of zinc borate to the epoxy resin.

Nir *et al.* [40] have studied the mechanical properties of brominated flame retarded and non-brominated epoxy (tris-(hydroxyphenyl)-methane triglycidyl ester)/graphite composites. While the incorporation of bromine did not change the mechanical properties within ±10% of those of the non-brominated resin, it helped in decreasing water absorption and increasing environmental stability, thereby indicating that this is an easy method to flame retard without unduly influencing the impact strength of graphite-reinforced composites.

Intumescents

In our earlier review, we considered the potential for inclusion of intumescents within the composite structure and noted at that time (1998), very little interest had been shown [3]. Kovlar and Bullock [41] have reported using an intumescent component as an additive in a phenolic matrix and developed a formulation with phenolic resin and intumescent in 1:1 ratio, reinforced with glass fabric. Upon exposure to fire the intumescent composite panel immediately began to inflate, foam, swell and char on the side facing the fire, forming a tough, insulating, fabric-reinforced carbonaceous char that blocked the spread of fire and insulated adjacent areas from the intense heat. These intumescent-containing samples showed marked improvement in the insulating properties than control phenolic or aluminium panels.

Most work since that time has been undertaken in our own laboratories based on a patent awarded in 1995 [42] in which novel combinations of intumescent and flame-retardant fibres are described. These yield complex 'char-bonding' structures when heated, which demonstrate unusually high fire and heat resistance compared with individual component performance. Work at the University of Bolton since 1998 has extended this concept into composite structures where the flame-retardant fibre component may become part of the reinforcement and the FR fibre-intumescent system interacts positively with the otherwise flammable resin component present. From this work, a number of publications and a second patent have arisen [21, 43–49].

We have studied the possible interaction between resin (polyester, epoxy and phenolic) with a phosphate-based intumescent and FR cellulosic fibre (Visil, Sateri) with thermal analytical techniques [21]. Studies of different components and their mixtures in different combinations indicated that, on heating, all components degrade by physically and chemically compatible mechanisms, resulting in interaction and enhanced char formation. This led to the preparation of composite laminates, where these components were added either as additives in pulverised form or fibre interdispersed with intumescent as a fabric scrim for partial replacement of glass fibre [46, 47]. The composite series in Table 9.3 were prepared to investigate the effect of thickness and nature of glass reinforcement (random matt versus woven) (PS1–PS6) for polyester resin-based composites. A similar set of epoxy-based composites, EP1–EP3 also comprising combinations of intumescent and FR celluosic were fabricated. The intumescent comprised melamine phosphate and glass fabrics were typically 300 g/m^2

LOI results for polyester sample PS1–PS3 are only slightly increased with intumescent presence (PS2, 22.6%) and is unaffected by additional presence of Visil (PS3, LOI = 22.6%) and these suggest that the composites will still burn in air in spite of the intumescent (Int) and FR viscose (Vis) presence. However, for epoxy composites, the presence of intumescent raises LOI from 27.5 (EP1) to 35.2% (EP2) and with additional Visil, to 36.2% (EP3), in both cases composites will not ignite and sustain burning in air.

Cone calorimetry (under 50 kW/m^2 external heat flux) behaviour is typified by the heat release rate (HRR) properties shown in Figs 9.21 and 9.22. Generally, the addition of intumescent and/or FR viscose has little effect on

Table 9.3 Physical and LOI properties of composite laminates [46, 47]

Sample No.	Sample details	Mass fraction (%)				Thick-ness (mm)	LOI (%)
		Glass	Resin	Visil	Intume-scent		
Polyester (PS) laminates with four layers of random mat glass							
PS 1	Res	39.9	60.1	–	–	2.7	19.3
PS 2	Res + Int	29.5	64.2	–	6.3	3.8	22.6
PS 3	Res + Vis + Int	25.8	62.0	6.1	6.1	4.6	22.6
Polyester (PS) laminates with four layers of woven roving glass*							
PS 4	Res	62.7	37.3	–	–	1.0	–
PS 5	Res + Int	57.2	38.8	–	3.8	1.2	–
PS 6	Res + Vis + Int	49.6	42.2	4.1	4.1	1.5	–
Epoxy laminates with four layers of woven roving glass*							
EP 1	Res	55.0	45.0	–	–	1.9	27.5
EP 2	Res + Int	53.0	42.3	–	4.7	2.0	35.2
EP 3	Res + Vis + Int	50.0	40.0	5.0	5.0	2.3	36.2

Note: *Woven roving glass fabrics used here have plain weave structures.

the time to ignition and extinction times, but does reduce the peak heat release rate values, which in a real fire, is the measure of the ability of a fire to grow in intensity. For polyester-based composites PS1–PS3, Fig. 9.21(a) shows this effect for samples of increasing thickness and hence fuel load and PHRRs decrease from 314 to 246 kW/m^2. Where composite thickness is almost constant, however, the effects of intumescent and intumescent-treated FR viscose fabric are less with peak heat release values reducing from 477 to 387 kW/m^2 (see Fig. 9.21b). These effects are particularly noticeable in the epoxy-based composites (see Fig. 9.21c) where PHRR values reduce from 385 to 262 kW/m^2, reflecting behaviour of LOI results.

Since the char retained after burning a polymer is also a measure of its flammability, the mass loss curves accompanying heat release data give insight into the fire performance of the samples. For polyester samples PS1–PS3, mass loss curves as a function of time and for all samples the effect of additives on the residual char retained after 300 s (360 s for samples PS1–PS3) are given in Fig. 9.22. Figure 9.22(a) shows that presence of intumescent (sample PS2) and Visil-intumescent (sample PS3) makes the samples more thermally stable than resin only (PS1) for about 240 s by slowing down volatilisation and burning. But after complete combustion, residual chars for these samples are less than control sample as can be seen from Fig. 9.22(b).

Mass loss curves for epoxy samples showed that the presence of intumescent alone and with Visil fibre increases the residual mass at any time [47] and even after complete combustion after about 300 s, as can be seen from Fig. 9.22(b). This supports our earlier thermal analytical results [21] that these components promote char formation of the resin.

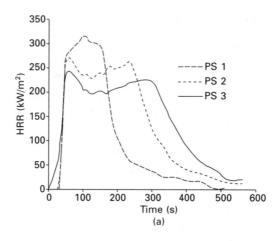

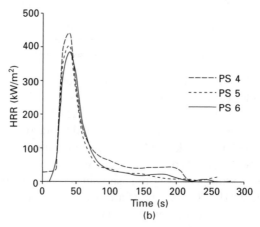

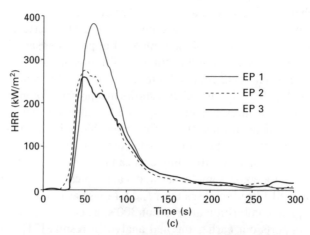

9.21 HRR versus time curves of (a), (b) polyester and (c) epoxy composite laminates at 50 kW/m² heat flux.

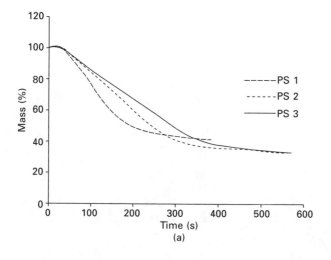

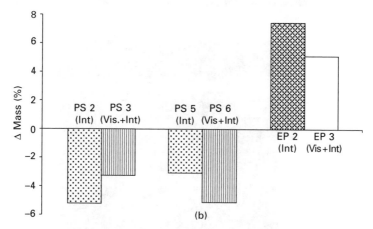

9.22 (a) Mass loss versus time curves of PS1-PS3 samples at 50 kW/ m² heat flux and (b) change in residual mass (Δ mass%) of PS2, PS3, PS5, PS6, EP2, EP3 samples compared with respective control PS1, PS4 and EP1 samples.

Smoke production appears to increase from PS1 to PS3 measured during cone calorimetry, although when measured using the standard 'NBS Smoke Chamber' according to ASTM E662, a reduction is seen [45]. However, for epoxy resin-based samples, a progressive decrease in smoke generation is seen for EP1, 2 and 3 samples. These results again illustrate the apparent synergy between the intumescent system and the epoxy resin matrix.

The charred epoxy samples EP1–EP3 left from cone-calorimetric tests were examined for changes in appearance by taking photographs with a digital camera. These samples were also examined under an optical microscope for finer details on the surface and through cross-sections of the laminates. Results are shown in Figs 9.23 and 9.24; in the latter, cross-sectional micrographs show evidence of delamination.

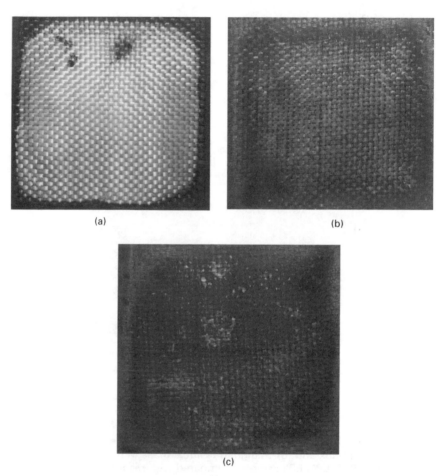

9.23 Images with a digital camera of samples (a) EP1 (b) EP2 and (c) EP3 after cone experiments.

Figure 9.23(a) for the control sample (EP1), which contains glass fibre and resin only, shows that after cone exposure all the resin has burned away. Charred residue on the edges is due to the shielding effect of the sample holder edge during the cone experiments. The cross-sectional view in Fig. 9.24(a) also shows that most of the resin has burned and only glass fibre is seen in the first five layers. The effect of intumescent additive (sample EP2) on the burning behaviour of resin is clearly seen in Fig. 9.23(b), where charred residues are seen on the surface and within layers of glass fabric in

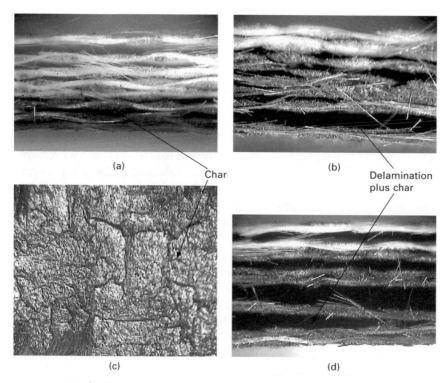

9.24 Optical microscopic images of samples EP1–EP3 after cone experiment.

cross-sectional view (Fig. 9.24b). However, when both Visil and intumescent are present as additives (sample EP3), the char formed is higher in quantity as seen on the surface (Fig. 9.23c and 9.24c) and between layers of glass fibre (Fig. 9.24d). The char on the surface of sample EP3, when seen under the microscope, shows the complexity of the charred structure, the chemical nature of which has been discussed in detail elsewhere [21].

9.4.3   Use of external coatings and outer protective surfaces

Another way of flame-retarding or fire-hardening composite structures is to use flame-retardant (usually intumescent based) paints or coatings. Intumescent systems are chemical systems, which by the action of heat evolve gases and form a foamed char. This char then acts as an insulative barrier to the underlying structural material against flame and heat. One very effective intumescent coating is fluorocarbon latex paint [50].

Tewarson and Macaione [51] have evaluated the flammability of glass/resin composite samples treated with intumescent and ceramic coatings by

FMRC (Factory Mutual Research Corporation) 50 kW-scale apparatus (discussed in ref. 51) methods. As expected, the calculated thermal response parameter (TRP) values showed that ceramic and intumescent coatings are quite effective in improving fire resistance. The intumescent coatings were superior on the vinyl ester and phenolic resinated composites while the ceramic coating was best on the epoxy composite.

Sorathia et al. [52] have explored the use of integral, hybrid thermal barriers to protect the core of the otherwise flammable composite structure. These barriers function as insulators and reflect the radiant heat back towards the heat source, which delays the heat-up rate and reduces the overall temperature on the reverse side of the substrate. Treatments evaluated included ceramic fabrics, ceramic coatings, intumescent coatings, hybrids of ceramic and intumescent coatings, silicone foams and a phenolic skin. The composite systems evaluated in combination with thermal barrier treatments included glass/vinyl ester, graphite/epoxy, graphite/bismaleimide and graphite/phenolic combinations. All systems were tested for flammability characteristics by cone-calorimetry and PHRR values at 75 kW/m^2 cone irradiance are plotted in Fig. 9.25. Without any barrier treatment, all composites failed to meet the ignitability and PHRR requirements, whereas all treated ones passed. Ceramic/intumescent hybrid coatings seem to be very effective. More recently Sorathia et al. have conducted an investigation of different commercial protective intumescent coatings for potential use on ships for the US Navy [53]. All coatings failed the fire performance criteria necessary to meet the US Navy requirements for high-temperature fire insulation in accordance with draft

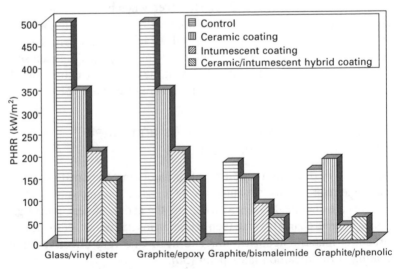

9.25 Peak heat release rate (PHRR) values for composite materials with thermal barrier treatment at 75 kW/m^2 cone irradiance [52].

military standard DRAFT MIL_PRF_XX 381. This led them to conclude that intumescent coatings are not sufficient to protect shipboard spacings during a fire and are not equivalent when used alone as direct replacement for batt or blanket-type fibrous fire insulation (e.g. mineral wool, StrutoGard®) installed aboard ships. However, some of these coatings, when used combined with blanket-type fibrous fire insulation, were effective in meeting fire resistance criteria.

While the use of mineral and ceramic claddings is quite popular for naval applications [41] in preference to flameproof conventional composite hull, deck and bulkhead structures, the main disadvantages are that they occupy space, add significant weight and can act as an absorbent for spilled fuel or flammable liquid during a fire. When this occurs, extinguishing the fire will be more difficult and the insulating property of the ceramic wool is lost. However, if the mineral cladding as a fibrous membrane is incorporated as the final layer within the composite structure then this last problem is overcome and a 'fire hardened' face may be introduced to the otherwise flammable core. Such a system overlaps with the intumescent developments made in our own laboratories [43–48] as well as those developed by Sorathia [52, 53].

One recent UK commercial example of this sort of product is the Technofire® range of ceramic wet-laid, nonwoven webs produced by Technical Fibres Ltd in the UK; these are available with a number of different inorganic fibres, including glass and rock wool either with or without an associated exfoliated graphite present. They are designed to be compatible with whatever resin is used in composite production.

9.5   Conclusions

Fire performance of fibre reinforced composites is becoming increasingly important not only as composite usage increases generally, but also as regulations in construction and transport sectors, especially, become more stringent in regard of increased fire safety. Within military circles, composites are replacing metals in traditional naval and armoured vehicle applications where fire performance is as important as mechanical and ballistic properties. From the above discussion it can be concluded that choices of resin and fibre are crucial in determining the flammability properties of the whole structure. Because resins make up a significant fraction of all composites and these are organic in character, then they are the prime sources of fuel when composites are heated. We have demonstrated that relative fuel-generating tendencies are wholly dependent upon respective resin chemistries and that the presence of inorganic fibres such as glass and carbon does not help in reducing overall flammability. If flame-retardant chemicals, which are compatible with both fibres and resin matrix, are selected, resulting effects can be synergistic both towards improved fire performance but also positively or negatively towards smoke suppression. Within the most common composite markets where the cheaper and more flammable polyester resins are used, smoke generation, already large, and its consequent suppression is an equally significant fire performance factor; it is here that the traditional antimony–bromine formulations are weak and considerable interest lies in finding alternatives with comparable flame-retarding properties but with enhanced smoke reduction. Use of intumescents and the work in our own laboratories has shown encouraging results to date and confirms that char-forming agents are the best way forward. The role of nanoclays and nanocomposite structures within the macrocomposite itself is currently being addressed by a number of research teams, including our own. The next few years promise much excitement in the discovery of novel fire-resistant systems that will have superior performance to present systems and will be based on both present and developing fire science understanding.

9.6   References

1. Kandola B.K. and Horrocks., 'Composites' in Horrocks A.R. and Price D. (Eds), *Fire Retardant Materials*, Cambridge, Woodhead Publishing, 2001.
2. Jones F.R., 'Glass Fibres', in Hearle J.W.S. (Ed), *High-performance Fibres*, Cambridge, Woodhead Publishing, Chapter 2, 2001.
3. Kandola B.K. and Horrocks A.R., 'Flame retardant composites, a review : the potential for use of intumescents', in Bras M.L., Camino G., Bourbigot S. and Delobel R. (Eds), *Fire Retardancy of Polymers – The Use of Intumescence*, Cambridge, The Royal Society of Chemistry, 1998.
4. *Toxicological Risks of Selected Flame-Retardant Chemicals*, Sub-committee on Flame-retardant Chemicals of the United States National Research Council, Washington, DC; National Academy Press, Washington, 2000.
5. Horrocks A.R., 'Textiles' in Horrocks A.R. and Price D., (Eds), *Fire Retardant Materials*, Cambridge, Woodhead Publishing, Chapter 4, p. 128, 2001.
6. Levchik S.V., 'Thermosetting Polymers', in Bourbigot S., Le Bras M., Troitzsch J. (Eds), *Fundamentals in International Plastics Flammability Handbook*, 3rd edition, in press.
7. Brown J.E., Loftus J.J. and Dipert R.A., *Fire Characteristics of Composite Materials – A Review of the Literature*, Report 1986, NBSIR 85-3226.
8. Kourtides D.A. *et al.*, 'Thermochemical characterisation of some thermally stable thermoplastic and thermoset polymers', *Polym. Eng. Sci.*, 1979, **19(1)**, 24–29.
9. Kourtides D.A., *et al.*, 'Thermal response of composite panels', *Polym. Eng. Sci.*, 1979, **19(3)**, 226–231.
10. *Fire Safety Aspects of Polymeric Materials, Vol 1 – Materials State of Art*, Chapter 6, A Report by National Materials Advisory Board, National Academy of Sciences, Technomic Publ. Washington, 1977.
11. Learmonth G.S. and Nesbit A., 'Flammability of polymers V. TVA (Thermal volatilization) analysis of polyester resin compositions', *Br. Polym. J.*, 1972, **4**, 317.

12. Das A.N. and Baijaj S.K., 'Degradation mechanism of styrene–polyester copolymer', *J. Appl. Polym. Sci.*, 1982, **27**, 211.

13. Pal G. and Macskasy H., *Plastics – Their Behaviour in Fires*, Chapter 5, Amsterdam, Elsevier, 1991.

14. Vasiliev V.V., Jones R.M. and Man L.I. (Eds), *Mechanics of Composite Structures*, Washington, Taylor and Francis, 1988.

15. Phillips L.N. (Ed), *Design with Advanced Composite Materials*, London, Springer-Verlag, 1989.

16. Bishop D.P. and Smith D.A., 'Combined pyrolysis and radiochemical gas chromatography for studying the thermal degradation of epoxy resins and polyimides. I. The degradation of epoxy resins in nitrogen between 400° and 700 °C', *J. Appl. Polym. Sci.*, 1970, **14**, 205.

17. Paterson-Jones J.C., 'The mechanism of the thermal degradation of aromatic amine-cured glycidyl ether-type epoxide resins' *J. Appl. Polym. Sci.*, 1975, **19**, 1539.

18. Paterson-Jones J.C., Percy V.A., Giles R.G.F. and Stephen A.M., 'The thermal degradation of model compounds of amine-cured epoxide resins' *J. Appl. Polym. Sci.*, 1975, **17**, Part I, 1867–1876, Part II, 1877–1887.

19. Lo J. and Pearce E.M., 'Flame-retardant epoxy resins based on phthalide derivatives', *J. Polym. Sci. Polym. Chem. Ed.*, 1984, **22**, 1707.

20. Mikroyannidis J.A. and Kourtides D.A., 'Fire resistant compositions of epoxy resins with phosphorus compounds', *Polym. Mat. Sci. Eng. Proc.*, 1983, **49**, 606.

21. Kandola B.K., Horrocks A.R., Myler P. and Blair D., in Nelson G.L. and Wilkie C.A. (Eds), *Fire and Polymers*, ACS Symp. Ser., American Chemical Society, Washington, DC, 2001, 344.

22. Troitzsch J., *International Plastics Flammability Handbook*, Chapter 5, Munich, Hanser Publ., 1990.

23. Eckold G., *Design and Manufacture of Composite Structures*, Cambridge, Woodhead Publishing, 1994.

24. Mikkola E. and Wichman I.S., 'On the thermal ignition of combustible materials', *Fire Mater.*, 1989, **14**, 87–96.

25. Babrauskas V. and Grayson S.J., *Heat Release in Fires*, London and New York, Elsevier Applied Science, 1992.

26. Scudamore M.J., 'Fire performance studies on glass-reinforced plastic laminates', *Fire Mater,* 1994, **18**, 313–325.

27. Brown J.E., Braun E. and Twilley W.H., *Cone Calorimetric Evaluation of the Flammability of Composite Materials*, Report 1988, NBSIR-88–3733.

28. Brown J.R. and St John N.A., 'Fire-retardant low-temperature-cured phenolic resins and composites', *TRIP*, 1996, **4(12)**, 416–420.

29. Gabrisch H.-J. and Lindenberger G., 'The use of thermoset composites in transportation: their behaviour', *SAMPE J.*, 1993, **29(6)**, 23–27.

30. van Krevelen D.W., 'Some basic aspects of flame resistance of polymeric materials', *Polymer*, 1975, **16**, 615–620.

31. Gilwee W.J., Parker J.A. and Kourtides D.A., 'Oxygen index tests of thermosetting resins', *J. Fire Flamm.*, 1980, **11(1)**, 22–31.

32. Kourtides D.A., 'Processing and flammability parameters of bismaleimide and some other thermally stable resin matrices for composites' *Polym. Compos.*, 1984, **5(2)**, 143–150.

33. Brown J.R., Fawell P.D. and Mathys Z., 'Fire-hazard assessment of extended-chain polyethylene and aramid composites by cone calorimetry', *Fire Mater.*, 1994, **18**, 167–172.

34. Horrocks A.R., Eichhorn H., Schwaenke, Saville N. and Thomas C., 'Thermally resistant fibres,' in Hearle J.W.S. (Ed), *High Performance Fibres*, Cambridge, Woodhead Publishing, Chapter 9, 2002.
35. Hshieh F.Y. and Beeson H.D., 'Flammability testing of flame-retarded epoxy composites and phenolic composites', *Proc. 21st Int. Conf. on Fire Safety*, San Francisio 1996, **21**, 189–205.
36. Stevart J.L., Griffin O.H., Gurdal Z., Warner G.A., 'Flammability and toxicity of composite materials for marine vehicles', *Naval Engn. J.*, 1990, **102(5)**, 45–54.
37. Georlette P., 'Applications of halogen flame retardants' in Horrocks A.R. and Price D. (Eds), *Fire Retardant Materials*, Cambridge, Woodhead Publishing, Chapter 8, 2001.
38. Morchat R.M. and Hiltz J.A., 'Fire-safe composites for marine applications', *Proc 24th Int. SAMPE Tech. Conf.* Toronto, 1992, T153–T164.
39. Morchat R.M., *The Effects of Alumina Trihydrate on the Flammability Characteristics of Polyester, Vinylester and Epoxy Glass Reinforced Plastics*, Techn. Rep. Cit. Govt. Rep. Announce Index (US), 1992, **92(13)**, AB NO 235, 299.
40. Nir Z., Gilwee W.J., Kourtides D.A. and Parker J.A., 'Rubber-toughened polyfunctional epoxies: brominated vs nonbrominated formulated for graphite composites', *SAMPE Q.*, 1983, **14(3)**, 34–38.
41. Kovlar P.F. and Bullock D.E., 'Multifunctional intumescent composite fire barriers' in Lewin M. (Eds), *Proc. of the 1993 Conf. Recent Advances in Flame Retardancy of Polymeric Materials*,Vol IV, BCC, Stamford, Conn., 1993, 87–98.
42. Horrocks A.R., Anand S.C. and Hill B., Fire and heat resistant materials. UK Pat. 2279084B, 21 June 1995.
43. Kandola B.K. and Horrocks A.R., 'Complex char formation in flame-retarded fibre-intumescent combinations – III Physical and chemical nature of the char', *Text. Res. J.*,1999, **69(5)**, 374–381.
44. Kandola B.K. and Horrocks A.R., 'Complex char formation in flame-retarded fibre-intumescent combinations – IV. Mass loss and thermal barrier properties', *Fire Mater.*, 2000, **24**, 265–275.
45. Horrocks A.R., Myler P., Kandola B.K. and Blair D., Fire and heat resistant materials. Patent Application PCT/GB00/04703 December, 2000.
46. Kandola B.K., Horrocks A.R., Myler P. and Blair D., 'The effect of intumescents on the burning behaviour of polyester-resin-containing composites', *Composites Part A*, 2002, **33**, 805–817.
47. Kandola B.K., Horrocks A.R., Myler P. and Blair D., 'New developments in flame retardancy of glass-reinforced epoxy composites', *J. Appl. Polym. Sci.*, 2003, **88(10)**, 2511–2521.
48. Kandola B.K., Horrocks A.R., Myler P. and Blair D., 'Mechanical performance of heat/fire damaged novel flame retardant glass – reinforced epoxy composites', *Composites Part A*, 2003, **34**, 863–873.
49. Neininger S.M., Staggs J.E.J., Hill N.H. and Horrocks A.R., 'A study of the global kinetics of thermal degradation of a fibre-intumescent mixture' paper presented at the 8th European Conference on 'Fire retardant polymers', 24–27 June 2001, Alessandria, Italy; *Subsequently Published in Polym. Deg. Stab.*, 2002, **77**, 187–194.
50. Ventriglio D.R., 'Fire safe materials for navy ships', *Naval Engn. J.*, October 1982, 65–74.
51. Tewarson A. and Macaione D.P., 'Polymers and composites – an examination of fire spread and generation of heat and fire products', *J. Fire Sci.*, 1993, **11**, 421–441.

52. Sorathia U., Rollhauser C.M. and Hughes W.A., 'Improved fire safety of composites for naval applications', *Fire Mater.*, 1992, **16**, 119–125.
53. Sorathia U., Gracik T., Ness J., Durkin A., Williams F., Hunstad M. and Berry F., 'Evaluation of intumescent coatings for shipboard fire protection' in Lewin M. (Eds), *Recent Advances in Flame Retardancy of Polymeric Materials*, Vol XIII, Proc. of the 2002 Conf, BCC, Stamford, Conn., 2002.

10
Cost analysis

M D   W A K E M A N   and   J - A   E   M Å N S O N,
École Polytechnique Fédérale de Lausanne (EPFL), Switzerland

10.1   Introduction

Textile composites offer a diverse range of properties suited to an equally wide range of applications, offering the design engineer opportunities for many end-uses. Applications vary significantly in size, complexity, loading, operating temperature, surface quality, suitable production volumes and added value. The expanding choice of raw materials, in terms of reinforcement type (concentration and fibre architecture) together with matrix material (subsets of both thermoplastic and thermoset polymers), followed by many subsequent final conversion processes, gives impressive flexibility. These variables often interact to create for the uninitiated an often confusing material and process 'system'. The properties of the final moulded item are hence controlled by the initial choice of fibre and matrix type, together with the subsequent processing route.

The particular route selected through the choice of fibre type, resin system, processing technique, finishing operations and assembly sequence will not only affect the part performance, but importantly the part cost. For example, a manual-based lay-up process will be suited to low-volume components, whereas for higher manufacturing volumes considerable investment in processing equipment and automation can be made. Material scrap must be minimised. During the design of a component, the processing technique must be selected not only to give the desired geometrical complexity, but also to suit the cost structure of the component under consideration. Notably, the processing technique should be chosen to suit the manufacturing volume over which tooling costs and plant are to be amortised. Hence, for the full implications of any use of textile composites to be considered, both the benefits and costs must be quantified. This is shown schematically in Fig. 10.1.

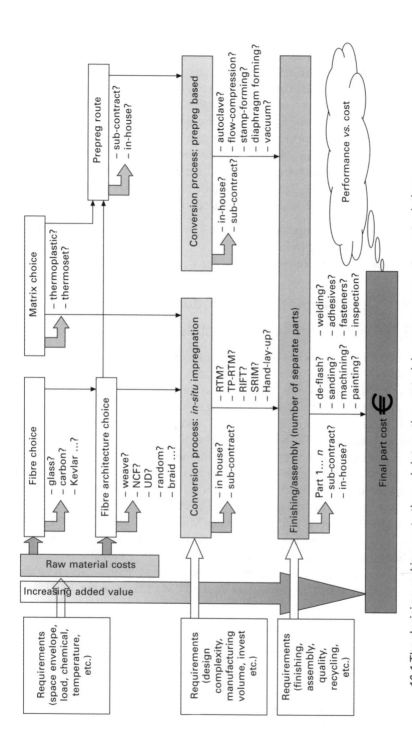

10.1 The decision making route through interacting material, process, property, cost relations.

10.1.1 Outline of this chapter

In order to illustrate different factors affecting the cost of textile composites, a cost modelling tool is described such that the process of cost estimation can be understood. Cost build-up in textile composite applications is then discussed, commencing with an examination of raw material prices. Typical process machine and tooling costs are discussed, followed by a summary of the effect of manufacturing volume on cost and the typical manufacturing volume of textile composite conversion processes. Assembly costs form important cost portions of a module and the effect of parts count reduction on cost is shown. Two case studies are then presented that show the build-up of system cost for components made of textile composite structures:

1. Chassis brace components produced by thermoplastic sheet stamping and over-injection moulding.
2. Carbon epoxy aft fuselage panels for the Airbus A380.

10.2   Cost estimation methodologies

10.2.1 Cost estimation approaches

Cost modelling approaches for composite manufacturing have been reviewed previously[1-3]. Beyond 'rule of thumb' approaches that use experience-based estimating, these can be summarised as comparative techniques[4-8], process-oriented cost models[9-11], parametric cost models[12-15], relational databases[16], object-oriented system modelling tools[17] and process flow simulations[18-23]. An understanding of how these techniques work and of their general suitability is of importance for the realistic modelling of a particular process.

As the basis for examples presented throughout this chapter, a cost prediction tool is described, which will drive an understanding of how cost is built-up in textile composite applications[3]. This parametric technical cost model (TCM) is based in MS Excel^{TM}, interacting with MS Visual Basic^{TM}. The TCM can be coupled with a discrete event-based process-flow simulation (PFS) tool that dynamically represents the interactions between the different manufacturing operations. Hence, an averaging effect is gained over a statistically significant period, thereby generating input data for the TCM and increasing the accuracy of the cost calculation.

10.2.2 Technical cost modelling

Parametric models offer flexibility together with easy manipulation of process and economic factors for sensitivity studies. Activity-based costing (ABC) accountancy attributes direct and overhead costs to products and services based on the underlying activities that generate the costs. However, as ABC is based upon historical data, it is of limited use when new processes are considered. In cases where detailed information is not available to define overhead costs, not all variable costs will be activity-based and volume-based approximations are applied (for example, a ratio of direct to indirect labour). Hence, TCM methodologies are related to ABC but use engineering, technical and economics characteristics associated with each manufacturing activity to evaluate its cost[24]. The technical cost modelling approach is shown in Fig. 10.2.

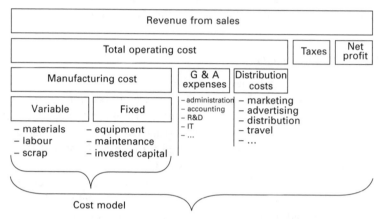

10.2 Technical cost modelling approach.

TCM commences with the identification of the relevant process steps required to manufacture a particular component. The approach is designed to follow the logical progression of a process flow. In this manner, the process being modelled is divided into the contributing process steps. Each of these operations contributes to the total manufacturing cost as resources are consumed. As such, each operation is modelled and the respective total manufacturing cost is divided into contributing cost elements. Hence, the complex problem of cost analysis is reduced to a series of simpler estimating problems. The contribution of these elements to the part manufacturing cost is derived from inputs including process parameters and production factors, e.g. production rate, labour and capital requirements, and production volume. These elements are calculated based on engineering principles, economic relationships and manufacturing variables.

Fixed and variable costs

A natural segregation of cost elements is between those that are independent and those that are dependent on the manufacturing volume within a given time frame. Variable costs are independent of the number of parts produced on a per piece basis. For example, the raw material cost is generally independent of the number of parts made. Additionally, labour, energy and sub-contracted costs remain constant regardless of production volume.

Conversely, fixed costs are capital investments that are necessary for the manufacturing facility. They are labelled as fixed because they are typically a one-time capital expenditure. These costs are distributed over the number of parts produced and the fixed costs per piece vary according to the production volume. As the production volume increases, fixed costs are reduced because the investment can be amortised over more parts. Machine, tooling, maintenance, cost of capital and building costs are typically fixed costs.

Input data and assumptions

The cost model input data are derived from the PFS output together with additional data concerning materials (e.g. weight fraction, costs), equipment (cost, area, energy, lifetime, maintenance, cost of capital), labour (number of workers, number of shifts, working area) and overheads (consumables, storage, scrap and reject). Input data is entered through a MS Visual Basic^{TM} user interface or directly into the spreadsheet.

The model will predict either the manufacturing costs occurring during continuous production, or the total cost including general and administrative (G&A) overheads (Fig. 10.2). This depends on factors such as the indirect to direct labour staff ratio. Development costs, production tests and machinery installation can be included with 5–10% of the initial machine purchase price per year (covering installation and planned maintenance schedules), or excluded. Machinery is normally assumed new and depreciated linearly, with a typical life of seven and ten years for three and two shift patterns respectively. Production periods of yearly increments are considered. Where applicable, a resale value can be applied to the equipment if production ceases before the defined life. Tool life is defined by a number of parts and hence costs are amortised over the life of the part. Labour costs comprise both direct labour and social costs. An augmented reject rate is used and material costs consist of the product, waste, rejected products and any internal recycling cost or benefit.

Model structure

Figure 10.3 illustrates the structure of the model where multiple materials and machines can be modelled. For example, several manufacturing cells can be modelled, with separate finishing and assembly steps. This would correspond to a manufacturing plant divided into cells, such as shown in Fig. 10.4. The manufacturing plant is described further in section 10.4 to illustrate the method of breaking down the overall estimation into the individual steps.

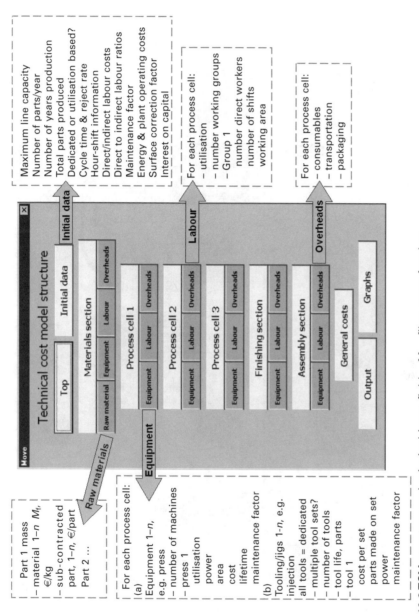

Initial data

- Maximum line capacity
- Number of parts/year
- Number of years production
- Total parts produced
- Dedicated or utilisation based?
- Cycle time & reject rate
- Hour-shift information
- Direct/indirect labour costs
- Direct to indirect labour ratios
- Maintenance factor
- Energy & plant operating costs
- Surface correction factor
- Interest on capital

Labour

For each process cell:
- utilisation
- number working groups
- Group 1
  number direct workers
  number of shifts
  working area

Overheads

For each process cell:
- consumables
- transportation
- packaging

Raw materials

- Part 1 mass
  - material 1–n M_f,
    €/kg
  - sub-contracted
    part, 1–n, €/part
- Part 2 ...

Equipment

For each process cell:
(a)
- Equipment 1–n,
  e.g. press
  - number of machines
  - press 1
    utilisation
    power
    area
    cost
    lifetime
    maintenance factor
(b)
- Tooling/jigs 1–n, e.g.
  injection
  - all tools = dedicated
  - multiple tool sets?
  - number of tools
  - tool life, parts
  - tool 1
    cost per set
    parts made on set
    power
    maintenance factor

10.3 TCM structure and principal input fields. M_f = fibre mass fraction.

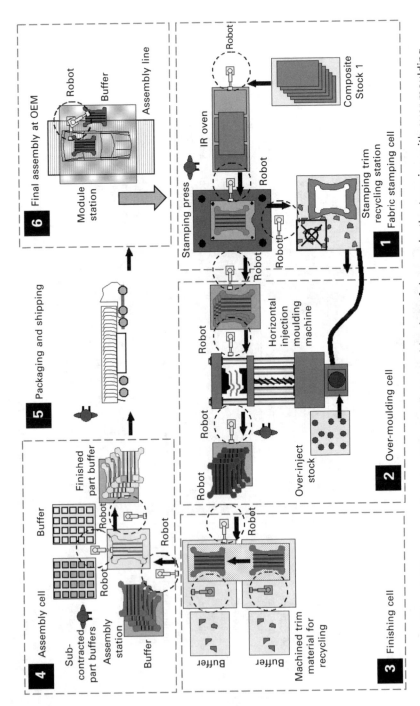

10.4 Schematic of a composite manufacturing line: combination of thermoplastic stamping with over-moulding, trimming, module assembly, shipping and body-in-white (BIW) assembly operations.

The TCM developed enables the amortisation of plant costs to be approached in two ways. First, a whole line could be dedicated to one product where all of the fixed plant costs are amortised over the number of parts produced for the total years of production. Cost against volume graphs can be generated simply by assuming that the full plant costs are spread over the parts produced, with strongly increasing costs at lower volumes. In the second case, only a fraction of either a line capacity or a plant would be assigned to one product while the remaining capacity would be sold to a second client. Fixed plant costs are amortised as a fraction of utilisation and the number of years that the plant is used, effectively giving a charge rate per minute for a manufacturing line.

Cost model output

The TCM enables factors, including equipment cost, depreciation and operating power, to be defined and sensitivity analyses to be performed for the whole line or sub-units of a manufacturing cell. The model predicts the manufacturing cost and cost segmentation as a function of volume. Cost versus volume relations are given together with segmentation of the total production cost into:

- material;
- direct labour;
- overheads (indirect labour and plant costs);
- depreciation, interest and maintenance;
- energy;
- consumables;
- tooling;
- transportation;
- sub-contracted costs.

Additionally, for each process step, costs can be further segmented into the above categories, and the material costs itemised. Other outputs of the cost model are global scrap and reject rates, production cycle time, production time, production rate, plant area and energy requirements.

10.2.3 Process-flow simulation tool

While cost calculations do not require input from process-flow simulations, a limitation of the parametric technical cost modelling approach is the assumption that each step in the manufacturing process operates independently from the others from a temporal standpoint of part flow in the manufacturing line. This assumption often results in an underestimation of the manufacturing cost[23]. This is overcome by PFS, which dynamically represents the interactions between the different manufacturing operations, gaining an averaging effect over a statistically significant period (depending on the number of parts made), rather than a simple 'static' representation. PFS includes commercial codes such as Witness™. Benefits include the ability to predict the cycle time and the capacity of the process, in addition to the manufacturing cost. These tools can also aid process improvement studies, identify bottlenecks and achieve a better distribution of personnel and raw materials on the shop floor.

Again, as a basis for the examples presented in this chapter, the PFS tool developed is briefly described. It is a discrete-event simulation tool written in MS Visual Basic™, consisting of a hierarchical structure in the way that a commanding finite state machine controls other finite state machines, which represent the objects in the line. Such simulations are used to simulate components that normally operate at a high level of abstraction[25]. Discrete-event simulation is relatively fast while still providing a reasonably accurate approximation of a system's behaviour.

PFS input and operation

The tool consists of a workspace where the user places all the objects (machines, buffers, robots, workers, etc.) that have an influence on the production flow of a given manufacturing line (e.g Fig. 10.4). The modules are, for example, machines, robots or buffers. Machines are preceded and followed by a transfer system, such as a robot. Graphic objects are descriptive symbols that are not part of the process flow. The end category is the final part buffer and the end of the simulation. Failure incidents can be simulated, where the user defines a failure probability and the time needed to fix the problem. The program randomly generates breakdowns according to these data. Both operators and workers can be considered. Workers are fully engaged in the manufacturing line and are part of the actual manufacturing process. Operators perform maintenance and problem-solving tasks. Their occupancy level is predicted by the failure generation function, enabling optimisation of operator allocation. Both convergent and divergent material or part flows are possible using different combinations of objects, with the following categories included:

- graphic object;
- processing machine;
- buffer;
- materials stock;
- end;
- continuous conveyor;
- indexed conveyor;
- single transfer;

- multiple transfer;
- unloading task;
- loading task;
- jig.

All these objects need to be defined. Figure 10.5 gives examples of PFS input data for a buffer, process machine and a transfer device for a section of the line in Fig. 10.4.

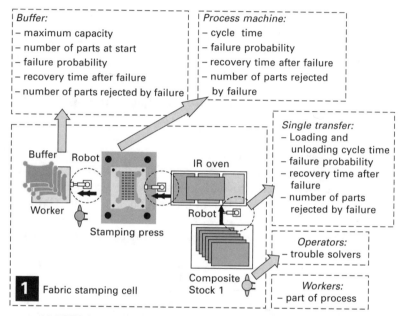

Buffer:
- maximum capacity
- number of parts at start
- failure probability
- recovery time after failure
- number of parts rejected by failure

Process machine:
- cycle time
- failure probability
- recovery time after failure
- number of parts rejected by failure

Buffer   Robot

IR oven

Single transfer:
- Loading and unloading cycle time
- failure probability
- recovery time after failure
- number of parts rejected by failure

Worker

Robot

Stamping press

Operators:
- trouble solvers

1 Fabric stamping cell

Composite Stock 1

Workers:
- part of process

10.5 PFS input data for buffers, transfer devices and processing machines.

Either a time goal or an output goal (a number of parts) can be selected which defines the end point of the simulation run. Input data are stored in a MS Access^{TM} database before the program processes the data. The program moves forward in time steps (simulated seconds) through each object in a user-defined sequence. Events (incidents that cause the system to change its state in some way) can occur only during a distinct unit of time during the simulation and not between time units. According to the category or type of an object, a function is run that virtually checks the objects and adjusts their simulation variables. When the simulation goal is reached, values of the simulation variables are written into a MS Excel^{TM} template.

PFS output

The results are processed by MS Excel™ macros giving information about: the global production scenario (e.g. production cycle time), the detailed object (e.g. number of parts processed or loaded, machine utilisation, buffer start/finish size), the number of produced parts, the time needed for production and the average cycle time of the line. Furthermore, process and throughput time are calculated considering any scheduled maintenance. Occupancy represents the relative time an operator is performing maintenance due to unscheduled stoppage. These results can then be transferred to the TCM. Hence, as a first step in a cost calculation, a PFS can be performed to generate input data for a subsequent TCM, thereby increasing the accuracy of the cost calculation.

10.3   Cost build-up in textile composite applications

10.3.1 A materials perspective

Following the sequence of the TCM (Fig. 10.3), the quantity and cost of the textile composite raw materials are required. Thus, defining the raw material cost is one of the steps needed towards calculating the system cost. The data in Tables 10.1 and 10.2 are intended for use as input into a full TCM and *not*

Table 10.1 Typical composite raw material costs: un-impregnated textiles and polymers

Reinforcement	€/kg	Matrix	€/kg
Glass	1.6	Polypropylene (PP)	0.7
Carbon (80k–12k)	15–17.5	Polyethylene terephthalate (PET)	3.5
Kevlar	23	Polyamide 66	2.5–4
		Polyamide 12 (PA12)	8.4
GF weave (1200 tex, 300 g/m²)	10	Polyetherimide (PEI)	17.6–22
Kevlar weave (300 g/m²)	47	Polyetheretherketone (PEEK)	68–77
CF weave (HS 12k CF, 300 g/m²)	78	Unsaturated polyester	1.5–1.8
CF weave (IM 12k CF, 300 g/m²)	124	Vinylester	2.5–3.5
GF NCF (100" wide, 1000 g/m²)	3	Epoxy	2.2–55
Commercial 12k CF NCF (100" wide, 1000 g/m²)	17–30	Phenolics	1.65–5
Aerospace 12k CF NCF (100" wide, 1000 g/m²)	45	Cyanate esters	62
GF biaxial braid	11–15	Polyurethanes	5.5–14
CF biaxial braid	31–90	Bismaleimides (BMI)	78

GF, glass fibre; CF, carbon fibre; NCF, non-crimped fabric.

Table 10.2 *Typical* textile composite raw material costs: semi-finished products

Thermoplastic-based textile composites			Thermoset-based textile composites		
Material form	€/kg	Example of supplier	Material form	€/kg	Example of supplier
CF/PA12 sheet	50–54	Schappe Techniques	GF/epoxy, woven prepreg, 720 g/m², 1 m × 50 m roll	26	SP systems, also: Hexcel Cytec
GF/PA12 sheet	13–17	Bond Laminates TEPEX	CF/epoxy unidirectional (UD) prepreg, 476 g/m² (CG carbon), 1 m × 150 m roll	29	SP systems, also: Hexcel Cytec
CF/PA66 sheet	30–50	Bond Laminates TEPEX			
CF/PA6 sheet	7–11	Bond Laminates TEPEX	CF/epoxy, UD prepreg, 461 g/m² (HM carbon), 1 m × 150 m roll	91	SP systems, also: Hexcel Cytec
GF/PET sheet	4.6–7.5	Vetrotex	CF/epoxy, UD prepreg, 461 g/m² (HM carbon), 1 m × 150 m roll	91	SP systems, also: Hexcel Cytec
GF/PP dry fabric	3–4.5	Vetrotex			
GF/PP sheet	3.5–5.5	Vetrotex			
GF/PP sheet, GMTex	3.5–5.5	Quadrant Plastic Composites	Aramid/epoxy UD prepreg, 545 g/m², 1 m × 150 m roll	50	SP systems, also: Hexcel Cytec
GF/PP UD tape	4.9–6.4	Plytron	CF/epoxy, woven prepreg, HS carbon), 517 g/m², 1 m × 50 m roll	59	SP systems, also: Hexcel Cytec
PEI/GF & PPS/GF	60	CETEX sheet (Ten Cate)	CF/epoxy, woven prepreg, HS carbon), 517 g/m², 1 m × 50 m roll	59	SP systems, also: Hexcel Cytec
PEI/CF & PPS/CF	140	CETEX sheet (Ten Cate)	Closed cell styrene acrylonitrile (SAN) core material, 5 mm, 50 kg/m³	10/m²	ATC
CF/PP tape	16–29	GuritSuprem/ Flex composites	Closed cell styrene acrylonitrile (SAN) core material, 30 mm, 50 kg/m³	41/m²	ATC
CF/PA tape	20–30	GuritSuprem/ Flex composites			
CF/PET tape	20	GuritSuprem/ Flex composites			

as a cost-based materials selection guide. Costs will vary with: oil price, polymer/fibre price, exchange rates, the application dimensions (*vs.* sheet or roll dimensions), and weight of material sold (20 boat hulls *vs.* an order for 200 000 automotive components/year for seven years). Suppliers are given as examples only and costs should be taken as commonly accepted representative guidelines rather than formal prices of any supplier.

10.3.2 General input data

The second step in the sequence of the TCM (Fig. 10.3) is to define the general input data for the manufacturing plant. Such data obviously vary with industrial sector, the manufacturing processes used and the planned production volume. This typically includes the following factors, which are determined on a case basis:

- maximum line capacity/yr;
- number of parts/yr desired;
- number of years production;
- reject rate;
- part cycle time;
- working days/year;
- number of shifts and hours/shift;
- combined direct and indirect labour costs including social;
- ratio indirect/direct labour staff;
- energy costs;
- plant operating cost per unit area;
- equipment maintenance;
- interest on capital.

10.3.3 Effect of manufacturing volume

The industry sector considered has an important role in composite manufacturing process selection. If it is assumed that the manufacturing line does not normally exist (compared with the highly established steel and aluminium manufacturing industries), then costs must be calculated on a dedicated basis. The effect of this is discussed further in section 10.4.5.

The fixed costs associated with setting up a manufacturing line need to be considered against the number of parts that need to be produced. Components produced at higher annual volumes will justify the use of automated equipment and robotic transfer systems, for example, the stamping of thermoplastic composites for automotive applications. In contrast, components produced at lower annual volumes, such as for niche marine products, often use increased manual labour (typically €30/h in Europe) rather than automation as high fixed costs would be amortised over uneconomic volumes.

Figure 10.6 shows an approximation of the parts produced per year from one tool set for different thermoplastic and thermoset-based textile composite processes. As the maximum manufacturing volume increases, the fixed costs also tend to increase. While high fixed cost processes can be used at lower volumes, it may not be the most economic approach. An exception to this would be the thermoset-based automated tape placement (ATL) and automated fibre placement (AFP) processes, where the fixed costs are high and the volumes low, justified by reduced material scrap of expensive aerospace grade carbon prepreg tape.

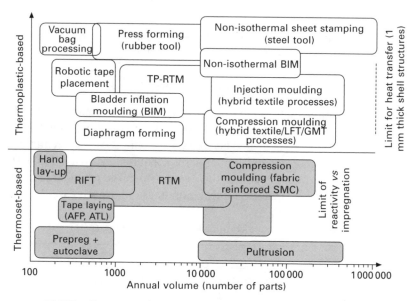

10.6 Textile composite annual manufacturing volumes (adapted from Månson *et al.*[26]).

Assuming 235 to 250 working days per year, and 90% efficiency, the effect of process cycle time on the number of parts produced per year from one tool set can be modelled for: one shift (7.5 h/day), two shifts (15 h/day) and three shifts (22.5 h/day). Figure 10.7 shows that processes requiring long cure times in ovens or autoclaves will be limited to low volume, high added value, applications. In contrast, thermoplastic composites can offer material and process combinations with cycle times of 1 (stamping glass fibre reinforced polypropylene (GF/PP) systems) to 5 min (rubber forming of carbon fibre reinforced polyetherimide (CF/PEI) sheet), giving low machine and tool costs per part.

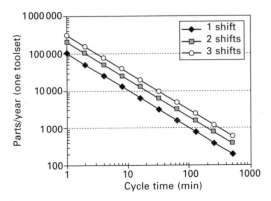

10.7 Effect of process cycle time versus maximum parts produced per year, for one-, two- and three-shift patterns.

10.3.4 Typical process machine cost

Any cost calculation requires identification of the process equipment needed, including machines, transfer devices and buffers, such as those shown in Fig. 10.4. Relations can be established between, for example, press size in tonnes and press cost to give general guidelines for equipment costs. Table 10.3 includes examples of textile composite equipment costs, which are size and application dependent. Specific information is used for each case.

Table 10.3 Typical textile composite process machine costs

Equipment	Cost
Braiding machine (172 carriers)	€250–350k
100″ (2.5 m) warp knitting machine	€1500k
1500 tonne hydraulic press	€900k
IR oven	€150k
Vacuum pump and tank (1–2 m² part, 30/day for 1 yr)	€6.1k*
RTM injection unit (high volumes)	€170k (€40k for lower volumes)
Autoclave, small	€230k
Automated fibre placement (AFP)	€5000k

* Not consumables.

In order to show typical utilisation-based equipment costs (part and size specific), a range of machine costs per minute has been calculated. This assumed a three-shift pattern with full utilisation and a seven year production period, with inclusion of plant surface costs (€85–115 m²/y), energy costs (€0.05–0.12/kW h), and cost of capital for the machines. Direct operators and indirect overheads were excluded. The results are given in Fig. 10.8.

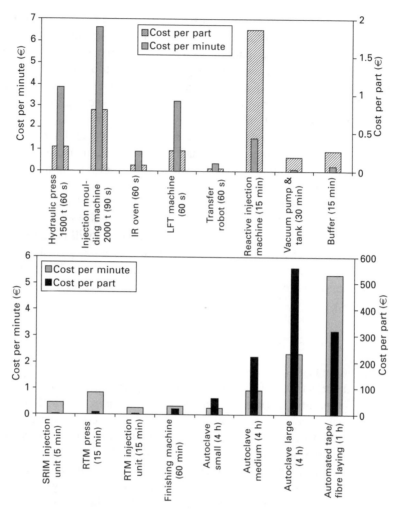

10.8 Examples of textile composite utilisation-based equipment cost (part and size specific), machine cost/min based upon: three-shift pattern, full utilisation, seven year production period, including: plant area cost, energy cost, cost of capital for the machine, *excluding* direct operators and indirect overheads.

10.3.5 Tooling costs

Tooling costs are an important issue for composite processing. Many textile composite techniques, notably for thermoset-based materials, have cycle times of several minutes (structural reaction injection moulding, SRIM) to several hours (autoclaving of prepreg) and tooling cost is a significant fraction of the total. The use of low-cost tools (e.g. nickel shell composite tools) can reduce costs compared with steel tools, especially for lower manufacturing volumes. For example, tooling costs of €920k for a resin transfer moulding (RTM) floor pan amortised over 13 000 parts/yr (the maximum from one tool set on a two shift pattern with a 15 min cycle time) for five years production (total parts = 65 000) would give a tooling cost of €25/part (including tool maintenance). A lower-cost, epoxy-based tool at €250k would reduce the tooling cost to €6.9/part. However, if the cycle time for the steel tool were reduced to 10 min, such that 20 000 parts were made per year, then the tooling cost per part would be €17/part. A general increase in cost competitiveness occurs when the manufacturing volume from one tool set is increased. For example, thermoplastic textiles can be stamp-formed with over 200 000 parts/yr from one tool set (€500k), giving a tooling cost of €0.9 (seven years' production). In comparison with steel stamping processes that have multiple tool sets, such composite processes offer lower tooling costs. High volumes with thermoset-based processes will need multiple tool sets and handling equipment. While multiple moulding cells using multiple tool sets are used in production, generally a process should be used that is adaptable to the manufacturing volume to avoid large moulding plants with many tool sets.

10.3.6 Effect of process scrap

An important cost issue with textile composites is that of material scrap. This is not the reject rate of the final conversion process, for example RTM, but that of preparing the preform. Taking the example of a RTM floor plan (16 kg), carbon fibre preforms could be taken from non-crimp fabrics (NCFs) to produce the structure. Even when using computer-optimised nesting patterns, NCF waste fractions of 30% can occur. With NCF costs forming 63% of the total floor pan cost, at an NCF cost of €30/kg, this can represent €80/part of waste. Reduction of fabric scrap, ideally before moulding, or maximising the value of the material post-process, are key to the economic use of non net-shape textile preform composite processes.

10.3.7 Assembly costs

Composite materials enable many features to be integrated into a single composite component. A 10:1 component consolidation is achievable but the added complexity in operations such as blank placement (for stamp forming) and preform construction (for RTM) should be assessed to evaluate the effect on cost (reduced parts count with a complex part *vs.* increased parts count with simple parts)[27]. Parts consolidation minimises tooling and parts count and the associated investment, inventory, tracking, and assembly effort and space. The reduced count of tools, jigs and fixtures can both increase the flexibility of assembly operations, simplifying the process of assembling model variants on the same line, and reduce the number of assembly steps.

Simply modelling a composite component cost and comparing this with an existing (metallic) product often results in an underestimation of the potential cost saving for a composite system compared with a steel system. The TCM approach consists of two sections (Fig. 10.9), where part 1 focuses on component costs and part 2 on the assembly cost to give a system cost. This requires definition of assembly scenarios, often based on a modular approach, to fully develop the system cost and the final weight-saving implications.

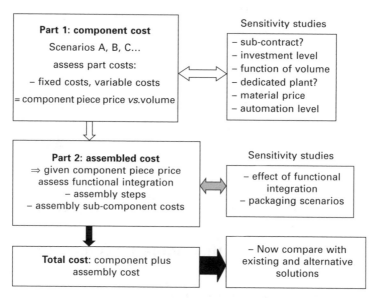

10.9 Cost modelling approach: Part 1 = component cost, Part 2 = module cost.

In a high-volume assembly line a reduction in the line length or the number of assembly workers allocated to a particular task can reduce the system cost. Comparison of a conventional steel system with a modular composite system can show considerable cost savings at the assembly stage (Fig. 10.10). As an example, a robot could be used to mount a one-piece composite moulding in a simple operation rather than the four stations and workers required to fit a metallic-based part. For a two-shift pattern assembling 500 modules per day, a comparison between the alternatives illustrates the point of considering the system cost. With typical line investment costs of €5000/m² and associated surface costs, with four workers included (and associated indirect costs), the costs of manual assembly would be considerable at €7.1/part. By assembling a drop-in module via robot on such an assembly line, assembly costs could be reduced to €1.1/ part.

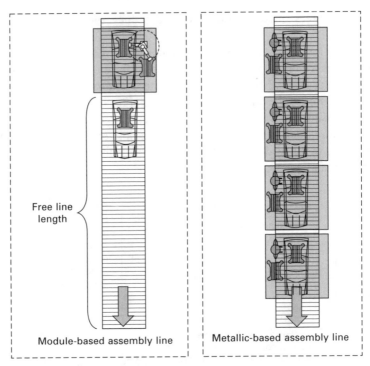

Free line
length

Module-based assembly line   Metallic-based assembly line

10.10 Assessment of assembly costs: cost reduction by functional integration, parts count reduction, and modular assembly.

10.4   Case study 1: thermoplastic composite stamping

10.4.1 Ford Thunderbird X-brace component

Using the TCM approach described in section 10.2, comparative cost estimations were made for X-brace components used to stiffen cabriolet body-in-white (BIW) assemblies, produced from different *candidate* composite materials. The prime application was the Ford Thunderbird, with a target manufacturing volume of 20 000 units per year. The existing steel X-brace formed baseline cost, assembly cost and mass values (Table 10.4).

Table 10.4 Steel X-brace system details

Item	Weight (kg)
Steel X-brace	15.6
Steel panel	3.6
Assembled to BIW: system mass and cost/part	19.3

Figure 10.11 shows the Ford Thunderbird vehicle and the steel X-brace assembly, where a steel tube-based brace mounts onto suspension towers. A vertical steel panel spanning the width of the vehicle attaches to the shock towers. The candidate system examined in this cost study is a composite replacement (Fig. 10.11) that could combine the X-brace and the vertical panel, but would use the existing steel suspension towers. A fibre-reinforced thermoplastic sheet would first be stamped to give a component of complex double curvature. To maximise the stiffness of the component, a ribbed structure would be over-moulded onto the stamped sheet, using the same polymer as the fibre-reinforced sheet matrix[28].

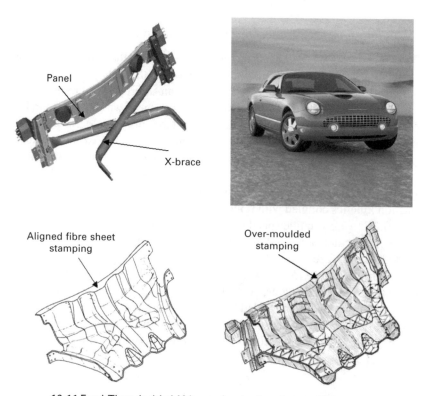

Panel

X-brace

Aligned fibre sheet stamping

Over-moulded stamping

10.11 Ford Thunderbird X-brace in steel and a *candidate* thermoplastic composite solution.

10.4.2 Thermoplastic composite material systems

Three material systems were examined, all consisting of thermoplastic composite prepreg sheets processed via non-isothermal stamping, followed by an over-injection moulding process incorporating closed loop recycling of stamping scrap. GF/PP sheet, commercially available as Twintex™, formed a benchmark system. In order to decrease the component weight and offer higher operating temperatures potentially compatible with steel E-coat and paint line temperatures, a glass fibre reinforced polyethylene terephthalate (GF/PET) sheet material (PET Twintex[TM]) was studied. The third material, offering the greatest weight-saving potential, was a carbon fibre reinforced polyamide 12 sheet product (CF/PA12). The over-moulding phase, for all three material variants, used recycled stamping waste diluted with virgin polymer, giving fibre mass fractions of 30–40%. Typical sheet material costs are shown in Table 10.2 and over-moulding polymer costs in Table 10.1. As a reference, automotive grade sheet steel is €0.9/kg.

Sheet cost estimates for a reactive impregnation route

Costs of existing CF/PA12 fabric grades were greater than could be justified for this particular application, but the properties that CF/PA12 could offer were still considered of interest and hence an alternative material supply route was invested using TCM techniques. This study was based around the ability to batch pre-impregnate sheets at a laboratory scale via a reactive thermoplastic RTM process using an anionically polymerised laurolactam system (APLC12)[29–31]. Commercialisation would require a continuous reactive impregnation unit to produce sheet material. Line speeds for a novel prepreg line were predicted, based upon experimental results, and modelling of the reaction kinetics coupled with impregnation phenomena (including capillary forces). With a given line speed (>3 m/min), these results were coupled with the TCM to calculate the sheet cost, based upon an estimated line cost of €2000k, the surface area and labour needed, and the energy costs. Automotive grade carbon fibre (€15/kg), weaving costs and APLC12 material costs were used as raw material inputs, with the line running costs and an additional profit factor that would be added by any commercial producer of such sheet material. Depending of the machine utilisation, a material cost of €22/kg was predicted, which offered cost reductions compared with existing CF/PA12 systems.

10.4.3   Weight saving assumptions

Prior to a full design study and finite element analysis (FEA) modelling, weight-saving assumptions (Table 10.5) were made compared with the existing steel X-brace system, to determine the initial cost case for a composite X-brace. From these assumptions, the average part thickness was calculated, assuming a final ratio of stamped material to over-moulded material of 60:40. Note that local increases in section thickness occur, for example to facilitate load introduction. The part thickness is an important variable in the later manufacturing cost prediction, where an excessive thickness will

Table 10.5 Weight savings from thermoplastic composite material systems

Composite solution	CF/PA12	GF/PET	GF/PP
Stamped sheet thickness (mm)	1.7	2.0	2.8
Average over-moulded polymer thickness (mm)	1.2	1.6	2.4
Composite part weight (kg)	7.7	12.5	13.5
Weight saved (%)	60	35	30
Raw material cost ($\in$)	153	58	44

require longer in-mould cycle times to accommodate shrinkage and reduce warpage.

The stamping process uses a blank-holder system, and as a base line, 30% of the initial stamped sheet was assumed to be scrap material. With closed-loop grinding of this stamping scrap for over-injection moulding, the additional virgin polymer fraction mass has been calculated to give the required thickness. The raw material costs (i.e. before the manufacturing processes of stamp-forming and over-moulding) for the X-brace component in different material systems are hence given in Table 10.5.

10.4.4 Manufacturing process

Figure 10.4 shows an envisaged manufacturing plant for producing the X-brace component. This cost comparison examines steps 1–3 and 5. Production was planned for a five-year period, using a three-shift pattern. A stamping cycle time of 50 s was assumed (3 mm average sheet thickness, 40 s in mould, 2×5 s transfer). The over-injection moulding cycle time was assumed as 90 s (allowance for locally thicker sections, 80 s in mould, 2×5 s transfer), giving a line cycle time of 90 s per part. The maximum number of parts/yr from one set of tools per machine is hence 205 000.

Starting from *composite stock 1*, pre-impregnated thermoplastic sheets are taken from a buffer by a *robot* and placed into an *infrared oven*. The materials are heated above their melting point before rapid transfer to a fast-acting *hydraulic press*, again via *robot*. The press closes and the hot thermoplastic prepreg is shaped to the steel tool profile, in cycle times of 10–30 s for thin shell structures. A blank-holder is used to hold the stamping material. The component is punched out of the overall shaped blank, giving scrap material. The scrap is removed via a *robot* and placed into a chopping and *grinding cell* that processes this scrap prepreg into pellets. This high-volume fraction material is combined with virgin injection moulding pellets to give the desired material for the over-injection stage. The stamped preform is hence transferred via *robot* into a *warm buffer*, which is to ensure that the average temperature at the interface of the stamped material and the over-injected material is above the composite matrix system T_m. Transfer by robot into the *over-injection moulding machine* enables over-moulding of the complex features. The net-shaped component is removed from the tool by *robot* and loaded onto a *conveyor system* that transports the part to the *trim and machine stations*, where any limited finishing of the net shape part is made, as required, and the finished parts are loaded by *robot* into a *buffer*. This can then be followed by assembly of subcontracted modules as required (this is not modelled in this example).

Quotations were received for the process machines including purchase price, maintenance details, plant floor area, operating power and reject rates where appropriate. As a summary, the stamping (including the recycling cell), over-injection and trimming cells represented investments of €1250k, €3365k and €200k, respectively. Plant areas and powers were: (233 m^2, 259 kW), (229 m^2, 530 kW), and (50 m^2, 100 kW) respectively for the three cells. Tooling quotations for the stamping and over-injection moulding tools were €130k and €500k respectively. All tooling costs were dedicated, while the manufacturing line was used on a cost per minute basis where it was assumed that the remaining 90% line capacity for the five-year production period and the full capacity for the remaining period of plant life were filled by a different client and product, but using nominally the same moulding process. As section 10.4.6 shows, dedication of an automated plant with heavy processing equipment with low utilisation levels results in substantially increased costs. Manual workers are shown in the plant diagram. Transportation costs of €3/part and subcontracted steel load introduction washers at €1/part were included.

10.4.5 Effect of material type

Using the material weight and cost given in Table 10.5, X-brace component cost in the three material types was calculated. Conversion costs will typically be higher for engineering polymers, such as PET, compared with PP. In contrast, the increased wall thickness required for GF/PP, particularly compared with CF/PA12, would increase cycle times for the GF/PP material.

Figure 10.12 shows the effect of material type on part cost. For each material type, a high, medium and low cost boundary is shown. The high case considers increased equipment investment and higher raw material costs while the lower case considers lower equipment and raw material costs. As Fig. 10.12 shows, the lowest part cost was for a GF/PP X-brace structure, at €70/part for 20k parts/yr. Materials costs formed 63% of the total (for the median cost case), and tooling 16%, with 11% for equipment. As the structural performance of the raw materials increased (and temperature rating for PET), so did the part cost (in the materials category), with part prices of €83 and €179 for the GF/PET and the CF/PA12 systems respectively.

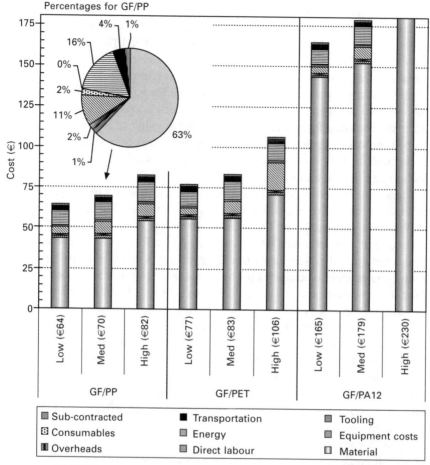

10.12 X-brace part cost segmentation for GF/PP, GF/PET and CF/PA12 material solutions.

An increased weight saving corresponded to an increased cost. The euros needed to save 1 kg of weight (over GF/PP) were €6/kg for PET and €23/kg for CF/PA12 variants. The total investment per part, ranged from an additional (compared with GF/PP) €13/part for GF/PET, to €109/part for CF/PA12. The large total cost increase for CF/PA12 is due to the high absolute mass saved versus GF/PP of 5.8 kg.

10.4.6 Plant utilisation: dedicated vs. utilisation based

The overall strategy of a manufacturing plant is an important assumption to set at the outset of any TCM study, with the effect often influencing other manufacturing parameters. The assumption used in this case study is that X-brace production would occupy a percentage of an existing plant capacity, with the X-brace product hence paying on a cost per minute basis. This enables high-throughput thermoplastic stamping and over-injection moulding processes to be used for a medium volume (20 000 parts/year) application, which would be uneconomic if a manufacturing line capable of over 200 000 parts/yr were used at 10% utilisation. This assumption holds for existing manufacturing processes or where exclusivity for a particular technology is neither held nor desired. However, if a new technology has been developed that requires a new integrated plant to be built to gain maximum cost benefits, then the product must pay for the whole line cost. This also applies if exclusivity is desired, such that the supplier is not permitted to use the same process technology for another client.

Hence the effect of either a *percentage line utilisation* or *amortising the full line costs* is shown to illustrate the importance of this effect.

Where a percentage of the line is attributed, the plant costs (initially assuming all new equipment) are calculated as a function of the use rate. It is assumed that the plant has a life of seven years (three-shift pattern). The remaining production capacity for the years that the X-brace is produced and the full remaining capacity for the final years of the plant life are assumed to be fully utilised by different products by the same or a different client.

Where the full line costs have been considered, X-brace production is assumed to last for five years. Hence the plant life is set to a period of five years. Even for lower production volumes, such as 20k/year, the full plant cost is attributed to the part. Labour costs are also fully attributed to the one product, as the workers are hired to run the plant and cannot be reallocated (assuming a dedicated stand-alone plant). However, energy costs are obviously calculated based upon the running time of the machines needed to produce the total number of parts. At the end of X-brace production, here set as five years, all plant has zero value. An obvious consideration is that a dedicated plant set-up to produce 20k parts/yr would not have the same shift pattern and labour levels as a plant set-up to produce 200k parts/yr. Hence for up to 70k parts/yr, a one-shift pattern has been used, a two-shift pattern used from 70k to 135k parts/yr, and a three-shift pattern used from 135k to 200k parts/yr.

Using common input data for both dedicated and utilisation-based scenarios, but changing the plant life, percentage plant utilisation and shift information, comparisons were made for processing the commingled GF/PP weave, followed by over-injection moulding.

Utilisation-based plant

Figure 10.13 shows cost versus volume curves for five plant assumptions. Costs for a utilisation-based plant are shown, together with a dedicated plant, with the volume portions of the three different shift patterns. The maximum parts/yr that can be made with one-shift is 70k, so for higher

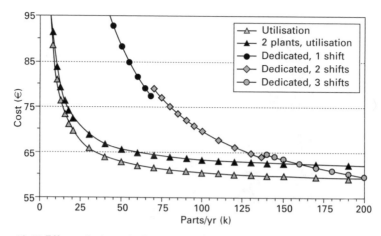

10.13 Effect of plant dedication or utilisation scenario on part cost versus volume.

volumes a two-shift pattern is needed, increasing cost at 70k, but quickly giving cost reductions when more than 80k parts/yr are made up to the two-shift pattern limit of 135k. From here on, a three-shift pattern can be used to the maximum of 200k parts/yr. However, a utilisation-based plant always gives a lower cost.

The utilisation-based cost versus volume curve in Fig. 10.13 for GF/PP shows that part costs for this hybrid moulding process reduce steeply to 20k parts/yr (€70/part). From 20k to 50k per year (€63/part), part costs reduced by 10%. From 50k to the maximum one-tool volume of 200k units per year (€59/part), a further reduction of 7% occurred. With only a 7% cost shift between 50k and 200k parts/yr, the hybrid moulding process is suited to a wide range of manufacturing volumes. If higher volumes than 200k parts/yr were required, additional processing cells (utilisation-based) and tool sets (dedicated) would be needed. Higher costs at lower volumes were principally due to the amortisation of tooling costs over the lower manufacturing volume, with all equipment costs on a percentage utilisation basis.

Dedicated plant scenario

Figure 10.14 compares cost breakdowns for volumes of 20k and 200k parts/ yr, for both dedicated and utilisation-based operations. Where 20k parts/yr are produced, the cost per part increases by 151% for a dedicated plant used for five years (one-shift pattern) compared with a utilisation-based plant at 200k. Cost increases occur for equipment and tooling cost categories, with smaller increases for direct and indirect labour costs. Comparison of dedicated and utilisation-based plants running at 20k parts/yr shows a reduced but still significant cost increase of 115% for the dedicated operation. Comparison of

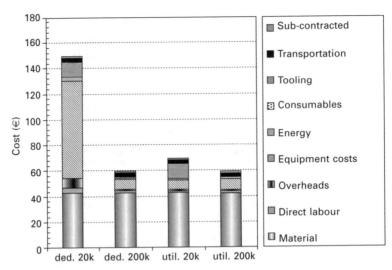

10.14 Effect of plant dedication or utilisation scenario on part cost segmentation versus volume.

utilisation-based plants at 20k and 200k parts/yr shows the same cost levels for all cost categories, with the exception of tooling cost that is always dedicated. Production of 200k parts/yr in a dedicated plant (five years, three-shift pattern) showed only a marginal cost increase of 0.5% versus a utilisation-based line at the same volume. Therefore, dedicated plants should ideally only be considered where close to the production capacity would be used.

Utilisation-based two-plant scenario

In practice, it is difficult to assume an optimised plant layout and still claim a utilisation-based line, because for the remaining 90% capacity a different product would need to be produced that may require a different physical plant layout or dynamics. A two-plant scenario is therefore examined where the stamped fabric part is made in a first factory, which is then shipped (with additional transport and labour costs) to a second factory. The stamped part is then over-moulded and any final trimming or finishing performed, again with additional labour assumed. The stamping cycle time of 50 s enables 307k parts/yr to be produced (three shifts), and hence plant costs for 20k stampings are amortised over a high volume if the stamping sub-supplier is fully utilising the stamping line. As standard machine layouts do not need to be altered to produce the X-brace in two such steps (1st stamping plant and 2nd over-moulding plant), it is particularly suitable for lower volumes and in fact gives lower costs than a dedicated (three-shift) plant up to volumes of 160k parts/yr (Fig. 10.13). From here on, the higher transportation and labour costs of a two-plant scenario are outweighed by the optimised, dedicated plant. Hence subcontracted two-plant set-ups are the lowest cost alterative for lower manufacturing volumes, while (for this case) dedicated plants are only justified at above 80% plant utilisation.

All sensitivity studies from this point forward consider a one-plant utilisation-based scenario.

10.4.7 X-brace production sensitivity studies

The following section examines the effect of the following examples on GF/PP X-brace production costs (20k parts/yr, three-shift pattern, utilisation-based):

- percentage weight saving *vs.* steel;
- high *vs.* low automation levels;
- high *vs.* low equipment investment;
- 60 s *vs.* 120 s line cycle time;
- material cost (−10% and +10%);
- 2% to 20% reject rates;
- material scrap: hot drape forming, 10% sheet scrap;
- sheet material recycling strategy: 30% scrap with and without recycling;
- labour level: two *vs.* six workers;
- tooling cost (−10% and +10%).

Figure 10.15 plots the baseline X-brace cost and the cost change over the factor range studied in each case. Rather than plotting a simplistic ±% difference to all factors, the difference in each factor that could be expected in reality was compared.

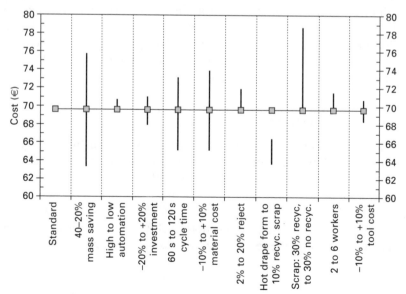

10.15 Cost variation for X-brace components produced at 20 000 per year in GF/PP.

Effect of weight-saved assumption

An important assumption behind TCMs of textile composites is the weight saving that could be expected from using a polymer and composite system, here to replace the current 19 kg steel system. A baseline weight saving of 30% for a GF/PP system was assumed for this study. Full design and FEA simulations would be needed to verify this estimation and hence sensitivity studies have been made of 20–40%, assuming that all the other TCM parameters are unchanged. This change in weight-saving assumption corresponds to costs of €76 and €63 respectively. Figure 10.15 shows that this gives the largest cost variance and that this estimate is hence an important figure, justifying further analysis before any decision for full-scale development. Here the material cost percentage in the overall process was 63%. A lower proportion would reduce the effect of the weight-saving assumption, for example including all the steps in Fig. 10.4. A 30% weight saving has been assumed through the remainder of this study.

Effect of automation level

The effect of the degree of automation was studied by comparing the standard high automation level (two workers and nine robots) with a lower automation level (seven workers and four robots). The lower level increased part cost by 2%. Caution should be used in interpreting these results because the extra hidden set-up costs and higher production engineering overhead costs of the high automation case have not been specifically included. For high-volume industries, the choice of a high or low automation level will depend on the length of a contract from an OEM (original equipment manufacturer) to a Tier 1 supplier such that the capital invested can be recovered in the total production period. Lower automation, with corresponding lower capital costs, and the ability to relocate labour, may create a lower risk for the Tier 1 supplier despite the marginally higher part cost.

Effect of plant investment

The effect of equipment cost on part price was studied by investigating a −20% and +20% variation. Cycle times and scrap rates were assumed constant. This would reflect any estimates made in the necessary equipment capacities or to cover the addition of extra items that were not initially included. This 40% difference caused a 4% shift in part cost, which is a larger effect than a changing automation level. However, the +20% increase in plant cost had a smaller effect than increasing the number of workers, as discussed below.

Effect of cycle time

Increasing the cycle time from 90 s to 120 s decreased the maximum parts/ yr from 205k to 153k. With plant costs now amortised over a decreased total number of parts, the part cost increases, with additional increases in direct and indirect labour costs per part. If the plant were running at maximum capacity, the tooling costs per part would also increase. Decreasing the cycle time to 60 s increased the maximum annual production rate to 307k parts/yr. Between 120 s and 60 s a cost reduction of 11% (€7.2) occurred, showing that cycle time is a parameter of key importance towards reducing part cost, provided that scrap and reject rates are not affected.

Effect of raw material price

The effect of raw material cost (both sheet material and over-moulding material) on part price was studied by investigating a –10% and +10% variation. A smaller difference was assumed compared with plant cost due to increased confidence in the quotations. As expected for a process where material costs form 63% of the total cost, changing raw material costs had an important effect, with the part price ranging from €65.2 with a –10% material cost to €73.8 at a +10% material cost (a 13% change). Hence, material cost for such textile composite processes is an important issue, notably if higher added-value sheet materials were used.

Effect of reject rate

Reject rate refers to parts at the end of the production line that cannot be sold. If a part becomes rejected at an earlier stage in the process, less value is lost than if rejected at the end of the line. When a part is rejected, it is assumed that new material and additional machine time in all process steps preceding the reject decision are needed to replace it. Rejected part costs are hence compounded through the process steps. Rework of rejected parts is not assumed. Reject parts can be sold as material for recycling, where payment is received, or a cost may be incurred to pay for removal. Here rejected parts are assumed cost neutral in terms of sale or disposal. While high-volume thermoplastic forming and over-moulding processes are relatively stable (when established), other lower-volume textile processes (with reduced plant investment) may have higher reject rates and hence the effect of 20% rejected parts is shown, increasing part cost by 3%. As Fig. 10.15 shows, higher reject rates have an important effect, such that if the +20% plant investment referred to above were to reduce reject rates, then the increased plant costs would be justified.

Recycling strategy

Owing to the high stamped sheet cost percentage, the effect of the percentage sheet that is trimmed from the stamped part before over-moulding is shown. Scrap results when:

- the developed part surface is not rectilinear (the difference between the developed shape and the rectilinear blank is waste, Fig. 10.16a);
- material is held by a blank-holder during the stamping process that is not part of the final component.

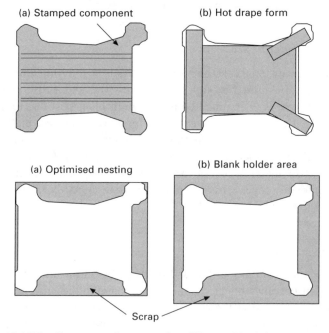

(a) Stamped component   (b) Hot drape form

(a) Optimised nesting   (b) Blank holder area

Scrap

10.16 Textile composite scrap for different blank layouts.

The standard stamped sheet waste allocated here is 30% (Fig. 10.16d). Additional stamped sheet is therefore specified to give that required for the part while leaving an additional 30% scrap. The scrap is directly reused, with additional virgin granules to adjust the fibre fraction, in the over-moulding process. While a closed loop recycling solution is proposed here, the stampable sheet is more expensive than virgin injection moulding pellets.

The first comparison compares: a scenario where the stamping scrap is not recycled but can be essentially eliminated, and a low stamping scrap percentage of 10% (recycled in-line). Stamping waste could be eliminated if the sheet area were inside the part edges (Fig. 10.16b). Pre-consolidated sheet would be heated and then hot draped to the tool geometry using a lower-cost shaping tool, and placed locally in the mould where it is needed. Hence a separate stamp-forming stage would not be used, reducing tool and press costs. A limitation would be that structural sheet would not span the whole component, notably at non-linear edge regions. A 10% stamping scrap scenario would represent optimised 'nesting', here shown as an example with the equivalent rectilinear blank minus the final part shape (Fig. 10.16c). Both the hot-drape forming and 10% scrap scenarios reduced costs by 9% and 5% compared with the standard 30% scrap. The 10% scrap scenario would keep full fabric placement freedom and therefore a choice between the processes will be driven by a combination of part design and achieving the lowest manufactured part price.

To show the effect of closed-loop recycling that is possible with the GF/PP stamped textile sheet, costs were compared for 30% stamping scrap (giving material area for a blank-holder that controls fabric deformation to the mould) both with and without recycling. Equipment costs were marginally reduced without recycling while less over-moulding material was needed where the grinding machine was used. Elimination of in-line recycling increased part cost by 13% (€9.2). In this case, in-line recycling is clearly advantageous.

Effect of operator level

The standard layout allocated two workers to run the line. Without detailed knowledge of the process, gained through experience and process-flow simulations, specification of the exact number of operators is difficult. Hence the operator level was increased to six, with a corresponding cost increase of 3%. This was greater than the effect of increasing the plant cost by 20%. Similar attention should therefore be paid to controlling the operator levels as to optimising plant investment. A larger effect would occur in a dedicated plant running at below the maximum capacity for any given shift pattern.

Effect of tooling cost

The final comparison made is that of tooling cost. If, for example, sliding cores are needed, or a choice of materials/surface treatments (chrome plated *vs.* polished steel *vs.* surface texture) must be made, then tooling costs may vary during the development of a component. Here a −10% to +10% variation gave a cost difference of 3% for a manufacturing volume of 20k, decreasing to a difference of below 0.5% at 200k parts/yr. For lower manufacturing volumes, tooling costs should be more carefully controlled, while for high manufacturing volumes tooling costs are amortised over an increased number of parts and tooling should be optimised for performance.

10.5   Case study 2: composites for the Airbus family

10.5.1 Introduction

The demand for weight saving in aerospace applications, with a lower sensitivity to production rates and material costs, has led to the development of composite processing techniques that can achieve both cost and weight reduction when a system approach is taken. The first composite primary structure to enter production on a commercial aircraft was the A300/A310 CF/epoxy prepreg composite rudder[32]. This replaced its metal counterpart without design changes to the aircraft, reducing 2000 parts (including fasteners) for the metal system to fewer than 100 for the composite system with a 20% weight saving, and an overall cost saving, despite the higher raw material cost. Combined with other design changes, the composite rudder and vertical fin lead to reduced fuel consumption.

Airbus and Fokker have demonstrated cost reductions through using thermoplastic textile composites. For example, giving a total of 1000 kg of textile thermoplastic composites per aircraft, the Airbus A340-500/600 incorporates: engine pylon panels, keel beam ribs and profiles, lower wing access panels, the inboard fixed wing leading edge and aileron ribs. Such textile materials (CF/PPS, polyphenylsulfone), used for example in rib manufacturing (for 16 ribs), have shown similar costs to metallic structures with €95 for aluminium versus €125 for CF/PPS produced by rubber stamp-forming. However, a weight reduction for CF/PPS compared with the metallic structure was valued at an additional cost of €285 for the metallic structure, giving an effective cost saving for the CF/PPS parts. CF/PPS showed a 90% cost reduction compared with prepreg materials and autoclave processing (€245), for the same overall weight[33].

10.5.2 Composite material systems for A380 fuselage panels

The second case study, using data supplied by Airbus[34], focuses on the Airbus A380 that in passenger configuration seats 550–650 people (mixed-class) by using an oval cross-section with a double-deck cabin configuration (full aircraft length). To minimise airport infrastructure modifications, the wingspan and aircraft length were required to fit within an 80×80 m^2 box. Maximum outer dimensions of the fuselage are 7.2 m wide by 8.6 m high. The following candidate carbon fibre reinforced epoxy composite material systems are compared for A380-family fuselage panels aft of the rear pressure bulkhead, as shown in Fig. 10.17[34]:

- hand lay-up of pre-impregnated woven fabrics;

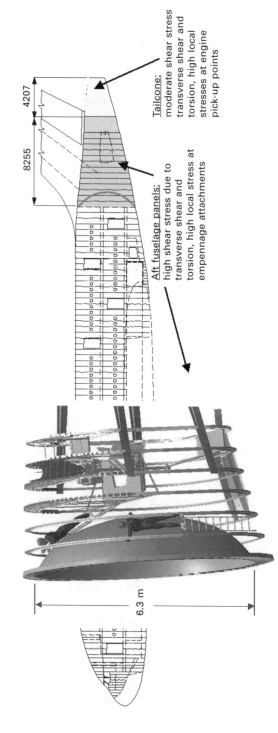

Tailcone:
moderate shear stress
transverse shear and
torsion, high local
stresses at engine
pick-up points

Aft fuselage panels:
high shear stress due to
transverse shear and
torsion, high local stress at
empennage attachments

4207

8255

6.3 m

10.17 Airbus A380 structural design for aft fuselage (adapted from Hinrichsen[34]).

- hand lay-up of dry fabrics (for resin film infusion);
- automated tape laying (ATL) (pre-impregnated tapes);
- automated fibre placement (AFP) (pre-impregnated slit tapes/tows).

Panel size and the number of joints needed for fuselage shells are important parameters that determine both weight and manufacturing costs. The large size of the A380 fuselage requires materials and manufacturing processes for longer and wider panels at twice the average thickness compared with smaller aircraft (A320 and A340). 'Design for maintainability' (repairs after tail-strike events) requires that the panel arrangement and additional frame joints allow exchange of lower fuselage structure elements using spare part kits, independent of material and manufacturing process selection. Another parameter for the selection process is the complex aerodynamic shape of the aft-fuselage, affecting the number of panels and joints. Restrictions of metal stretch-form operations for strongly double-curved geometries limit panel sizes and consequently increase the number of panel joints compared with a composite design solution. The aft-fuselage shape also has a strong impact on manufacturing process selection for composite panels.

10.5.3 Material costs

The four composite material systems were compared for the aft-fuselage panels based upon the material costs associated with the manufacture of 1 kg of flying structure. The material cost consists of the material purchase price and the material overhead costs (which in this study is 5% covering all costs linked to purchase activities and acceptance control on delivery and storage). The material cost is segmented into the allocated material overhead costs, material waste costs and the flying material cost, as shown in Fig. 10.18. The lowest material cost occurred for hand lay-up of dry fabrics. It can be seen that waste is an important fraction of the material costs, which is lowest for AFP. Waste occurs principally during pre-form cutting, resulting in unusable pieces of pre-impregnated woven fabrics, and during final edge trimming of the cured lay-up. The waste assumptions made here reflect the large fuselage panel application in combination with state-of-the-art materials handling and different buy-to-fly (B/F) ratios would apply for different geometries and applications.

10.5.4 Manufacturing process performance

With a target annual production rate of 48 aircraft/year, and with six panels per unpressurised fuselage, the annual panel production rate is 288 per year, representing a lower annual rate compared with the case study in section 10.4. The panel arrangement for a composite solution achieves cost savings

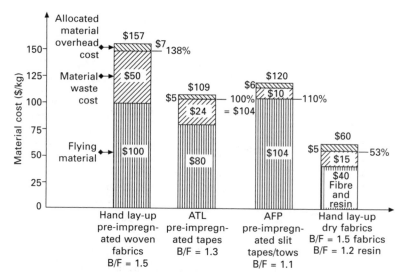

10.18 Airbus A380 aft fuselage material cost comparison (adapted from Hinrichsen[34]).

primarily through a reduction in the number of joints. Simulation of lay-up for the four processes revealed that ATL was not suitable due to the degree of double curvature. The required split into four large panels was feasible for both AFP and hand lay-up. Both hand lay-up processes had adequate access to the female mould for the worker, with deposition rates of 1.6 kg/h, but eventually this would be constrained by the mould size. AFP is less constrained by the size of the male tool and for skin manufacture is suited to 2–4.7 kg/ h rates. However, AFP is also constrained by subcomponent size as the uncured skin is transferred into a female tool for stringer placement prior to the cure cycle, where stringer positioning requires equivalent accessibility as for hand lay-up.

Using the above deposition rates for each process and introducing costs for labour and equipment, scatter-bands for 'process cost' versus 'complexity' were established, as shown in Fig. 10.19. A charge rate of $500/h was assumed to cover the costs related to the AFP-machine at average utilisation, including supervision. For the hand lay-up processes, a charge rate of $80/h was used, including all costs for workforces and the use of shop-floor facilities. All costs are recurring costs, including all work linked to the delivery of uncured *skins* for fuselage panels ready for positioning stringers on the delivered lay-up, which occurs in a further process. Hence, materials and manufacturing costs are defined as recurring cost per kg flying structure.

Figure 10.19 enables the manufacturing process cost to be determined versus degree of complexity, depending on the deposition rate. The triangular symbols indicate average deposition rates based upon the aft-fuselage panel

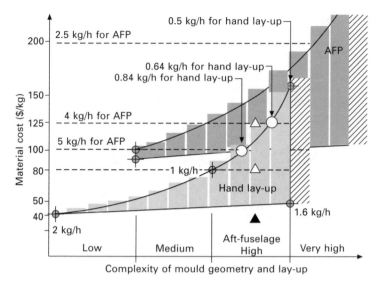

10.19 Process cost comparison for Airbus A380 aft fuselage (adapted from Hinrichsen[34]).

complexity and hence $/kg cost values for hand lay-up and AFP processes. As hand lay-up of dry fibres (e.g. NCFs for the resin film infusion process) requires additional time compared with hand lay-up of pre-impregnated fabrics, a lower kg/h rate is used, indicated by circular symbols. This is due to the difficulties of placing dry fibres with the right accuracy, and the fact that resin film has to be arranged between the fibre layers.

The combined material and process costs, valid for the delivery of uncured skins, are summarised in Table 10.6. Hand lay-up of prepregs and AFP end up at equal cost, whereas the lay-up for the RFI process yields savings in the order of 25%, despite the fact that labour costs are 50% higher compared with hand lay-up of prepregs. It can be seen that, except for the resin film approach, material costs form 50% of the total cost linked to the delivery of laid-up skins. For hand lay-up of woven fabrics, half of these material costs are due to waste, showing the prime driving force for AFP.

Table 10.6 A380 aft-fuselage panel material and process costs

	Hand lay-up, pre-impregnated fibres ($/kg)	AFP ($/kg)	Hand lay-up, dry fibres ($/kg)
Material	157	120	60
Labour	80		120
Machine	–	125	–
Total	237	245	180

10.5.5 Assessment on an aircraft section level

A comparison of panel materials requires comparison over an aircraft section. The conventional all-metal design with Al2524 panel skins, Al7075 stringers and Al7050 milled frames, served as a reference. Composite materials were assumed to replace both aluminium panels and sheet metal frames, but with both metallic and composite panel solutions sharing 36% weight common metal parts.

Composite materials reduced the total weight by 15%, while also decreasing the panel count from 16 aluminium panels to four carbon fibre reinforced polymer (CFRP) panels. This reduced assembly costs from 17% total for the metallic solution to 12% total for composite. Cost per unit mass was estimated by considering the uncured panel cost, the curing costs, finishing costs and assembly costs. For finished (cured) CFRP parts substituting metal parts, this was estimated as $300/kg, the common metal parts as $220/kg, and for the metal parts competing with CFRP, $250/kg. Relative costs and weights are compared in Fig. 10.20 for the two materials. Costs for metal panels were 11% higher than composite panels due to high waste fractions for each of the 16 panels resulting from stretch-forming, the high price of advanced aluminium alloys, and the subsequent assembly of 16 metallic panels with the associated labour-intensive forming and heat-treatment operations.

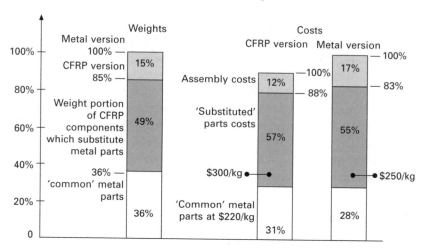

10.20 Airbus A380 aft fuselage cost assessment on an aircraft section level (adapted from Hinrichsen[34]).

Consideration of the system cost hence gives an 11% cost reduction for CFRP using either hand lay-up or AFP, together with a 15% weight saving. The weight saving will reduce fuel usage thereby decreasing aircraft operational costs during the product life cycle, while achieving a lower initial construction cost.

10.6   Conclusions

The TCM approach has been shown to offer a route to assess the cost of textile composite structures to include the effect of fibre type, matrix resin, prepregging route, conversion technique, scrap and waste management, finishing and assembly. Where the system cost is considered, textile composite structures were shown through two case studies to offer cost reductions compared with metallic references. The reduction of scrap in textile composite processing was shown to be of high importance. Tooling cost was also shown to be an important issue, notably where metallic tools are needed for lower production volumes. The strategic consideration of manufacturing plant utilisation or dedication was shown to be a key assumption in cost calculations in situations where the manufacturing supply base is still being established. Materials and processing techniques should be considered together from the outset to suit the desired manufacturing volume and the cost structure of the application market sector.

10.7   Acknowledgements

The authors wish to acknowledge funding from the many anonymous research projects that have enabled the creation of modelling tools and case study data that have been generalised into the material in this chapter. Colleagues at the EPFL and IMD including F. Roduit, C. Menzl, A. Mahler, H. Hermann, Dr P. Sunderland, N. Weibel, N. Mona, K. Becheler and Prof. T. Vollmann are also thanked for their research collaboration. Industrial contacts who have provided information specifically for this chapter include: Dr P. Blanchard (Ford Motor Company), Mr J. Hinrichsen (Airbus), Mr W.V. Dreumel (TenCate Advanced Composites), Mr D. Cripps (SP systems), Mr P. Lucas (Vetrotex International), Mr M. Schrief and Mr. T. Bischoff (Saertex), and Mr I. Toll (Aeroform Ltd). The authors also wish to thank the many anonymous industrialists who have contributed to the cost modelling approach and the data presented in this chapter.

10.8   References

1. Wakeman M.D., Bonjour F., Bourban P.E., Hagstrand P.O. and Månson J.A.E., 'Cost modelling of a novel manufacturing cell for integrated composite processing', *23rd International SAMPE Europe Conference*, Paris, France, 9–11 April 2002.
2. Bernet N., Wakeman M.D., Bourban P.E. and Månson J.A.E., 'An integrated cost and consolidation model for commingled yarn based composites', *Composites Part A*, 2002, 33, 495–506.
3. Wakeman M.D., Sunderland P.W., Weibel N., Vollmann T. and Månson, J.-A.E., 'Cost and implementation assessment illustrated through composites in the automotive industry – part 1: methodology', in preparation, 2005.

4. Weibel N.D., Sunderland P.W., Wakeman M.D., Månson J.A.E., Vollmann T.E. and Bechler K., 'Beyond a cost model: assessing implementation of new materials technologies', *10th International Conference on Management of Technology IAMOT*, Paper 231NW, Lausanne, 2001.

5. Weibel N.D., Sunderland P.W., Wakeman M.D., Vollmann T.E. and Månson J.A.E., 'Beyond a cost model: assessing implementation of new composite technologies', *J. Cost Management*, 2002, 16/3, 21–29.

6. Silverman E.M. and Forbes W.C., 'Cost analysis of thermoplastic composites processing methods for spacecraft structures', *SAMPE J.*, 1990, 26(6), 9–15.

7. Walls K.O. and Crawford R.J., 'The design for manufacture' of continuous fibre-reinforced thermoplastic products in primary aircraft structures', *Compos. Manuf.*, 1995, 6(3–4), 245–254.

8. Bader M.G., 'Materials and process selection for cost-performance effective laminates'. *Proceedings of the Eleventh International Conference on Composite Materials*, Gold Coast, Australia, 1997, 621–629.

9. Gutowski T., Henderson R. and Shipp C., 'Manufacturing costs for advanced composites aerospace parts', *SAMPE J.*, 1991, 27(3), 37–43.

10. Karbhari V.M. and Jones S.K., 'Activity-based costing and management in the composites product realization process', *Int. J. Mater. Prod. Technol.*, 1992, 7(3), 232–244.

11. Mayer C., Hartmann A. and Neitzel M., 'Cost-conscious manufacturing of tailored thermoplastic composite intermediates using a double belt press', *Proceedings of the 18th International SAMPE Europe Conference*, Paris, France, 1997, 339–351.

12. Wang E. and Gutowski T., 'Cost comparison between thermoplastic and thermoset composites', *SAMPE J.*, 1990, 26(6), 19–26.

13. Foley M. and Bernardon E., 'Thermoplastic composite manufacturing cost analysis for the design of cost effective automated systems', *SAMPE J.*, 1990, 26(4), 67–74.

14. Gutowski T., Hoult D., Dillon G., Neoh E.T., Muter S., Kim E. and Tse M., 'Development of a theoretical cost model for advanced composite fabrication', *Compos. Manuf.*, 1994, 5(4), 231–239.

15. Kang P.J., 'A technical and economic analysis of structural composite use in automotive body-in-white applications', MSc Thesis, Department of Materials Science and Engineering, Massachusetts Institute of Technology, 1998.

16. Marti H.G., 'The cost modeling of automotive body-in-white assembly using relational databases', BSc Thesis, Department of Materials Science and Engineering, Massachusetts Institute of Technology, June 1997.

17. Kirchain R.E., 'Modeling methods for complex manufacturing systems: studying the effects of materials substitution on the automobile recycling infrastructure', PhD thesis, Department of Materials Science and Engineering, Massachusetts Institute of Technology, Feb. 1999.

18. Lee D.E. and Hahn H.T., 'Virtual assembly production analysis of composite aircraft structures', *Proceedings of the 15th International Computing Engineering Conference and the 9th ASME Engineering Database Symposium*, Boston, USA, 1995, 867–874.

19. Jones S.C., 'Profit Cue™ (process-fitted cost and quality evaluator)', CompositeTechBrief N° 103, Center for Composite Materials, University of Delaware, Newark, USA, 1997.

20. Li M., Kendall E. and Kumar J., 'A computer system for lifecycle cost estimation and manufacturability assessment of composites', *Proceedings of the Eleventh*

International Conference on Composite Materials, Gold Coast, Australia, 1997, 630–639.

21. Olofsson K. and Edlund A., 'Manufacturing parameter influences on production cost' submitted for publication at the, *International Conference on Advanced Composites, SICOMP Technical Report* 98–009, 1998.

22. Evans J.W., Mehta P.P. and Rose K., 'Manufacturing process flow simulation: an economic analysis tool', *Proceedings of the 30th Int. SAMPE Technical Conference*, San Antonio, USA, 1998, 589–595.

23. Kendall K., Mangin C. and Ortiz E., 'Discrete event simulation and cost analysis for manufacturing optimisation of an automotive LCM component', *Composites Part A*, 1998, 29A(7), 711–720.

24. Clark J.P., Roth R. and Field F.R., 'Techno-economic issues in materials selection', in *ASTM Handbook, vol. 20*, Materials Selection and Design, 1997, 256–265.

25. Rampersad H.K., *Integrated and Simultaneous Design for Robotic Assembly*, John Wiley & Sons, Inc. 1994.

26. Månson J.-A.E., Wakeman M.D., Bernet N., 'Composite processing and manufacturing – an overview', in *Comprehensive Composite Materials*, Ed. Kelly A. and Zweben C., Vol 2, 577–607, Elsevier Science, Oxford, 2000.

27. Johnson C.F. and Rudd C.D., in Kelly A. (ed.), *Comprehensive Composite Materials*, Elsevier, 2000, Cht. 2.32, 1049–1072.

28. Månson J.A.E., Bourban P.E. and Bonjour F., 'Process and equipments for the manufacture of polymer and for composite products', European Patent Office, EP 0825 922 B1, 17 May 1996.

29. Luisier A., Bourban P.E. and Månson J.A.E., 'Time–temperature-transformation diagram for reactive processing of polyamide 12', *J. Appl. Polymer Sci.*, 2001, 81, 963–972.

30. Wakeman M.D., Zingraff L., Kohler M., Bourban P.E. and Månson J.A.E., 'Stamp-forming of carbon fiber/PA12 composite preforms', *Proceedings of the Tenth European Conference on Composite Materials*, 3–7 June 2002, Brugge, Belgium.

31. Zingraff L., Bourban P.E., Wakeman M.D., Kohler M. and Månson J.A.E., 'Reactive processing and forming of polyamide 12 thermoplastic composites', *23rd SAMPE Europe International Conference*, Paris, France, 9–11 April, 2002.

32. Anon, 'A brief look at composite materials in Airbus commercial aircraft', *High Performance Composites*, March/April 1999, 32–36.

33. Mr. Dreumel W.V., 'Personal communication', TenCate Advanced Composites, November 2002.

34. Hinrichsen J., 'A380 – the flagship for the new century', Proceedings of JISSE-7, Japan, 2001.

J L O W E, Tenex Fibres GmbH, Germany

11.1   Introduction

The drive within the aerospace composites field over the last decade has
been to reduce cost, increase component performance and reduce component
weight. Composites have now gained an accepted position in aircraft design,
while carbon fibre reinforced materials have become the mainstay of secondary
components such as wing movables (flaps, spoilers, rudder, etc.) and have
found their way into primary structural components such as complete horizontal
stabiliser and vertical stabiliser structures.

This chapter is solely related to the use of carbon fibre reinforced textile
materials, since the other widely used composite reinforcement fibres, glass
and aramid have gained relatively limited use in the aerospace sector, owing
to weight and lack of stiffness with regards to glass and to the problem of
moisture absorption with respect to aramid fibre. Similarly applications
concentrate on the use of epoxy resin systems, as these dominate the aerospace
composites sector.

At the present time pre-impregnated materials are utilised for nearly all
composite components for aerospace structures. The largest volume of materials
use carbon fibre fabric, though in comparison carbon unidirectional prepreg
is growing at a stronger pace. This trend is, however, beginning to change
with the advent of reliable processing routes with resin transfer moulding
(RTM) or resin film infusion (RFI) and a variety of other processing routes
connected to these basic methods (Fig. 11.1). This trend has come about due
to the textile processing improvements of using high tow count carbon fibres
(i.e. larger 6K carbon fibre tow) for the preparation of textile fabrics and
preforms. Applications for textile composites are usual where part consolidation
is possible and overall sub-assembly construction cost is lowered against
metallic components.

Lifetime costs for components are also a driver for these types of textile
composites where the part cost may be similar or more expensive than metal
components and the consideration of much higher fatigue life and lower in-

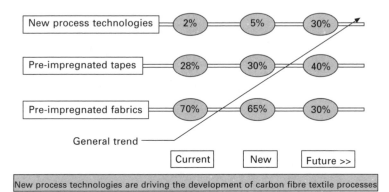

11.1 Expected change in usage of aerospace composites.

service cost is an overriding factor. All the major airframe manufacturers accept that higher component costs may be accepted if lower flying weight and the associated fuel savings can be achieved.

It has to be understood from the general process of manufacturing carbon fibre yarn that the production cost decreases with increasing tow count for the same type of yarn. In general the cost difference between a 7 µm diameter, 240 GPa modulus, 4000 MPa strength yarn can be approximated as follows: 3000 filaments, 3K (100%) > 6000 filaments, 6K (75%) > 12 000 filaments, 12K (50%). Carbon fibre yarn cannot be defined as a low-cost raw material and for this reason the textile development described in this chapter tends to describe material structures which have been developed to move away from the traditional carbon fibre fabric materials, based upon 3K yarns, towards more cost-effective 12K and 24K yarns.

In many applications the high cost of prepreg materials available from only a selected range of suppliers is driving composite textile development. At the present time, however, the lower properties of textile materials compared with prepreg restrict widespread use. The general material trends for the future will include the development of resins and other components to complement the carbon fibre textile properties and improve mechanical properties such as impact, compression and interfacial strengths, since these properties still lie below those attainable with advanced carbon fibre prepregs.

11.2   Developments in woven fabric applications using standard prepreg processing

The main composite material used in the aerospace industry is pre-impregnated carbon fibre fabric. The development of these materials is evolving to accept the challenges of lower cost and improved or comparable material properties. For secondary sandwich structures, owing to the reasons of coverage and thickness (ply drop-offs), the industrial norm in has always been to use a minimum two-ply carbon fabric layer over honeycomb core (to be able to seal the structure). These layers have been traditionally 3K fabric prepreg in ply thicknesses of 0.2 mm (plain weave: 200 g/m^2), 0.25 mm (5-harness satin 285 g/m^2) or 0.4 mm (8-harness satin: 370 g/m^2).

In recent years the trend has been for the industry to search for lower-cost woven fabric solutions, moving from 3K to 6K prepreg fabrics. Owing to improvements in weaving techniques and technology it has been possible for the same fabric quality with the same coverage to be achieved moving from 3K to 6K yarns (Fig. 11.2). Most recently this trend has continued with a movement towards 12K fabric prepregs.

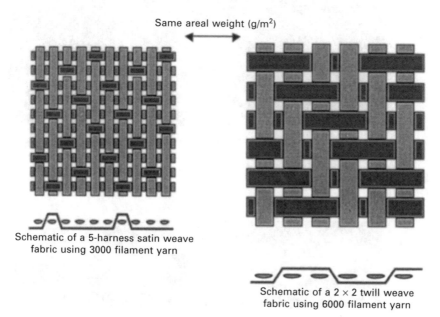

Same areal weight (g/m^2)

Schematic of a 5-harness satin weave
fabric using 3000 filament yarn

Schematic of a 2 × 2 twill weave
fabric using 6000 filament yarn

11.2 Achieving the same fabric weight and thickness with different carbon tow counts.

One of the first examples of has been achieved in Japan by Kawasaki Heavy Industries, where a 380 g/m^2 2 × 2 twill carbon fabric using Tenax HTS 5631 12K yarn using a special weaving technique is able to achieve complete coverage and the same ply thickness as fabrics woven with 3K or 6K carbon fibre yarns (Fig. 11.3). The prepreg developed using this fabric also exhibits self-adhesion to the honeycomb core, thus removing the requirement for an adhesive film to create filet joints between the honeycomb and the prepreg laminate. Several applications have already been found for this prepreg, including the Embraer ERJ 170 Inboard flaps, Embraer

11.3 Prepreg component manufacture using 12K carbon fibre fibric.

ERJ190 Outboard flaps and wing stubs and the Boeing 737-300 winglets (Fig. 10.4).

11.4 Boeing 737–300 winglet programme for Aviation Partners Boeing. Supplied by Kawasaki Heavy Industries (Japan); manufactured using KMS6115 style, Tenax HTS 12K fabric prepreg.

11.3   Carbon fibre multiaxial fabric developments

Multiaxial (non-crimp) fabrics using carbon fibre have been selected as they are seen to offer a textile solution to meet material properties similar to unidirectional prepreg while offering lower cost than established fabric solutions. During the late 1990s multiaxial fabric using carbon fibre yarn was the subject of many different development programmes, including the AMCAPS programmes funded by the UK's Department of Trade and Industry, several elements of the European-funded TANGO programme and the well-documented NASA composite wing programme (started by the Douglas Aircraft Company – now Boeing). These development programmes initiated and drove the improvement of the textile quality of the carbon multiaxial fabrics.

Initially the multiaxial fabrics manufacturing looms (two types are used within the industry, Liba and Karl Mayer-Malamo) were hydraulically driven, which caused the placement of the filaments to be inexact and the general movement of the laying-down heads to be clumsy and damaging to the brittle carbon fibre yarn. As a result, the earliest fabrics manufactured using this technology tended to exhibit poorly placed filaments, gaps and uneven stitching (Fig. 11.5a). Driven by the quality concerns of the aerospace industry, Liba in particular improved its multiaxial looms to use electronic controls and electrically driven motors. This has improved the accuracy of filament placement and has acted to reduce the damage of the carbon fibre yarn by controlling acceleration forces and using improved fibre handling mechanisms (Fig. 11.5b). This generation of machines, known as MAX3 Liba looms (Fig. 11.6), are the basis on which all present carbon fibre fabrics are produced for aerospace application and development programmes.

11.3.1 Fabric development

Nearly all the developments of carbon fibre multiaxial fabrics have concentrated on the use of 12K carbon fibre yarn, owing to the cost implications of using lower tow counts. During this time the major carbon suppliers have also realised that these developments require carbon fibre yarns that are supplied as ribbon-like (flat) tows, able to spread and provide easy coverage once the tows are knitted through with polyester yarn to create the multiaxial fabric. For this reason, carbon fibre yarns such as Tenax HTS 5631 12K or Toray T700 12K carbon have been developed, which are supplied as a flat tow without any false twists than can spread to provide gap-free fabrics.

Many problems still need to be overcome in the development of multiaxial fabrics. The mechanical properties of laminates produced using multiaxial textile fabrics are still lower than equivalent unidirectional prepreg laminates and the variation in properties is higher, dictating that the design properties are lowered. Limitation of low layer weights from multiaxial fabrics reduces design flexibility (due to large ply drop-offs) since at the present time individual layer weights are limited to approximately 200 g/m^2 (from conventional 12K carbon fibre yarn). The higher layer weights for thick constant section components are attractive, since the issue of high fabric thickness and the ability to apply triaxial or quadriaxial fabrics quickly is of great advantage when compared to unidirectional prepreg structures.

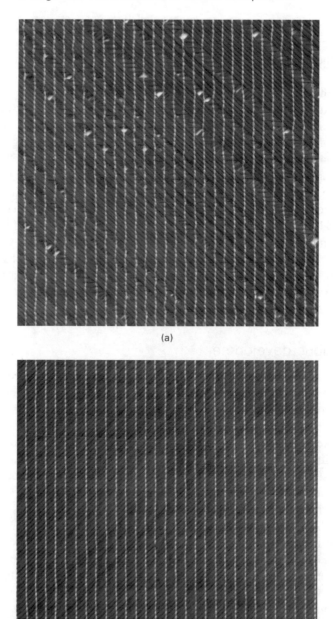

(a)

(b)

11.5 400 g/m^2 biaxial fabrics (±45°) Manufactured using (a) older and (b) the Liba multiaxial technology.

11.6 Liba multiaxial MAX3 machinery.

11.3.2 Stitching thread

An issue of great interest for the future is the development and improvement of stitching thread technology. The polyester stitching used as a processing aid in holding the fabric imparts an inbuilt flaw into the microstructure of the fabric. When the structure of the fabric is viewed closely, the carbon fibre filaments bundled by the stitching thread can be seen. This fibre bundling creates uneven fibre distribution at a microscopic level and is believed to be the cause of lower material properties in certain orientations. As such, the larger the stitching fibre used, the greater the fibre disruption, bundling and 'fish eyes' within the microstructure of the fabric.

Presently, the standard in the industry is to use a non-sized 80 dtex polyester multifilament yarn. Improvements are constantly underway, and recent technology available from Teijin Polyester Filament (Japan) allows the application of an unsized 33 dtex multifilament polyester yarn. The first fabrics being developed with these finer stitching yarns show promising improvements.

11.3.3 Quality control

One of the most problematic issues for multiaxial fabrics is quality control. Many of the manufacturers of multiaxial fabrics have been used to supplying industrial or marine industries, which have much lower-quality requirements. The aerospace industry has the requirement for the highest quality control and product release specifications. The raw material standard, prepreg materials, have tended to require mechanical property testing (e.g. interlaminar strength and tensile strength) from specimens taken from random samples within a production run.

With multiaxial fabrics this type of quality control would be too difficult to organise, owing to the variety of processing routes and resins that may be used. As such, there are developments underway to implement quality detection using computer graphical detection methods. Images of randomly sampled fabric samples could be digitised, analysed and compared against an accepted standard (Fig. 11.7). In this manner it is hoped that by controlling the yarn orientation, coverage, stitching density, etc., the fabric can be released easily for aerospace fabrication without the need for costly mechanical testing.

11.3.4 Lighter-weight fabrics

One of the most restrictive issues with carbon fibre multiaxial fabrics is to be able to produce very light-weight layers (i.e. lower than 150 g/m^2) using lower-cost 12K or 24K filament tows. In order to be able to spread the larger tows to create closed, gap-free unidirectional layers, Liba has modified the multiaxial loom (named MAX5) such that the yarns are spread and laid onto the fibre bed. The spread tows are clamped (Fig. 11.8) and held in place with a newly devised system instead of the usual pins around which tows are usually wrapped.

This type of multiaxial loom can also to pre-spread tows of carbon filament yarn. Using pre-spread tows manufactured either from 12K or 24K yarn, layer weights as low as 50 g/m^2 can be achieved. Widely spread carbon fibre tows are available from a variety of sources, most of these being in Japan. As such the Liba MAX5 is capable of manufacturing quasi-isotropic (quadriaxial) fabrics with areal weights as low as 200 g/m^2 (0.2 mm ply thickness).

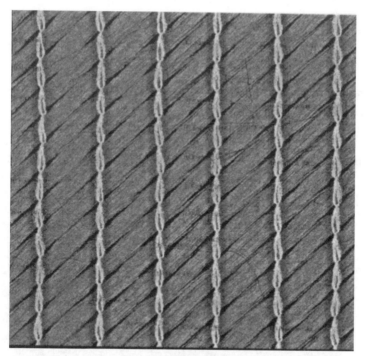

11.7 Digital graphical control of multiaxial fabrics. Digital image shows excessively large 'fish eyes' created by high knitting yarn tension.

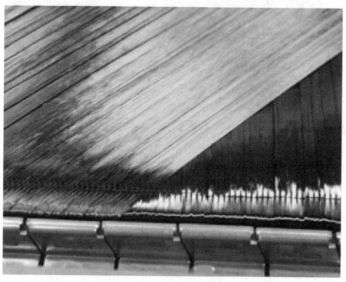

11.8 MAX5 Liba multiaxial technology is able to clamp unidirectional tapes in position before assembling the fabric.

Example 1: *Multiaxial fabric components – using interleaved RFI*

GKN Aerospace has started series manufacture of lower wing access panels for the Airbus A380 (Fig. 11.9) using a newly developed technology that involves interleaving layers of fabric with resin film (M36 epoxy film supplied by Hexcel Composites). The components use a 440 g/m² biaxial carbon fibre (Tenax HTS 5631 12K) multiaxial fabric (±45°) manufactured by Saint-Gobain (UK), stiffened by a honeycomb core. The manufacturing method requires no autoclave, merely vacuum and heat for the fabrication of these components. Also, owing to the viscosity of the resin film, no adhesive film is required to bond the laminate to the honeycomb. Because of the lower cost materials, reduction in lay-up times and the need for lower cost processing tools (i.e. no autoclave), the component manufacturing cost is estimated to be 30% less than a conventional fabric prepreg manufacturing route.

11.9 Airbus A380 lower wing access panels manufactured by RFI using biaxial (±45°) multiaxial HTS 5631 12K fabric supplied by Saint-Gobain (UK).

Example 2: *Multiaxial fabric components – RFI, using resin film blocks*

The pressure bulkhead developed for the Airbus A380 has been devised using carbon fibre (Tenax HTA 5131 12K) biaxial (0/90°) multiaxial fabrics manufactured by Saertex GmbH and is fabricated using resin film infusion with 977-2 resin film supplied by Cytec Engineered Materials (Fig. 11.10a). The bulkhead uses a fabric preform supplied by Saertex, which is assembled and stitched to form an assembly. This preform assembly is rolled onto a core unit, that is supplied to Airbus for component fabrication (Fig. 11.10b). The preform is rolled out onto a convex form that is pre-covered with the resin film. This assembly is vacuum bagged in the conventional manner and cured in an autoclave.

(a)

(b)

11.10 Rear pressure bulkhead or the Airbus A380. (a) Assembled multiaxial dry fabric preform and (b) finished component.

11.4   Improvement in standard fabric technology for non-prepreg processing applications

11.4.1   Unidirectional fabrics

Unidirectional (UD) fabrics are commonly used for adding stiffness in structures such as wing skins, spars or stiffeners (gliders, small general aircraft). UD fabrics also allow the lay-up of complex laminates in a similar fashion to UD prepreg tape structures. UD fabrics manufactured using Liba multiaxial looms tend to allow the filaments in the 0° direction to wander. Hence the resulting mechanical properties have been low in comparison with UD prepreg. Other UD materials have also been evaluated, such as dry UD tape held together with a fibre mesh backing, which is usually coated with an epoxy resin or a thermoplastic binder. The main problems with these types of UD tapes is that they have very poor impregnation properties, especially through-thickness.

To solve these issues and to be able to offer a UD fabric structure that exhibits repeatable quality, easy impregnation without excessive pressure and high mechanical properties, a new generation of fabrics has been developed using 12K carbon fibre yarns. In particular, advanced UD weaves using glass or thermoplastic (polyester) fibres in the weft direction to hold the carbon fibre yarns in place can be used as tape laid like products in similar manner to UD prepreg (Fig. 11.11). The scaffolding structure of the support yarns allows for good permeability of the fabrics for vacuum infusion processes

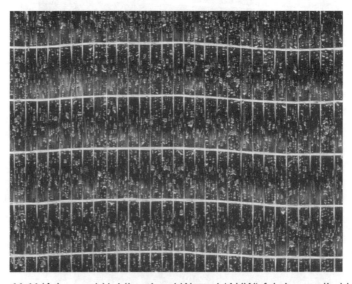

11.11 'Advanced Unidirectional Weave' (AUW) fabric supplied by Cramer (Germany) using Tenax HTS 5631 12K carbon fibre yarn. Epoxy powder binder is applied onto the fabric surface.

and the fact that the carbon filaments are not crimped allows the fabrics to exhibit high mechanical properties.

Since these types of fabric are manufactured on a conventional weaving loom, issues such as quality control and repeatability are well understood and accepted by the aerospace community. There are numerous development programmes presently underway to use the opportunities offered by these new types of materials.

11.4.2 Non-crimped biaxial fabrics

The type of non-crimp woven fabric described above is also available in biaxial form. As such there are development programmes and projects underway to qualify and use these materials in new component forms.

11.5 Braided materials

Braided carbon structures have been used for many years in aerospace applications. The best known applications are propeller blades (see Fig. 11.12) by Dowty-UK and Ratier-Figac-F. In recent years, braided materials have also been used in shear web stiffeners for applications such as landing flaps for the Bombadier-Shorts CRJ, in development programmes such as TANGO or for reinforcement of support struts for Airbus components

11.12 Over-braiding of a propeller fabric preform using Tenax HTA 5131 12K carbon fibre yarn.

(manufactured by SARMA, Lyon). In the USA the company Fibre Innovations manufactures complete cruise missile casing preforms in a complex braiding procedure.

Bombardier Aerospace has developed a process to manufacture outboard wing flaps for the Canadair regional jet (Fig. 11.13). The components replace metal outboard wing flaps, and are manufactured in a single injection process using HTA 5131 6K carbon fibre fabric in the skins, stiffened by braided ±45° socks folded to create shear webs.

11.13 Outboard wing flaps manufactured by RTM using Hexcel RTM6, Hexcel Injectex Fabric, braid from 'A&P' using Tenax HTA 5131 6K carbon fibre.

11.6   Tailored fibre placement

Since 1995 there have been several well-documented development programmes to use textile embroidering technology to place carbon fibre tows, then stitch the tows into position using fine polyester monofilaments to produce complex preform structures (Fig. 11.14). The most successful of these developments has enabled the company Hightex (Germany) to offer complex preforms for specialist applications. Hightex has used this technology to construct a complex preform structure for Eurocopter. The first components to be selected are hoop-fuselage frames for the NH90 troop transport helicopter.

For these frames the preform is built up from a backing structure consisting of specially woven band fabrics onto which the tow is placed. The components are manufactured using the RTM process in a two-part matched metal tool.

11.14 Tailored fibre placement (TFP) uses a single carbon fibre yarn which is stitched in place to create a complex preform. Several placement heads are able to run in parallel.

Even though the preform manufacturing process is expensive, the raw material cost is low and the waste material is estimated to be less than 5%. It is understood that the metal component that this composite piece replaced was 30% heavier and was more expensive to manufacture.

11.7   Preforming

11.7.1 Textile assembly

The company FACC (Ried Austria) has used new material and preforming developments from Cytec Engineered Materials, to develop a demonstration composite replacement part for a highly stressed aluminium spoiler centre fitting on the Airbus A340-400 (Fig. 11.15). Owing to the complexity of the component RTM was chosen. The high fibre volume content and the complexity of the preform meant that a low viscosity resin which could quickly and thoroughly impregnate the dry fibre preform was required.

However, for the high compression strengths necessary in this application, toughened epoxy resins containing high levels of thermoplastic additives are required. The thermoplastic additives increase the resin viscosity dramatically. This increased viscosity means that it is very difficult to permeate a preform

11.15 Spoiler centre fitting manufactured by RTM using a complex perform produced using Cytec Engineered Materials Priform Technology. Moulding by Fischer Advanced Composite Components.

without high injection pressures and temperatures. The 'Priform' system developed by Cytec Engineered Materials is said to overcome this problem by adding the thermoplastic toughener into the preform in the form of fibres, rather than including them in the resin system. The resin, a development of a widely used and aerospace-qualified toughened epoxy resin (977-2) is known as 977-20 and is a single pot system including hardener and accelerator agents.

The polymer toughening agent is spun into continuous fibres and is co-woven and added as a 3D stitching fibre into the fabric preform. The thermoplastic fibres are said to dissolve into the resin during cure at temperatures above 100 °C. The preform uses a standard aerospace type 370 g/m^2 HTA 5131 6K carbon fibre fabric. The component is manufactured using a matched metal RTM mould. Injection temperatures are initially low to avoid the thermoplastic component dissolving too soon, which would create variations in the thermoplastic contents throughout the moulding. This component, which is now due to be applied into full series production, weighs 30% less than the original metal component and has passed all required certification testing for aerospace application.

11.7.2 Preforming using heat-activated binder-coated fabrics

Thermoforming is the most well-known and accepted preforming assembly method either with the use of heated vacuum forming (Fig. 11.16) or heated press forming. In either case powder binders are applied to the fabrics to bond the layers together to produce a rigid preform. These binders are usually epoxy or thermoplastic based, and are compatible with the basic resin systems so that the binder does not adversely affect mechanical properties. Several

11.16 Heated vacuum performing facility used to form textiles coated with heat activated adhesive binders.

different binders are used, although each is coupled with its own resin system from specialised resin manufacturers (e.g. RTM6 from Hexcel Composites).

11.7.3 Preform assembly using stitching yarns

In very large preform structures such as the A380 pressure bulkhead, in order to be able to confection fabric textiles or to add extra localised reinforcement in certain highly loaded areas, specialised industrialised stitching/sewing machines have been developed and are used for a variety of aerospace preforming applications.

The choice of stitching thread materials are limited to a small selection of fibre types:

- Low tex carbon fibre (67 tex or 200 tex) yarn has been used for a variety of stitching applications. The major limitation with direct carbon fibre tow use is that the fibre has little knot strength and as such tends to filamentise when looped. Airbus is known to use direct carbon fibre tow to add reinforced stitching to preform stringers manufactured using RTM.
- Carbon stitching thread was supplied to the market up to the late 1990s by Toray, but is no longer available. However, new developments from the carbon fibre manufacturers could introduce a new carbon fibre sewing yarn into the aerospace market.
- Assembled carbon thread (stretch-broken assembled filaments from 12K/24K yarn). Thin assembled yarns are available for suppliers such as Schappe (France) who supply carbon yarns as low as 50 tex which can be used for stitching.
- Aramid threads or yarns are easily processed and give excellent out of

axis strength. Drawbacks include problems with ultrasonic inspection where the areas which have been stitched cannot be analysed since the signal range is too high. The second and most important factor for the lack of acceptance of aramid yarn for localised structural reinforcement is moisture absorption.

• Polyester yarn is used as an assembly aid. No strength is gained from the material. Saertex uses polyester yarn as a processing aid in the A380 preform (Fig. 11.17).

(a)

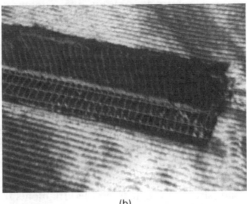

(b)

11.17 Preform assembly using stitching technology. (a) Stitching station used to assemble several layers of multiaxial fabric (Saertex, Germany); (b) stitched assembly of multiaxial fabric stringers onto a fabric skin.

11.8   Repair of fabric components

Repairing aerospace composite components has always been a hotly discussed topic. Through the mid-1990s to 2003 a programme involving all the major airframe manufacturers (Airbus, Boeing, Embraer, etc.) and many of the large aircraft users (British Airways, Air France, Fed Ex, United, American, etc.) devised design methodologies, repair techniques and qualification of material systems using a variety of repair methods. The system chosen was a plain weave fabric using 3K plain weave HTA 5131 carbon fibre and a two-part epoxy resin system Epocast 52A/B supplied by Huntsman.

The repair material is a 'wet lay-up' repair system that can be applied to a prepared damaged area using two laminating techniques, either the 'squeeze-out' or 'vertical-bleed' methods. Both methods use prepared layers of 'wet' resin impregnated patches to repair airframe structures, both for monolithic or sandwich laminates. The completed repairs are cured under vacuum conditions at 80 °C.

The repair system is now qualified to a universal specification and the certified data are available to all interested parties. As such there is now a material and laminating system that can be used in Aircraft repair stations worldwide. This fabric repair material can be applied to repair all epoxy-based aerospace components whether they are manufactured using prepreg or other material processing technologies.

12

Applications of textile composites in the construction industry

J C H I L T O N, University of Lincoln, UK and
R V E L A S C O, University of Nottingham, UK

12.1   Introduction

The use of textile composites offers attractive possibilities for the construction industry. Some potential advantages are the reduction of weight of construction, the (mass) production of complex form components, possible overall cost reductions thanks to industrialised off-site manufacturing processes, reduction in construction time, and production of multifunctional components. In addition, from an architectural point of view, textile composites offer variety of appearances: translucency, colour, surface texture, finish quality, etc.

Textile composites have been used in construction since the 1960s, and though they do not hold a prominent place compared with that of traditional construction materials, their use is on the increase. It is possible to categorise two different groups of textile composites used in construction: fibre reinforced polymers (FRP) (most commonly glass reinforced polyester (GRP), but also glass, carbon or aramid reinforced epoxides) and membrane structures (most commonly poly(vinyl chloride) (PVC)-coated polyester fabric and poly(tetrafluoro ethene) (PTFE)-coated (Teflon®) glass fabric but with other combinations such as silicone-coated glass, PVC-, PTFE- or silicone-coated aramid also being used).

12.2   Fibre reinforced polymers

The introduction and understanding of new materials in an industry sector is not a simple procedure. This is particularly true for the construction industry owing to its large segmentation. Nonetheless, there have been a considerable number of studies on applications and design guidance for FRP in construction. Even such early studies as those by Hollaway [1, 2] and Leggatt [3] are still very relevant today; it is normally suggested that architects or civil engineers not familiar with these materials should be aware of FRP composites having specific anisotropic properties, which means that existing design considerations may not be relevant. In addition, their mouldability and relatively low stiffness imply the use of structural surfaces and multifunctional elements. On one hand, the anisotropic nature of FRP, and the particularities of the manufacturing processes by which individual elements are produced, seem (on account of market constraints and building regulations) to call for levels of product development and industrialisation at which the non-standard building component cannot compete. On the other hand, their mouldability and the necessity (on account of relatively low stiffness) to adopt geometrical configurations of structural performance, confer a strong design character that makes it difficult for individual components to attain the flexibility that most traditional materials may offer.

Although these two main characteristics may appear contradictory from the traditional perspective of buildings as 'one-off' creations, there are plenty of opportunities for integration by adopting new approaches to the nature of the whole procurement process. One possibility would be to integrate adaptable design to production processes at early stages (e.g. mass customisation), thus avoiding the role of the architect as we know it. Perhaps these possibilities are the key to widespread use of FRP.

Even though FRP composites still do not hold a prominent position in building, they have attracted the attention of architects and engineers for many years. There was an initial rise in their application in construction during the 1960s and 1970s, a period during which a considerable number of buildings using FRP appeared. The reasons for the later reduction in their use are various, including the oil crisis (which importantly increased the price of resins), the low quality control exhibited by some manufacturers (which gave GRP a bad image) and the unsuccessful approach with which in early times the construction industry in general managed standardisation and prefabrication processes. In addition, because of the low structural requirements of the average component, textile FRP composites have only occupied a marginal place in the construction industry. The majority of FRP used have been chopped strand mat combined with glass reinforced polyester, using wet lay-up processes for (typically) cladding purposes. However, following advances in material technologies and owing to the economic need for industrialised production, new processes and applications have been found for composites in construction relying much more on the use of textiles as reinforcement, thus opening a completely new sphere for their use. Polymer composites are now employed to fulfil significant structural tasks such as concrete strengthening, concrete retrofitting and the production of entire lightweight structures such as pedestrian bridges.

Manufacturing processes such as pultrusion and pre-impregnation as well as the use of carbon and aramid fibres have gained significant importance in construction during the last decade. This has been especially the case in the area of infrastructure, where composites offer important advantages regarding durability (corrosion resistance) and ease of installation (light weight). These are crucial qualities at present, as a large proportion of the reinforced concrete infrastructure in the developed world is reaching the end of its life span. Examples of the various bridge decking systems using pultrusion processes are ACCS (Advanced Composite Construction System) from Maunsell Structural Plastics, ASSET (Advanced Structural System for Tomorrow's infrastructure) developed by a European consortium including Fiberline Composites and Mouchel Parkman, and Duraspan® developed by Martin Marietta Composites. An example of the use of pre-impregnated processes for strengthening and retrofitting is provided below as a case study. Another interesting field is the use of 3D textiles; a good example is Parabeam®, which uses velvet weaving techniques with E-glass, to produce an integral sandwich panel.

At the moment, the construction industry represents an important fraction of the total tonnage of composites being consumed, 11% being the figure for the UK [4] and 16.3% for the EU [5]. In both cases the construction industry represents the third major market for composites, but unlike other sectors where composites are used, in the construction industry plastics in general do not account for even 1% of the materials used [6]. Taking into account the size of the construction industry and the various current advantages of composites for building purposes, the above illustrates the enormous potential for a greater use of polymer composites in the construction industry. Accordingly, an increasing number of academic–industry projects and networks, such as Network Group for Composites in Construction (NGCC) and Polymer Composites as Construction Materials, on the use of composites in construction are being developed in the UK and abroad. The future of FRP composites in construction looks very promising.

12.3   Membrane structures

Coated textile membranes form a class of flexible textile composites that are used for small- to medium-span enclosures (usually roofs and air-halls), where their properties of lightness, translucency, high tensile strength, durability and fire resistance may be exploited. Here the textile is the primary load-carrying component of the composite while the flexible coating protects the textile from environmental degradation (e.g. the effects of UV light, moisture and pollutants) and creates a weather-tight enclosure.

Of the most commonly used yarn materials, polyester exhibits good tensile strength, flexibility and significant elongation before yield, while glass fibre shows brittle behaviour and low elastic strain [7]. Thus polyester fabric-based membranes have greater tolerance to dimensional errors in manufacturing, as they may be stretched more easily to fit on site, and they may be used in deployable membrane structures. Conversely, glass fabric-based membranes require more accurate manufacture and can be more easily damaged when folded.

The base fabric is generally woven in either basket or Panama weave, with the latter providing better mechanical properties [8]. One or more coatings are then applied, the most common being PVC for polyester fabric (usually with a top coat of acrylic lacquer, weldable or non-weldable poly (vinylidene fluoride) (PVDF) lacquer, or poly (vinyl fluoride) (PVF) film [9]) and PTFE for glass fabric (with, for example, a fluorinated ethylene propylene (FEP) topcoat to enhance impermeability, fungal resistance and weldability [10]). Other coatings are available, such as silicone, which is used primarily on glass fabric, giving a more flexible membrane than the more common PTFE-coated glass and offering greater translucency. Their perceived disadvantage has been a tendency to attract dirt more readily than the 'self-cleaning' PTFE material, although new surface treatments have overcome this, prompting renewed interest in their application.

Unlike the majority of conventional building materials, textile membranes display anisotropic, non-linear stress–strain behaviour, have negligible resistance to bending and low shear resistance. The structures for which they are used are geometrically non-linear under load. This requires a quite different approach to their design. To obtain some understanding of this difference, one might imagine the behaviour of a straight thread spanning between two supports. This is unable to resist loads applied to it (except loads applied in the direction of the thread) unless it deflects from its original straight profile[1] (i.e. the geometry changes). The profile of the deflected thread will also vary according to the applied load configuration. Equally, a textile membrane surface cannot remain perfectly flat if it is to resist loads applied out of the plane of the surface and will change its form according to the loading.

Snow (or rainwater) and wind are the most common external applied loadings in the case of textile membrane roofs. The former acts vertically downwards under gravity while the latter acts perpendicular to the membrane surface and can be either pressure or suction, depending on the wind direction and orientation of the surface. Both types of load are of much higher magnitude than the membrane self-weight, which is typically less than 1.6 kg/m^2. Thus to better resist the applied loads, textile membrane structures usually have double curvature and to reduce deformation under applied load the surfaces are prestressed. The curvature is anticlastic (Fig. 12.1) in the case of membranes tensioned externally (e.g. by masts, arches and edge cables) or synclastic, in the case of inflated and air-supported membranes, which are pretensioned by internal air pressure (Fig. 12.2). Hence, before carrying out an analysis to

---

[1] In theory, for a straight horizontal thread to resist vertical load (even its own weight), without deflecting from the original straight profile, would require an infinite tension to exist in the thread. The thread actually hangs in the shape of a catenary under its self-weight.

12.1 Simple anticlastic form of a small canopy, Millennium Garden, University of Nottingham, UK. The form has two high and two low points creating a classic saddle surface (photo: John Chilton).

12.2 Synclastic form of the air-inflated membrane tubes for the deployable roof of the Toyota Stadium, Toyota City, Japan (photo: John Chilton).

determine the stresses in the surface, the engineer must first determine the form of the membrane for the specified support (boundary) conditions. This process is known as 'form-finding' and is carried out using specialised software. Only once the membrane form is established can the stresses in the surface and forces in supporting elements be assessed.

A further peculiarity occurs because coated textile membrane material is produced in rolls of a specific width and length, the precise dimensions depending on the manufacturer. Therefore, as in the tailoring of garments, the cloth must be patterned and cut in such a way that individual planar pieces, when assembled, approximate to a double-curved surface. In the case of membrane roof structures this surface may be of considerable size. The pattern is normally arranged so that the warp and weft directions of the fabric follow the principal stress trajectories. Because the material is anisotropic and is also to be prestressed, it is essential that the correct allowance or 'compensation' is made in the cutting pattern for the stretch of the fabric under tension. Biaxial tensile tests are usually carried out to determine appropriate compensation factors for the material. An appropriate allowance must also be made for overlap at the seams. Therefore, all pieces are cut slightly smaller than established from the designed form and are then stretched to the required shape and size during erection, known as 'stressing-out'. This ensures that the form of the structure, as erected, is as close as possible to that originally designed and should avoid unsightly wrinkles. Owing to the translucency of the fabric, seam lines are a very prominent feature and the final pattern is often determined by both engineering and architectural considerations.

Extensive details of membrane material manufacturers, fabricators, designers and researchers and a wide-ranging database of projects, engineers and architects are provided by TensiNet [11], the EU-funded network for tensile membrane structures.

12.4   Case studies

12.4.1 Boots building, Nottingham (FRP strengthening)

The Boots Building in High Street, Nottingham (Engineer: Taylor Woodrow, Manufacturer: Advanced Composites Group; Fig. 12.3) is a historical building (Grade II listed) whose steel structure dating from 1921 had lost 30% of its section due to corrosion. For its recent refurbishment, the structure had to be able to carry loads superior to those specified by the original design. A particularly important problem was the flexural strengthening of the curved corner beams, for which (in two cases) FRP composites were used.

The solution chosen was a glass and high-modulus carbon fibre–epoxy low-temperature curing pre-impregnated system, maintaining fibre orientations at 0° and ±45° to the directionality of the beams. After installation (Fig. 12.4), the beams were vacuum bagged (Fig. 12.5) and heated to achieve a fibre volume fraction of 65%. The vacuum method also ensured a resin void content below 4%. The advantages of the solution include short installation time, no disruption to the refurbishment process, and minimal additional

12.3 Elevation view of the Boots Building, showing the curved corner section (photo: Advanced Composites Group Ltd).

12.4 Placement of a carbon prepreg layer around the flanges and web of a steel beam (photo: Advanced Composites Group Ltd).

12.5 A vacuum bag used to apply pressure on the composite material (photo: Advanced Composites Group Ltd).

weight or volume to the structure (about 5 mm in thickness). The latter allowed the position of the terracotta cladding to be maintained, thus complying with requirements from heritage authorities.

12.4.2 Inland Revenue amenity building, Nottingham (membrane structure)

Opened in 1995, the Inland Revenue complex (Architect: Michael Hopkins and Partners, Engineer: Ove Arup & Partners), in Nottingham, UK, has at its heart an amenity building (Fig. 12.6) which includes an indoor sports hall, covered by a single-layer, PTFE-coated glass fabric membrane. This is bounded on two sides by two-storey bar/restaurant facilities and by fully glazed façades to the north and south, partially shaded by the membrane. The sections of membrane are supported by lenticular windows suspended from four main masts and tensioned between the rigid edge frames of the windows and cables linking peripheral masts, (Figs 12.7 and 12.8).

12.5   Future developments

The built environment as an important part of and medium for cultural expression is now facing challenges of complexity to a hitherto unparalleled level. Digital computer-aided design (CAD) technologies, which allow for

12.6 Membrane roof at the Inland Revenue amenity building, Nottingham, UK (photo: John Chilton).

12.7 Stressing out the membrane at the Inland Revenue amenity building, Nottingham (photo: Alistair Gardner).

the development of 3D intricate configurations, along with the possibility of their direct linking to manufacturing processes (CAD/CAM), are now becoming standard in various industries. The implications of vast information availability, fast project development speeds and new techniques as mentioned, are already of importance to the way we think, and thus design and build. In this context, textile composites offer extensive opportunities due to high flexibility of physical configurations, and to the industrial nature of the processes involved in their production. One could easily imagine mass-customisation processes being employed by using for instance membranes, pre-impregnated composites

12.8 Cable and edge details, Inland Revenue amenity building, Nottingham (photo: Alistair Gardner).

or any other structural textile being automatically cut and sewn to form 3D structures that would comply with specific requirements. The textile reinforcements per se may have been tailored using different densities, directionalities and fibre types, to suit the individual structural and visual requirements for each part of the component they would form.

In the case of FRP composites particularly, there is a growing concern for the environmental impact of their production and end of life. There is significant interest in the development of biodegradable fibres and resins, very often based on the utilisation of natural substances. Recyclability issues have focused attention on the use of thermoplastic matrices in construction. However, taking a more realistic view of the immediate future, and understanding the issue of sustainability from a broader perspective, it is certain that present applications of FRP composites will continue to spread. Thanks to their efficiency, ease of transportation and assembly, they still will have much to contribute towards global economic growth and environmental welfare. The near future of construction is also likely to be marked by the emergence of new ways of using FRP composites, probably in combination with traditional materials (e.g. masonry, concrete, timber, steel and glass) to form hybrid structures. Furthermore, following an initial period of technical investigation and validation, a more liberal and creative use of the materials can be expected, where aesthetic values take foremost importance. An example of this is the F100 street station (Fig. 12.9), a glass–FRP hybrid structure being developed

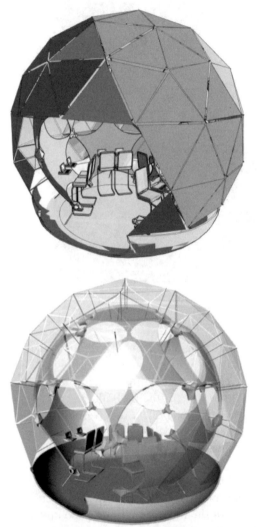

12.9 F100 street station (design and 3D image: Rodrigo Velasco).

within the Architectural Construction Research Group (ACrg) at the University of Nottingham.

Tensile fabric membranes have, until recently, have been used primarily for canopies and/or temporary enclosures where environmental performance is not a high priority. However, with the increased use of textile membranes for permanent or semi-permanent enclosures, for example the Inland Revenue amenity building described above and the Millennium Dome in London, there is a move to improve, in particular, their thermal and acoustic properties. To achieve this, one normally needs to provide at least a two-layer system with or without an intermediate thermal and/or acoustic insulation layer. However, this tends to adversely affect the desirable property of translucency. Consequently, textile membranes are currently being developed with additional coatings designed to improve thermal performance, such as low-emissivity (low-e) coatings and surface treatments that selectively reflect certain wavelengths of solar radiation. Some of these materials are being used in the New Bangkok Airport Terminal, currently under construction [12].

12.6   References

1. Hollaway L., (ed), *The use of Plastics for Load Bearing and Infill Panels*, Manning Rapley, Croydon, 1974.
2. Hollaway L., *Glass Reinforced Plastics in Construction: Engineering Aspects*, Surrey University Press, Guildford, 1978.
3. Leggatt A.J., *GRP and Buildings*, Butterworths, London, 1984.
4. Sims G., Bishop G., *UK Polymer Composites Sector: Foresight Study and Competitive Analysis*, NPL Materials Centre/NetComposites, 2001 (www.netcomposites.com)
5. Starr T., *Composites, A Profile of the Worldwide Reinforced Plastics Industry*, Elsevier Advanced Technology, Oxford, 1995.
6. Dufton P. and Watson M., *Materials in Use and Developments in Markets*, Composites and Plastics in Construction, BRE, Watford, 1999.
7. Houtman R., 'There is no material like membrane material', in *Designing Tensile Architecture*, eds Mollaert M. *et al*, Proceedings of TensiNet Symposium, Brussels, 2003, p 182.
8. Ibid, p 183.
9. http://www.architecturalfabrics.com (last accessed 28 December 2003).
10. http://www.4taconic.com Taconic Architectural Fabrics Division (last accessed 28 December 2003).
11. http://www.tensinet.com TensiNet (last accessed 28 December 2003).
12. Schuler M. and Holst S., 'Innovative energy concepts for the New Bangkok Airport', in *Designing Tensile Architecture*, eds Mollaert M. *et al*, Proceedings of TensiNet Symposium, Brussels, 2003, pp 150–167.

Textile reinforced composites in medicine

J G ELLIS, Ellis Developments Limited, UK

13.1 Splinting material

Splinting materials for the repair of broken bones are not only the largest medical market for textile reinforced composites, but the oldest. The use of plaster of Paris was certainly known by the 10th century as a support for fractures and other bone injuries of limbs. Because of the brittle nature of plaster of Paris, it is probable that it was reinforced with textile materials from a very early stage. However, the modern plaster of Paris bandage, still very extensively used, was not patented until 1851 by Antonius Mathysen, a Dutch military surgeon.

Many orthopaedic surgeons strongly believe that plaster gives a much better moulding to the limbs than other products. Plaster of Paris casts (Fig. 13.1), however, tend to have a high failure rate and have to be reapplied or repaired. Patient comfort is as important as having a light-weight, strong

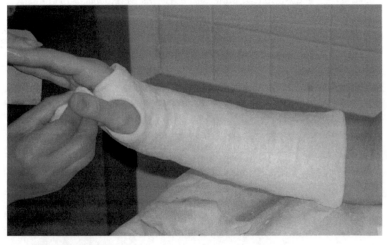

13.1 Plaster of Paris cast.

cast; not only does plaster of Paris fail in this regard but is also sensitive to water, which means that the patient cannot conveniently shower, and bathing may be almost impossible.

The requirement for a splinting material is a high degree of stiffness combined with conformability. This fabric reinforcement must be capable of being moulded around the awkward contours of limbs. The fabric is required to be of open construction to allow impregnation with a large quantity of plaster. These requirements can conveniently be met by a leno woven construction where the warp yarns cross over one another, locking the weft into place (Fig. 13.2). Such a fabric is also less inclined to fray when cut.

13.2 Woven construction of splinting material.

Similar characteristics can also be obtained with various warp knitted and some weft knitted constructions. A number of fibres have been used in conjunction with plaster, particularly cotton, although modern bandages on the market contain combinations of elastomeric yarn and polypropylene, polypropylene and polyester, or even glass. Glass has fallen into disfavour particularly among casting technicians, mainly because of the glass contained in the dust produced during cast removal. Other casting materials also include polyurethane-coated fibreglass and 100% polyester and polypropylene reinforcing fabrics.

In most modern casting products, the product is delivered to the plaster technician as a packaged product that requires dipping in warm water. This will then harden after wrapping around the limb of the patient within 3 or 4 min (Fig. 13.3). A cast is formed that is sufficiently strong, even capable of taking the patient's walking weight, within 20 min. Not only is this external support splinting system commercially valuable in its own right, but it has led to the technique of 'bandaging' bridge support pillars with textile reinforced materials when they are assessed as being too weak for anticipated loads.

13.3 Wetting the casting product.

13.2   Walking support frame

Walking support frames provide vertical support for patients who have dysfunctional lower limbs. They provide a rigid support to assist in both standing and walking. Conventionally, metallic frames have been used, but because of the comparatively low stiffness of metallic materials the weight of metal needed frequently inhibits use of the frame by the patient, who typically suffers from cerebral palsy. The reduction in both weight and volume of a carbon fibre reinforced orthosis has led to marked improvements in the continuity of walking as well as an obvious enhancement to the quality of life of the user.

Because of the greater stiffness and hence rigidity of the frame and its increased strength, larger patients are able to use composite orthoses more conveniently and continue to experience improved walking capability. Although carbon-based materials have brittle failure characteristics, the mode of failure of such devices is of progressive collapse and is therefore inherently safe. The use of composites in smaller orthoses, such as knee braces, is limited, because metallic knee braces are required to be available in only a limited range of sizes to reduce stock holdings. They are not customised to the same degree and metal can readily be bent into shape to fit the patient. Such a facility is not normally available with thermoset resin-based materials.

13.3   Bone plates

In orthopaedic surgery, bone plates and screws are often used to treat fractures, particularly of the long bones. Most bone plates are made of stainless steel, cobalt chromium and titanium alloys (Fig. 13.4). However, the elastic modulus of these metals is much higher than the elastic modulus of human bone (110–220 GPa in comparison with 17–24 GPa for human bone). Because there is a stiffness mismatch, the normal healing process of the bone tends to modify itself, changing the whole cycle of the formation of a callus around the broken section, the conversion of the callus to bone (in the process known as ossification) followed by the union of the broken bones. Therefore, development has moved on to consider the use of thermoplastic polymer-based composites and further to thermoplastic polymer composites for bone plate applications.

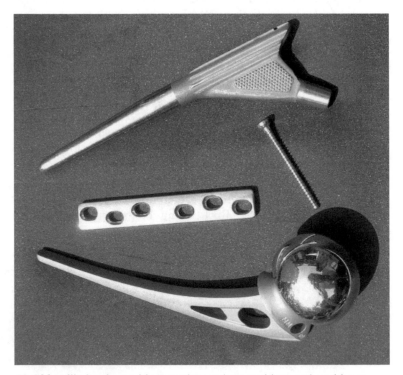

13.4 Metallic implants: hip prostheses (top and bottom) and bone plate and bone screw (middle).

Possible toxic effects (particularly from unreacted monomers) of all implanted polymeric products is of major concern. This, combined with the fact that thermoset bone plates cannot easily be bent to shape in the operating theatre, makes composites of limited attraction. Combinations of carbon/polystyrene, carbon/poly propylene, carbon/nylon, carbon/poly-(methylmethacry late) and carbon/poly (etheretherketone) for bone applications have all been considered. Manufacture has been by the conventional fabrication methods for thermoplastic textile composites, including film stacking (whereby layers of matrix and reinforcing fabrics are hot pressed to melt them together), co-mingling (where mixed yarns of reinforcing and matrix fibres are made into fabrics), hot pressing and the use of powders of matrix materials in between stacks of reinforcing fabrics.

One of the difficulties with composite bone plates is the necessity to provide screw holes within the plate to allow fixation to the bone. Drilling holes reduces the composite load-carrying capacity owing to the damage caused to the reinforcing fibres by drilling. However, if the hole is formed during the impregnation stage or earlier without significant damage to fibres, the loss of strength can be minimised. Computer aided design/manufacture (CAD/CAM) fibre placement technology, as is possible using conventional embroidery machinery, has been suggested for these applications (Fig. 13.5). Biodegradable materials have also been examined, but there are ongoing concerns particularly about the possible toxicity of degradation products as the materials are absorbed into the body.

Carbon hip prostheses have also been developed, which, being of composite, provide the potential to have controlled stiffness and stress distribution in the femur. This could result in significant property improvements over prostheses developed from conventional metal working methods, so that patterns of stress closer to those in the natural situation develop. This gives better regeneration of bone and reduces the potential for loosening. However, long-term success has yet to be proven, and such prostheses are not yet on free sale to surgeons.

External fixation devices made from iron were developed from horseshoes in Russia, now known as the Ilizarov ring fixator for fractures. The materials

13.5 Prototype carbon embroidered bone plate.

used have been developed and carbon is used with the main advantage of weight reduction and radio lucency (Fig. 13.6). They are not widely used, however, and many surgeons prefer to stay with steel devices, often on grounds of cost.

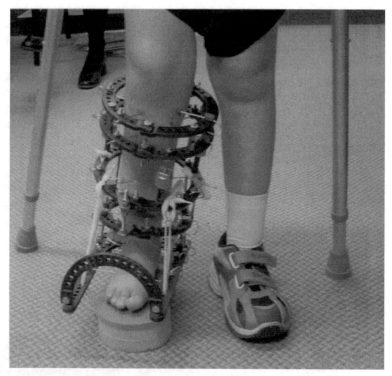

13.6 Illzarov ring fixators.

13.4   General application

There is potential for textile reinforced composites to be used in very many general applications within the medical industry, in common with many other areas described elsewhere in this book. For example, fabric reinforced printed circuit boards have applications in all walks of life including medical equipment. Specific applications include those requiring a low attenuation of radiation, in particular X-rays. With the increased use of minimally invasive surgical techniques requiring monitoring by X-ray during surgical operation, thin, non-metallic operating tables are useful in helping to reduce the radiation dose that is received by the patient and scattered to those working in the theatre. Carbon fibre operating tables have particular value in this area: they can be thin, yet retain the necessary rigidity. When covered with foam and a waterproof cover, a carbon fibre-based operating table will have all the performance characteristics needed.

13.5   Living composites

A number of textile materials can be used to reinforced natural tissues, as well as replace them. Textile surgical implants, such as suture threads, have been in use for many generations and over the past 40 years there has been extensive use of polyester woven and knitted tubes in the replacement of arteries. More recently, however, textile structures have been developed that will act as a scaffold for natural tissue re-growth, the textile remaining within the body for the rest of the patient's life and forming a living textile reinforced composite. One example is the Leeds–Keio artificial ligament for the replacement of the anterior cruciate ligament of the knee. The textile structure of this device is a mock-leno weave (Fig. 13.7) formed into a complex pocketed integrally woven tube (Fig. 13.8). During the operative procedure a bone plug is removed from a biomechanically appropriate site on both the tibia and the femur before a bone tunnel is drilled through the centre of the joint in the optimum position to join the two bone plug sites. The pocketed tubular woven fabric is threaded through the bone tunnel and the bone plug inserted within the pocket at one end of the textile structure. The ligament device is pulled tight through the bone tunnels and the second bone plug placed within a pocket formed at the other end of the artificial ligament.

13.7 Mock leno woven tape.

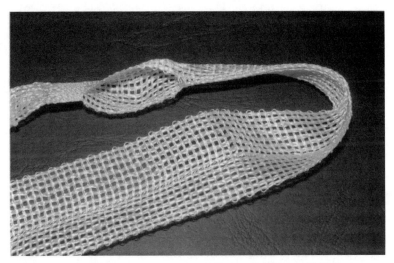

13.8 The woven pocket tape used in the Leeds–Keio artificial ligament.

Post-operatively, the bone will grow between the bone plug of both the femur and the tibial bones to lock the textile structure in place. Along the rest of the length of the polyester device natural tissue ingrowth will occur. As the knee is flexed and loaded, the tissue eventually orients itself and a natural composite of ligamentous connective tissue reinforced by polyester fibre is formed in a true tissue/textile composite material.

14

Textile composites in sports products

K VAN DE VELDE, Ghent University, Belgium

14.1   Introduction

Conventional composite materials generally consist of a reinforcing textile structure and a surrounding matrix with other mechanical properties. By combining two or more materials, the best properties of each material are combined. The reinforcing fibres can supply strength and stiffness in any wanted direction, while the (generally polymer-based) matrix protects the mostly brittle fibres against shocks and chemical agents. Whether or not mechanical stresses and strains are efficiently transmitted between fibres and matrix depends on the adhesion between the both of them. Using (physico-) chemically compatible components is thus necessary when designing a composite. Fibre/matrix compatibility is enhanced by applying sizings on the fibres, thus avoiding delamination between fibres and matrix.

The use of textile composites in sports gear is relatively new. In the earlier days, other materials were used. Calfee and Kelly[1] report that natural materials (e.g. wood) were initially used because of their good shock absorption, but these had many drawbacks. According to these authors – together with Jenkins[2] and Axtell et al.[3] – their anisotropic nature results in low perpendicular strength, and their large variation in properties and high moisture absorption result in unwanted deformations.

In the 1970s light-weight metals such as aluminium and titanium became popular[2]. These provided higher stiffness and a significant reduction in weight[1,2]. However, since aluminium has no fatigue stress limit, even small stresses contribute to fatigue[1]. Combined with its great flexibility, this led to overbuilt designs[1]. Other drawbacks include its high shock transmittance (resulting in, e.g., tennis elbow) and inherent isotropic nature, which leaves no freedom to meet mechanical demands in different directions (as opposed to composites)[1].

Later on, glass/epoxy composites, followed by carbon/epoxy and others, replaced metals[2]. These anisotropic materials allowed the 'insertion' of (mechanical) properties at certain places where extra strength is required[1, 2].

Miracle and Donaldson[4], Spry[5] and Mattheij et al.[6] mention that, by changing the amount, direction and type of reinforcement, one may vary properties along a certain cross-section, resulting in an optimal combination between performance and low weight. According to Jenkins[2], Spencer[7] and Chou et al.[8] composites provide higher specific stiffness, fatigue performance and shock damping than metals.

One of the latest developments is the combination of composites with other materials. Composite or metal baseball bats, for example, cause excessive ball speed (injuries), while a combination of wood with E-glass (or graphite) fibre reinforced composite results in an optimal combination of ball speed control and shock absorbance[3, 4]. Other good examples are modern skis that combine composites with metal and natural (and other) materials. Depending upon the desired properties, a combination of materials is made in such a way that optimal properties are achieved. According to Murphy[9], the combination of composites with other materials (e.g. in metal–composite combinations) can also be interesting given that when composites do fail, they generally fail spectacularly.

The usefulness of composites in sports gear depends upon the intended end-use. Some applications require good shock (and thus energy) absorption, while others require a minimal energy loss in order to generate high speeds. Most of the time, a balance between several more or less contradictory requirements has to be sought. The eventual properties of the product depend upon the materials used, the design and the production technology. The effects of changes in these factors, as well as resulting property changes, are discussed below, together with numerous examples.

14.2   Materials

14.2.1 Reinforcing fibres

Fibreglass is the most classical reinforcement. Its specific stiffness equals that of steel and it is more flexible and tougher than carbon[4, 5]. Its use in sporting applications however, is – according to Spencer[7] and Jacoby[10] – limited because of its poor fatigue performance and high vibration transmittance.

Carbon is five times stiffer than glass, but lighter than aluminium[5]. Together with its good fatigue resistance this results, for example, in light and strong bicycle frames[4, 7]. Carbon's vibration transmittance might be a drawback in some applications[10]. Graphite is even stiffer than carbon but also more brittle[4]. Its shock resistance makes it ideal for skis, bicycle frames and tennis rackets.

The 'sweet spot' transmits minimal vibrations when hit, and exists in equipment such as tennis rackets. Miracle[4] and Goode Snow Skis[11] report that the use of graphite broadens this spot, resulting in more 'forgiving equipment', ideal for inexperienced players and for avoiding injuries (such as tennis elbow).

Kevlar is a poly-aromatic amide that provides light-weight tensile strength and toughness, combined with good vibration damping and impact resistance[4, 10]. It is used in skis and also in many types of protective gear[4]. Boron is much stronger than carbon. Compression properties are especially good[4]. Combining carbon (tensile stiffness) and boron (compressive stiffness) results in a synergistic effect, i.e. the overall stiffness is better than could be predicted based on individual strengths[1]. Boron can, for example, be used as longitudinal reinforcement in golf clubs[4]. Celanese[12] produces light-weight Vectran® liquid crystalline polymer (LCP) fibres that are as strong as boron fibres. Their stiffness, however, is comparable to that of glass fibres. Their impact and shock resistance are outstanding, which leads to many possible uses, such as in golf clubs, bicycle wheels and tennis rackets.

Shock damping and fatigue resistance can further be increased by using poly(ethylene) (PE) fibres. A possible example is a carbon/PE reinforced bicycle[4]. Further weight reduction is also possible by using ultra-high modulus fibres. Such carbon and graphite fibres can be obtained by stripping off the outer fibre layer, leaving the stronger core[1, 13]. This list of possible fibres is by no means exhaustive.

14.2.2 Resins

Generally resins provide good vibration damping (far better than metal)[1]. Conventional thermoset resins (such as polyester and epoxy) are often combined with epoxy and rubber modifiers that lower vibration transmission[2]. One must, however, make sure that vibrations are not completely eliminated since the player needs 'information' from the impact in order to play a ball well.

Unsaturated polyesters are the most commonly used in the composites industry because of their good mechanical and chemical properties, in combination with their relative cheapness. They are mostly combined with conventional glass fibre reinforcement. Epoxies, on the other hand, are more expensive but provide better wetting out of the fibres. Their strength and corrosion resistance is better than those of polyester resins, and they are mostly used in combination with high-performance reinforcements (carbon, graphite, etc.) or with high glass contents. Vinyl esters have properties that lie between those of polyesters and epoxies. When considering sporting goods, epoxy is generally used as a matrix material because cost is not a major issue. Epoxy furthermore provides better adhesion and consequently has better resistance to harsh conditions such as water or moisture in general.

Thermoplastic resins offer very high toughness and durability. Compared with thermoset resins they have a higher damage tolerance, are 100% tougher, and are 600% more resistant to cracks and invisible damage. This makes them ideal for use in heavy-duty equipment such as (nylon/carbon) hockey sticks[14].

The following example illustrates the importance of the matrix. Vibration and shock transmission is higher in conventional carbon/epoxy tennis rackets than in those with a polyurethane-modified epoxy matrix, and certainly than in those with a thermoplastic polyamide-6 matrix. The latter absorbs more moisture but this only further reduces vibrations[8].

14.3   Design

14.3.1 General

In general, using different fibre angles (reinforcement shapes), different plies or thicknesses, different (combinations of) materials, etc., may all contribute to the eventual resulting properties[1]. But special techniques that are not specifically composite related, such as microbearings or overall shape modifications, may be also used to obtain the same effect[13].

14.3.2 Reinforcement shapes

Only a few examples of reinforcement shapes are discussed here. Braided reinforcements combine multidirectional reinforcement with an automated process, high uniformity, few seams and overlap, and good draping characteristics[3, 4]. Reinforcement angle and thickness can be greatly varied, resulting in products such as baseball bats that can appropriately be reinforced near the area of maximum stress, namely the handle[3]. Use of braids also reduces torsion – compared with unidirectional reinforcement – such as in ski 'torsion boxes' (see further on) or prepreg carbon braids in tennis rackets[4, 7].

Weaves have lower draping characteristics, and overlaps are often necessary to enable smooth load transfer between different plies. They may, for example, be used (as 0/90° weaves stacked at ±45°) in combination with unidirectional reinforcement in order to produce tennis rackets with high shear strength and stiffness (torque reduction)[2, 8].

The 'tailored fibre placement' (TFC) concept is worth mentioning. Fibre preforms with stress-aligned fibre orientations can be made, based on the embroidery technique that is also used for decorating purposes. This technique is suitable for lightweight parts with a complicated stress course, such as bicycle frames and brake boosters[6]. Completely unidirectional reinforcements can also be used. These result in excellent flexural and tensile stiffness and strength, but very limited torsion resistance. Generally, several forms of reinforcement are combined in order to obtain the desired combination of flexural, tensile and torsional properties.

14.3.3 Materials

It is important to choose a material according to the desired effect. Stiffer materials such as glass, carbon, graphite and aramide are needed when one, for example, wants to achieve high speeds. If not, some of the energy that has been put into the system may be lost in deformations. A good example is the outside of skates.

Less stiff materials, on the other hand, are needed when some or more degree of deformation is wanted in order to absorb shocks. Good examples include protection equipment, but also the inside of skates. The latter can, for example, be made from heat-mouldable foams mixed with carbon fibres. When the boots are preheated, one is expected to step in and the interior of the skate is consequently remoulded in order to fit perfectly to the feet[10].

14.3.4 Material partition and positioning

Differences in reinforcement and resin placement enable a shift in the centre of gravity. The ends of kayaks, for example, may contain more (light) PE than the rest of the boat, enabling the kayak to move more easily over the waves. Other examples include ski poles, bats and golf clubs; where shifting the centre of gravity can provide more speed and better equilibrum[4]. Pole vaults are another possible example. They are made stiffer at the butt end using a mixture of carbon and fibreglass in order to obtain optimal properties[5].

Combination with other non-composite materials may consist of using foams or honeycomb structures in order to reduce weight and increase (flexural) strength, e.g. closed cell foam is used between two layers of fibreglass in boating, while polyurethane (PUR) honeycombs or foams are used in the core of ski 'torsion boxes' (see Fig. 14.2 on page 454[4, 5]). Quaresimin *et al.*[15] mention the use of epoxy foam in between layers of carbon/epoxy for the production of bicycle cranks. Using lighter foam hardly affects the specific properties of the end-product.

Damping rubber or thermoplastics, on the other hand, may be used for vibration damping in tennis rackets. The 'ISIS' (impact shock isolation system) in tennis rackets, for example, consists of a separated graphite handle that is reconnected using graphite rods encased in PUR elastomer or thermoplastic (e.g. nylon) resin[13].

Overall, design features may also be effective in reducing torque. Stiffening both sides of the throat area and cross-bar with titanium/graphite in tennis rackets, for example, instead of using braided reinforcements will also reduce torque[13].

14.3.5 Some special features

Many features (not specifically related to composites) are possible, but only a few examples are given here. The insertion of thousands of microbearings in tennis rackets allows the racket to store energy as it swings backwards. This energy is released as the ball is hit, adding kinetic energy and thus speed to the ball. The increase of ball speed is combined with a reduction in shocks and vibrations[13]. Inclusion of piezoelectric materials, for example in skiing, is a possible future application. Its principle is the conversion of mechanical energy to electrical energy that is consequently dissipated, thus reducing vibrations[2].

14.4   Production technology

14.4.1 Continuous processes

Continuous processes are only suitable for (quasi) continuous cross-sections. Pultrusion, for example, is a relatively cheap way of producing profiles with a high level of unidirectional reinforcement. It involves pulling a profile through a heated die where it cures. Multidirectional reinforcement (mats), however, may also be inserted to provide transverse strength. Possible applications are ski poles and parts of carbon/nylon bicycle wheels[4, 7]. Suominen[16] and Spencer[7] also mention combinations with other techniques such as (filament-, co- or pull) winding or roll wrap. The latter enables the production of hollow shapes by working around a round mandrel. Using this combination of techniques leads to a combination of flexural and torsional resistance. Continuous processes result in low property variations and possible examples include golf clubs and fishing gear[4, 5, 7]. The latter can be produced as a combination of continuous carbon fibres that are spirally wound (torque resistance and transverse strength), unidirectional longitudinal glass fibres (vibraton transmittance) and longitudinal carbon fibres (flexural strength)[5].

14.4.2 Discontinuous processes

Discontinuous processes are generally slow and expensive, but they are necessary when cross-sections vary. Most sporting goods have complex shapes, thus necessitating the use of discontinuous processes. Tennis rackets may be made by resin transfer moulding or compression moulding (with internal bladder), while a special technique such as balloon moulding (that does not require rigid tools) may be suitable for baseball bats[3, 7]. Carbon/epoxy bicycles and carbon/glass/aramid surf paddles are generally integrally moulded[4, 17]. Again, many other examples are possible: the number of possible techniques is virtually unlimited.

14.5   Applications

Several examples are given here. As will be demonstrated, different applications require different combinations of materials, design features and production processes.

14.5.1 Pole vault

In pole vaulting, the early use of bamboo poles and of the aluminium poles in the 1960s have long gone. These poles have been replaced with carbon fibre composites. Froes[18] states that the ideal pole should be light and highly flexible but also stiff and torsion-resistant, and that energy loss should be minimal. Nowadays, a typical pole consists of three reinforcement layers. The outer layer contains epoxy reinforced with unidirectional carbon fibres, which provide high stiffness for low weight and good fatigue resistance, and return a maximum of energy. The intermediate and inner layers respectively consist of glass fibre webbing and wound glass filaments in an epoxy matrix, thus increasing the torsion resistance. The amount of glass and carbon can be varied in such a way that the pole is much stiffer at the butt end[5]. According to Bjerklie[19], the pole may thus be custom built to the vaulter's weight, take-off speed and hold technique.

14.5.2 Fishing gear

Fishing gear is somewhat similar in construction but requirements are different. Unidirectional carbon fibres provide flexural strength, while unidirectional glass fibres provide the necessary vibration transmittance. Torque (torsion) resistance and transverse strength is obtained through continuous carbon fibres that are spirally wound[5].

14.5.3 Bicycles

Bicycles have greatly evolved in the past few decades. The two major advances are in the frame and wheels. Aiming at minimal frame bending combined with minimal weight, carbon fibre composites are the materials of choice if there is no concern over cost. A light-weight race bicycle (used by professionals) with a carbon reinforced frame, front fork and seat post is shown in Fig. 14.1. In addition, frames have recently been produced from magnesium, aluminium, titanium and metal–matrix composites. Hybrid frames such as carbon fibre reinforced composites combined with titanium have also been produced[18]. Carbon has a relatively high vibration transmittance but good fatigue resistance, while the resin matrix has low vibration transmittance (good shock absorption), and titanium – as a metal – low fatigue resistance.

14.1 Lightweight race bicycle made of composite parts (courtesy: Giant, http://www.giant-bicycles.com)

Combining these materials may result in combining high fatigue resistance with increased shock absorption, while the overall weight remains low.

When going off-road, shock absorption becomes very important. Appropriately designed wheels can absorb a significant part of these shocks. Glass fibre reinforced nylon wheels have been produced to this end. Using a thermoplastic matrix such as nylon (polyamide) results in better shock absorption[4, 18]. Disc wheels (i.e. without the traditional spokes) made of aluminium alloys and carbon fibre reinforced composites have been developed for reasons of aerodynamics[18]. However, when crosswinds occur, these wheels make the bicycle difficult to control and so they become no longer beneficial in these conditions. A compromise might be a three- or five-spoke wheel, which besides making the bicycle more controllable, also cuts drag (by flattening the few remaining thin blades that slice through the air)[19]. Depending upon the track and weather conditions, other materials and designs may be necessary in order to obtain the best results.

14.5.4 Golf

Golf clubs are nowadays lighter, longer and have a bigger head (with equal mass) than before. The net results are greater club head speeds – because of the long arc – and straighter shots (because of the bigger sweet spot). The club shaft may be constructed from graphite reinforced epoxy – and even boron fibres may be used[14] – while the oversized hollow head is made of titanium[9, 18]. Changing material partitioning may also be helpful in shifting the centre of gravity, thus providing more speed[4]. These modern technologies result in an equalising effect (less gifted players perform much better owing to the reduced skill needed for playing golf), and change the game. This is why the US Golf Association has imposed a limit on the club head volume, as well as on the COR or 'coefficient of restitution'. This is the ratio of the speed of a ball before and after hitting the club and has been set at a maximum of 83%[9].

14.5.5 Baseball/softball

Aluminium baseball bats have recently been banned in the major American leagues because they resulted in excessive ball speeds, which led to more injuries upon impact, but also because they altered the game itself as the field became too small. However, after the ban in baseball, new double-walled aluminium and titanium softball bats were made. These provided a bigger sweet spot, combined with a greater ball velocity[18]. These light-weight metal bats, however, acted as 'equalisers': they turned average hitters into spectacular ones. Again, this type of bat was banned for security reasons and now a limit is set to what is called the BPF or 'bat performance factor'. For a conventional wooden bat, the BPF is set to 1.0. An aluminium bat that returns 10% more energy to the ball than the wooden one receives a BPF of 1.1 and bats that exceed 1.2 BPF are considered illegal. Designers have responded to these restrictions by making new types of bats, e.g. hybrid constructions with carbon fibre reinforced composite and honeycomb aluminium in a double-wall design[9]. The former provides strength and stiffness but is thought to reduce ball speed compared with aluminium, while the latter provides a weight reduction combined with increased flexural stiffness. Speed control is also possible by adjusting the partitioning of materials (e.g. reinforcing fibres), thus affecting the position of the centre of gravity[4]. This is an example of how composites may also be used for security and fair play reasons, instead of only for conventional design reasons (speed, vibration damping, weight, etc.).

14.5.6 Tennis

Tennis rackets have evolved from wooden and metal frames (the latter were introduced in the late 1960s) to the modern ones, which are made of monolithic metals, metal–matrix composites and carbon fibre reinforced composites. The goal in designing these modern rackets ranges from efficiency increase – i.e. accelerating the ball across the net – to damping the dangerous vibrations that can lead to tennis elbow. Accordingly, many types of rackets are possible. Increasing the rackets' sweet spot, which depends upon the stiffness of the frame and the size and shape of the racket handle and head, can reduce vibrations. Modern technologies have enabled the production of relatively large but still mechanically stable rackets, so the International Tennis Federation has now imposed a limit on the size of the racket[18]. An example of such a banned racket is a type that used elongated strings in order to create a larger sweet spot and generate more power and spin for less effort[9].

The use of carbon fibre reinforced composite frames results in high stiffness and corresponding efficiency. To reduce the high-frequency vibrations upon impact, racket handles may be constructed by wrapping multiple fibre reinforced layers around a soft core of injected PUR or a honeycomb construction[18]. An alternative way of damping vibrations may consist of using a separated graphite handle and reconnecting it by using graphite rods encased in PUR elastomer or a thermoplastic resin such as nylon[13].

A state-of-the art racket may, for example, be based on a urethane core, graphite fibres and – to a lesser extent – Kevlar fibres. The graphite provides strength and stiffness, and also prevents twisting of the racket head upon impact outside the sweet spot. The Kevlar fibres lead to additional strength and durability, and furthermore contribute to damping vibrations[18]. Using weaves of these materials and stacking them at ±45° (or braided reinforcement[13]), in combination with unidirectional reinforcement, results in rackets with high stiffness and high resistance to twisting[2, 8]. These state-of-the art rackets are lighter and stiffer, can be swung faster and give balls more rebound. Because of these high available speeds, modern tennis has shifted in favour of the fast servers[9]. A possible solution might lie in imposing energy-related limits as in golf and baseball.

Reduced torsion in tennis rackets may also be achieved by changing overall design features. Stiffening both sides of the throat area and cross-bar with titanium/graphite, for example, may eliminate the need for using braided fabrics in the frame. Other, not necessarily composite-related, design changes include the insertion of microbearings. These store energy, thus resulting in higher ball speeds (efficiency) combined with shock reduction[13].

Depending on the player level and his or her requirements a whole range of rackets is being made, using countless combinations of materials, material shapes, design features, etc. This explains the great variations in racket types.

14.5.7 Kayaks

Competition kayaks were once made of mahogany veneers but are now constructed of a combination of carbon fibre cloth, Kevlar and epoxy resins. This stiff design minimises the amount of energy wasted in flexing as the hull passes through the water, thus making this energy available for upholding the maximum possible speed. Weight reduction is also remarkable, although competition kayaks are now subjected to minimum weight requirements. Because of this, other parts such as foot brakes and seat supports are now being made of fibre reinforced composites in order to further reduce the overall weight[19]. Again, changing material partition may also improve the products' qualities: by introducing more – light – PE in the ends than in the rest of the hull, the kayak will more easily move over the waves[4]. Paddles were once made of solid poplar but are now also made of composite materials. Changing the paddle shape into what is called a spoon-shaped 'wing' paddle has further improved propulsion efficiency, although new paddling techniques are required[19]. This is again an example of how new technologies can change a sport.

14.5.8 Skis and snowboards

Figure 14.2 gives an idea of the possible build up of a K2 ski[20]. In fact, the same principle is also valid for water skis and snowboards. In the centre of the construction one can find the core. Aström[21] and others report that this flexible and lightweight part may be made of (PUR) foam, wood/foam composite or wood, or be simply empty[4, 10, 22]. Note the use of natural materials. Sometimes channels are introduced into the core in order to further reduce weight and increase flexibility[10, 23].

The core is wrapped with a composite material, resulting in the 'torsion box'. This torsion box is contoured to give the right amount of flex and

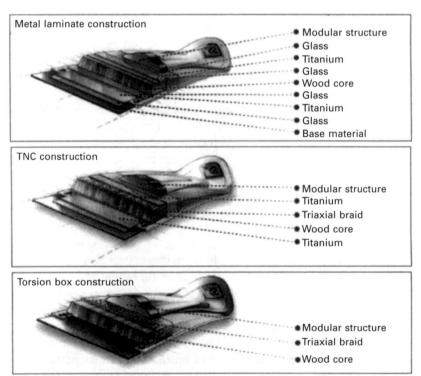

14.2 Possible ski constructions (courtesy: K2 Skis[20]).

spring and – especially – to eliminate twist[24]. This wrapping may consist of glass/carbon/epoxy prepreg (reinforced in different angles) or triaxial braids, for example three layers of pre-impregnated fibreglass[4, 10, 20, 22]. The latter may also be wet-rapped but this known to cause more mistakes. *Design News* magazine[25] reports that wrappings consisting of co-extruded acrylonitrile-butadiene-styrene (ABS)/fibreglass sheet, rubber and fillers may also be used.

This central section is then sandwiched between a top and precision-milled base layer(s). Unreinforced thermoplastics are generally used as base layer[21]. Moulding (and other) techniques are used for the construction of both layers. Multilayered skis may furthermore consist of sheets of glass/epoxy or reinforcing fibres such as carbon (strong, fatigue resistant), Kevlar (impact resistant), graphite (vibration resistant), titanium or others[10, 21]. On top of this construction, a secondary core is possible. This secondary core is based upon modular technology and works both as a suspension system as well as a mass damper[20].

Different styles of snowboards depending upon the snow type are now available. Deep, powdered snow requires more flexibe boards, while icier terrain requires stiffer boards. Because of their laminated construction, the boards in question require an epoxy adhesive that will keep the layers in place through all types of terrain. This adhesive provides excellent wet-out of the fibreglass and other reinforcing layers, resulting in good fatigue and thermal resistance. Increasing the amount of epoxy without changing the amount of glass fibre results in stiffer boards, which perform better on icier terrain. The bindings can be made of nylon-based composites. This thermoplastic matrix provides good impact resistance, even in sub-zero temperatures[25].

14.5.9 Ice hockey

Ice hockey is an application where shock resistance is of major importance. Heavy-duty equipment such as hockey sticks can be made of nylon/carbon composite. The thermoplastic matrix provides toughness, durability and shock resistance. Hockey skates may also be a combination of thermoplastic matrices and stiff fibres. The outside of these skates may be reinforced with stiff fibres such as glass, carbon, graphite or Kevlar. The stiffer the fibres used, the less energy will be wasted on deformations[10]. A heel stabiliser wedge made of an engineering thermoplastic elastomer, on the other hand, may improve the skater's efficiency by allowing more forward flexing than traditional skates while providing lateral and tendon support[25]. The inside of skates can be made from heat-mouldable foams mixed with carbon fibres. The boots are preheated and the user puts them on; the interior of the skate is consequently remoulded in order to fit the feet perfectly[10].

14.6   Conclusion

Considering the given examples, one can easily see that a virtually limitless number of material combinations are possible. Composite materials, but also other materials, are used for sports gear in order to obtain an optimal combination of properties for each possible application. By furthermore optimising design and production techniques, one may obtain a product that is suited for any possible combination of applications, conditions and player experience. Modern technologies also allow for the production of custom-built items.

In some sports these improvements have led to serious changes in the game itself, the required skill to play the game, security issues, etc. As a reaction to these technological changes, many sporting federations have imposed limits on sporting goods. Inserting composite materials into sporting goods has proved to be very useful – not only in improving overall performance – but also in controlling every possible property, thus leading to safer and fairer sporting. The future of composite materials in sporting goods probably consists of combining them with other materials, thus using the best properties of each constituent, and obtaining products that are suited for any given requirement.

14.7   Acknowledgement

The author would like to thank Stefaan Janssens for his contribution to this chapter.

14.8   References

1. Calfee C. and Kelly D., 'Technical White Paper – Bicycle Frame Materials Comparison with a Focus on Carbon Fiber Construction Methods', Calfeedesign, http://www.calfeedesign.com/calfee_TWP.pdf, 2002, 1–13.
2. Jenkins M.J., 'Good vibrations – materials swing into action', *Materials World*, 2000 **8**(6) 11–13.
3. Axtell J.T., Smith L.V. and Shenoy M.M., 'Effect of composite reinforcement on the durability of wood baseball bats', *32nd International SAMPE Technical Conference*, Boston, SAMPE Publishing, 2000.
4. Miracle D.B. and Donaldson S.L., *ASM Handbook – Volume 21 Composites*, Ohio, ASM International, 2001.
5. Spry W.J., (1998), 'Sports and Recreational Equipment', in Dostal C.A., *Engineered Materials Handbook – Volume 1: Composites*, Ohio, ASM International, 845–847.
6. Mattheij P., Gliesche K. and Feltin D., 'Tailored fiber placement – mechanical properties and application', *J. Reinf. Plast. Comp.*, 1998 **17**(9) 774–786.
7. Spencer B.E., (1998), 'Composites in the sporting goods industry', in Peters S. T., *Handbook of Composites*, London, Chapman & Hall, 1044–1052.
8. Chou P.J.C., Ding D. and Chen W.-H., 'Damping of moisute-absorbed composite rackes', *J. Reinf. Plast. Comp.*, 2000 **19**(11) 848–862.

9. Murphy M., 'Blast off!', *Popular Science*, 2002 (June issue, via http://www.popsci.com).

10. Jacoby M., 'Olympic science', *CENEAR*, 2002 **80**(5) 29–32.

11. 'Goode Snow Skis – Technology Overview', http://www.goode.com/sstech.html, 2003.

12. Celanese Acetate L.L.C., 'Vectran – Grasp the World of Tomorrow, Engineering Data' http://www/vectran.net, 2003.

13. Tennis Warehouse Recquet Technology Glossary, http://www.tennis-warehouse.com, 2003.

14. 'Falcon Sports Senior Hockey Sticks', http://falconsports.com/Products/Senior.html, 2003.

15. Quaresimin M., Meneghetti G. and Verardo F., 'Design and optimisation an RTM composite bicycle crank', *J. Reinf. Plast. Comp.*, 2001 **20**(2) 129–146.

16. Suominen M., 'High performance carbon fiber pultrusion profiles for the future', *10th European Conference on Composite Materials*, Brugge, 2002.

17. 'AT Paddles – Bentshaft Whitewater and Touring Paddles, AT2 series', http://www.atpaddle.com, 2003.

18. Froes F.H., 'Is the use of advanced materials in sports equipment unethical?', *J. Minerals, Metals Materials Soc.*, 1997 **49**(2) 15–19.

19. Bjerklie D., 'High-Tech Olympians', *Technol. Rev.*, 1993 (January) 22–30.

20. 'K2 Skis', http://www.k2skis.com/skis/technology.asp, 2002.

21. Aström B.T., *Manufacturing of Polymer Composites*, London, Chapman & Hall, 1997.

22. 'Evolution Factory Tour', http://www.evoski.com/factory_tour.html, 2002.

23. 'The Dynamic Ski Technology: Hyper Carbon, Texalium ...', http://www.skidynamic.com, 2003.

24. 'Surflight Hawaii Composite Controlled-Flex Technology', http://www.surflight.com/tech.html, 2003.

25. 'Plastics create a winter sports wonderland', *Design News*, 1995 (20).

Glossary

Anisotropic	Having properties that vary with direction within a material.
Autoclave	A closed and heated pressure vessel used for consolidation and curing of thermoset composites.
Binder	An agent (usually a thermoset or thermoplastic polymer) used to provide cohesion to fibres, yarns or fabric layers within a *preform*.
Blend	Mixture of two or more fibre types in a yarn.
Braid; plait	The product of braiding.
Braiding; plaiting	The process of interlacing three or more threads in such a way that they cross one another in diagonal formation. Flat, tubular or solid constructions may be formed in this way.
Commingled yarn	A yarn comprising intimately mixed reinforcement and matrix polymer fibres.
Crimp	The waviness or distortion of a yarn that is due to interlacing in the fabric.
Cure	Change of thermoset resin state from liquid to solid by the formation of crosslinks during polymerisation.
Delamination	Failure of a laminate by separation of layers, usually due to external stresses.
Denier	A measure of linear density, corresponding to the mass in grams of 9000 m of yarn.
Dobby	A mechanism for controlling the movement of the heald shafts of a loom. It is required when the number of picks in a repeat of the pattern is beyond the capacity of cam shedding (~8).
Drape	The ability of a fabric to conform to a complex geometry.
Ends	Another name for warp yarns.
Fibre	Textile raw material generally characterised by flexibility, fineness and high ratio of length to thickness.

Filament   A fibre of indefinite length.

Harness   A frame positioned across a weaving loom with a set of heddles mounted on it. This is used to raise or lower the warp yarns.

Heddle   A wire or thin plate with an eye through which a warp yarn passes.

Homogenisation   Determination of material properties from unit cell or representative volume analysis (RVE) for use in models at larger scales.

Intra-ply shear   Deformation mechanism for bidirectional fabrics, involving in-plane shear of the material via rotation of yarns at their crossovers.

Isotropic   Having properties that do not vary with direction within a material.

Jacquard   A shedding mechanism, attached to a loom, that gives individual control of up to several hundred warp threads and thus enables large figured designs to be produced.

Knitting   A process to manufacture textiles via looping of yarns around a series of needles.

Laminate   A composite material reinforced by a number of distinct textile layers (laminae, in this context).

Linear density   The mass of a yarn per unit length.

Locking angle   The maximum level of intra-ply shear that can be achieved for a fabric reinforcement before wrinkling/buckling occurs.

Matrix   In this context, the polymer resin in which textile layers are embedded, providing the medium for load transfer.

Maypole braider   A machine provided with three or more carriers that are driven by means of horn gears along tracks that cross at intervals, enabling the yarn drawn from the carriers to interlace to form a braid.

Monofilament   A yarn consisting of one filament only.

Net-shape   Referring to a manufacturing process whereby the component is produced in the correct dimensions and requires no trimming.

NCF   Non-crimp fabric; a fabric constructed from unidirectional crimp-free fibre layers, assembled together using a light stitching yarn. More properly referred to as a warp knit with inserted yarns.

Orthotropic   Related to properties with three perpendicular axes of symmetry.

Permeability   The property of a porous material (textile reinforcement here) allowing pressure driven fluid flow.

Picks   Another name for weft yarns.

Plain weave   The simplest fundamental weave pattern, where each warp yarn interlaces with each weft yarn.

Ply yarn   Several strand yarns twisted together.

Porosity   The volumetric fraction of voids within a body (here usually either a textile reinforcement or solid composite).

Preform   An arrangement of dry fibres in the three-dimensional shape of the final component for liquid moulding processes such as resin transfer moulding (RTM) and vacuum infusion (VI).

Prepreg   Pre-impregnated composites, comprising reinforcements combined (usually) with a partially cured thermosetting polymer.

RFI   Resin film infusion; a manufacturing process that involves consolidation of interleaved layers of thermoset resin films and fabric layers.

Roving   A flat yarn, normally sized, containing straight filaments

RTM   Resin transfer moulding; a form of liquid moulding whereby positive pressure is used to drive a liquid thermosetting resin into a mould cavity containing a dry fibre preform.

RVE   Representative (or repeating) volume element; the smallest element that can be used to describe the behaviour of textile reinforcements or composites. In this context this is normally equivalent to the smallest region that can be used to define the textile structure.

Satin   A fundamental weave characterised by sparse positioning of interlaced yarns, which are arranged with a view to producing a smooth fabric surface devoid of twill lines (diagonal configurations of crossovers).

Shed   A gap between warp yarns where a weft yarn may be inserted.

Shuttle   A weft insertion device, which is shut from side to side of a weaving loom.

Sizing   An agent, normally resin, added to yarn to keep fibres in a non-twisted yarn together and/or to improve adhesion to the matrix in composites.

Spinning   A process, or series of processes, used in production of yarns by insertion of twist.

SRIM   Structural reaction injection moulding; a liquid moulding process by impingement mixture of two or more reactive monomers, which impregnate a dry fibre preform.

Staple yarn A twisted yarn comprising fibres of predetermined short lengths.

Tex A measure of linear density, corresponding to the mass in grams per kilometre of yarn.

Textiles Originally woven fabrics, but now also applied to fibres, filaments and yarns, natural or synthetic, and most products for which they are the principal raw materials. This definition embraces, for example, threads, cords, ropes and braids; woven, knitted and non-woven fabrics; hosiery, knitwear and other garments made up from textile yarns and fabrics; household textiles, textile furnishing and upholstery; carpets and other fibre-based floor coverings; industrial textiles, geotextiles and medical textiles.

Thermoplastic A polymer material that is softened by heating and hardened by cooling in a reversible process.

Thermoset A polymer material that is hardened by an irreversible chemical reaction (curing).

Tow An essentially twist-free assemblage of a large number of substantially parallel filaments.

Twill A fundamental weave with yarns interlaced in a programmed pattern and frequency to produce a pattern of diagonal lines on the surface.

Twist Number of turns per unit length in a twisted yarn.

Twisted yarn A yarn with fibres consolidated by twisting.

Unit cell A representative unit of a textile or composite used in analysis (usually equivalent to a representative volume element, RVE).

VI Vacuum infusion; a liquid moulding process where the preform is enclosed between one solid tool and a vacuum bag, so that the driving resin pressure is one atmosphere. Also known as vacuum-assisted resin transfer moulding (VARTM).

Void A resin-free cavity within the composite (usually corresponding to entrapped air).

Volume fraction The volumetric proportion of a constituent in the composite (often an abbreviation for fibre volume fraction).

Warp The machine direction for a weaving machine/yarns in the machine direction.

Weave The pattern of interlacing of warp and weft in a woven fabric.

Weaving The action of producing fabric by the interlacing of warp and weft threads.

Weft (fill)   The direction across the width of a weaving machine/
                       yarns across the width of a fabric.
Wrinkle   A defect formed during composite forming, formed via
                       in-plane compression of the fabric/composite.
Yarn   A product of substantial length and relatively small cross-
                       section consisting of fibres and/or filaments with or without
                       twist.
Yarn count   The length of a yarn per unit mass.
(number)
Yield   A measure related to yarn count or linear density,
                       corresponding to length per unit mass (yd/lb).

Index